Vanished Islands
and Hidden Continents
of the Pacific

Vanished Islands and Hidden Continents of the Pacific

Patrick D. Nunn

 A Latitude 20 Book

University of Hawai'i Press | Honolulu

14 13 12 11 10 09   6 5 4 3 2 1

Library of Congress Cataloging-in-Publication Data

Nunn, Patrick D., 1955 –

Vanished islands and hidden continents of the Pacific /
Patrick D. Nunn.

   p. cm.

Includes bibliographical references and index.

ISBN 978-0-8248-3219-3 (hardcover : alk. paper)

   1. Oceania — Geography. 2. Pacific Ocean — Geography.
3. Islands of the Pacific — History. 4. Lost continents —
History. 5. Geographical myths — Oceania. 6. Geographical
myths — Pacific Ocean. I. Title.

DU18.5.N86 2008

001.94 — dc22   2008017180

Designed by April Leidig-Higgins

Printed by Edwards Brothers

For Roselyn

Out of monuments, names, words, proverbs,
traditions, private records and evidences, fragments
of story, passages of books . . . and the like, do save
and recover somewhat from the deluge of time.
— Francis Bacon, *The Advancement of Learning*

Contents

Acknowledgments xi

1 Introduction: A Personal Odyssey 1

2 The Earth's Dynamic Third: The Pacific Basin 7

3 Islands That Vanished Long Ago 35

4 Ancient Continents Hidden by Time 62

5 The Coming of Humans to the Pacific 71

6 Mythical Islands in Pacific Islander Traditions 86

7 Mythical Continents of the Pacific 108

8 Vanishing Islands: Processes of Island Disappearance
   Witnessed by Humans 130

9 Recently Vanished Islands in the Pacific 147

10 Vanished Islands of the Future 182

11 Vanished Islands and Hidden Continents in the World's Oceans:
   Last Thoughts 195

Appendix 1 201
Appendix 2 205
Notes 209
References 241
Index 265

Acknowledgments

This book is dedicated to my wife Roselyn, without whose unrelenting encouragement it could never have been written. The support of my children — Rachel, Warwick, and Petra — sustained me in bleak times. The University of the South Pacific granted me a year free of other responsibilities to focus on this book, for which I am grateful, and the Kagoshima University Research Center for the Pacific Islands gave me a base for four months to write the book. I wish to thank the staff of the center — Shinichi Noda, Tetushi Hidaka, Shunsuke Nagashima, Kei Kawai, and Hiroko Kusumoto — and Hiroshi Moriwaki for providing an environment conducive to writing this book.

For particular help, I am grateful to Bill Aalbersberg, Wal Ambrose, Dalsie Bani, Russell Blong, Edward Bryant, Rajesh Chandra, Marjorie Tua'inekore Crocombe, Ron Crocombe, Terry Crowley, Simon Day, William R. Dickinson, Anne Felton, Robert Fernau, Paul Filmer, Lawrence Foana'ota, James Goff, Gary Greene, Michael Heads, John Hosack, Kerry Howe, Toshio Kawana, Barbara Keating, Masaaki Kimura, John Lynch, Nic Maclellan, Bruce Masse, Sepeti Matararaba, Clare Morrison, Dieter Mueller-Dombois, Alain Nicolas, Robert Nicole, Ralph Regenvanu, Barry Rolett, Robert Schoch, Sandra Tarte, John Terrell, Marika Tuiwawa, Gerard Ward, Marion Ward, Michael White, Esther Williams, and Steve Wyatt.

The oral traditions reported for the first time in this book were collected over the past ten years as part of my research projects concerned with myths by Mary Baniala, Vicky Claude, Morris Harrison, Tony Heorake, Teena Kum Kee, Elia Nakoro, Alifereti Nasila, Bronwyn Oloni, Kori Raumea, Kellington Simeon, Esther Tegu, Samuela Tukidia, Sereana Usuramo, and Lysa Wini. A number of research assistants have worked on these projects, including Francis Areki, Roselyn Kumar, Tamara Osborne, Wainikiti Waqa-Bogidrau, and, most recently, Maria Ronna Pastorizo, Tammy Tabe, and Ledua Traill Kuilanisautabu. The help of Paul Geraghty is gratefully acknowledged. These research projects were funded mostly by the University of the South Pacific and the Government of France. I have been fortunate to work with a superb copy editor, Eileen D'Araujo, and with Keith Leber at the University of Hawai'i Press.

The following informants are acknowledged for sharing their oral traditions.

In the Cook Islands: on Mangaia, Mataora Harry, Tuaiva Mautairi; on Rarotonga, Ioane

Kaitara, Raumea Koroa, William Powell, Ngatoko Rongo, Sonny Tatuava, Tangianau Upoko, and Tony Utanga.

In Fiji: on Moturiki, Rusiate Balevusa, Timoci Boso, Timoci Bukasoqo, Eremasi Cokoisau, Jone Golea, Simione Kaikai, Manueli Lutu, Manu Malamala, Viliame Misipa, Samisoni Qereqere, Semi Ravuaceva, Savenaca Tagarekece, Apisai Taqaratu, Tevita Tui, Rusiate Tuinarabe, Marika Vakacakau, Viliame Vunalia, Peni Wainilagi, Josua Waqatabu; on Matuku, Ratu Osea Banuve, Ratu Jeremaia Coriakula (Tui Daku), Isireli Temo, Emitai Vakacegu; on Totoya, Rokosau Ratu Timoci Batila Lakeba, Ratu Netani Bolatolu, Rasalato Cama, Reverend Colati, Joni Dalutua, Josaia Daulako, Tukai Jesa, Ratu Viliame Lutu, Isikeli Moce, Temo Sobutilau, Cama Tiko, Inoke Tokalau, Suliano Tuicakau, Sogotubu Turaga, Telu Turaga, and Semi Vuniwaqasau.

In Kiribati: on Butaritari, Kakiaman Bebeia, Maria Ioane, Tinaua Ioane, Mareweata Kiata, Beniamina Kiboboua, Motirake Kiboboua, Tamuera Nakoro, Tekiniman Nataake, Kautu Rabangalii, Mariana Tabuaka, Ioteba Te Mauku; on Kiebu, Babo Mabubu, Kaireti Nakabobouea, Taobura Tamoa, Tim Taom; on Makin, Taenibeia Kabaua, Bakoko Kaitaba'a, Karianako Katia, Teata Koriri, Barbara NaKairoro, Kobebe Tanaua, Tim Tebakabo, Neneaki Tetabo, Moanteata Tetaeka, Taake Tikam; on Tarawa, Bwere Eritaia, and Captain Edwin Mange.

In Solomon Islands: on Maramasike, Houaniuwaha, Thomas Houkilo, Philip Hou'uwa, Olota, Phillip Masura'a, Andrew Suraru; on Santa Catalina, Rev. Matthew Manae, Gaius Mananga, Gordon Ringikoro, John Ringikoro, Chief Francis Sao, Joseph Sugu; on Ulawa, Isaac Aduadu, Michael Harara, John Palm Haununumanii'e, Georgina Saulusu Houniala, Hudson Houniala, Basil Poki, Cecil Sautehi, Jimmy Sautehi, Selwyn Sautehi, Rolland Titiulu, Thomas Titiulu, Ulute Titiulu, and Levi Uwomatawa.

In Vanuatu: on Ambae, Edward Banga, Oscar Leo, Willie Levu, Daniel Moli, Peter Tagaro, Johnstill Tari, Hadson Vugi; on Atchin, Hosea Andrew, Meltetamat John, Martino Metsal, Tony Rowsy, Boneface Taly, Marcela Tavunwo, Ruruvanu Terry; on Malakula, Tite Malvanu, Claude Telukluk, Early Telukluk, Iferem Telukluk; on Pentecost, George Boemaruru, Olivea Daruhi, Mark Gaviga, Lengas Rara; on Maewo, Gideon Bani, Erick Boe, Harold Finger, John Mark Sine, Chief Jonah Toakanase; and in Port Vila, George Goose, Cyriaque Metsan, Jean Marie Metsan, Edmond Rory, Georgy Bernard Rowsy, Paul Telukluk, and Reubennie Ulnaim.

Vanished Islands
and Hidden Continents
of the Pacific

Introduction

A PERSONAL ODYSSEY

There are few topics that have captured the imaginations of people within the last few centuries more than the idea of vanished islands. For myself—and, I would argue, for most schoolchildren with inquiring minds growing up in Europe in the second half of the twentieth century—the questions of whether the fabulous island Atlantis, described in exhaustive detail by the Greek philosopher Plato about 350 BC (Before Christ), ever truly existed and where it might have been located proved compulsive. For me at that time, such questions seemed to go straight to the fundamentals of existence in ways that the questions raised within prescriptive curricula did not. In adolescence, it seemed to me that proving the former existence of Atlantis was tantamount to proving the existence of God for it was self-evident that only in the wisdom of the Ancients, unpolluted by the complexities and crass materialism of the modern world, could the answers to such fundamental questions be found.

Naturally these views have since been significantly tempered but, decades later, I realize that numerous people had similar views, both before and after I held them. In the "Dedication" to his 1880 poetry collection *Ultima Thule,* H. W. Longfellow wrote

> But that, ah! that was long ago.
> How far, since then, the ocean streams
> Have swept us from that land of dreams,
> That land of fiction and of truth,
> The lost Atlantis of our youth!

Today, a whole new way of thinking—the new age—has embraced the concept of Atlantis, making it the cornerstone of an edifice that is chaotic, contradictory, and unapologetically unscientific. Yet, for the same reasons that I toyed with these ideas as a teenager, many people today get reassurance from such charismatic explanations of the world that they have struggled to find elsewhere. The fact that these explanations are fuelled by pseudoscience writers who bend, distort, and selectively cite scientific data and explanations to support their often ludicrous theories of natural phenomena and human history is less palatable to me. It is no coincidence that the rise of pseudoscience has coincided with a diminishing of legitimate scientific interest in issues like vanished islands and hidden continents, not because the associated questions have become any less intriguing or indeed valid scientifically but because they have become tainted. In a critical 1978 book about Atlantis, E. S. Ramage commented

> Perhaps it is already clear why those who are best qualified [scientists] to speak about Atlantis are satisfied with offering incidental criticism or else ignore the problem entirely. The one common denominator among all the various theories that have been put forward is the singular lack of detachment shown by the [pseudoscience] theorists. Instead of beginning with Plato, most begin with a hypothesis and develop their ideas with an enthusiasm that often verges on fanaticism.[1]

Some years ago I wanted to visit the island of Moturiki, some 10 kilometers off the east coast of Viti Levu, the largest island in the Fiji Group of the Southwest Pacific. I drove to Ucunivanua (the tip of the land), in eastern Viti Levu, to rendezvous with the boat from Moturiki. It arrived much later than I did, and I spent some of the interim wandering along the foreshore, exposed by the low tide, wondering at the vast numbers of pottery shards sticking up through the mud. Some of these shards were intricately decorated, made this way I supposed by the distant ancestors of the people now living in the area, people who today have no memory of pottery making.[2]

When the boat arrived, we headed out to sea, threading our way through the myriad reefs that fringe this coast, buffeted by swells driven into our path by the southeast trade winds. Our route was not directly to Moturiki; the boatman first had to collect some supplies from isolated Leleuvia Island, 5 kilometers south, where a backpackers' resort is located.

Leleuvia is one of the smallest and lowest islands I have ever been on, perhaps 100 meters in diameter and rising no more than 3 meters above the surface of the patch of coral reef on which it is built. It is a pile of sand and gravel, thrown together by successive storm waves, then colonized by the coconut palms and creepers that are ubiquitous along the low, sandy coastlines of the tropical South Pacific islands. While the boatman went to do business on Leleuvia, I wandered along the shore and there — to my surprise — buried in the sand spit at the eastern end of the island, I found more pottery shards. There is no clay on Leleuvia so the pottery could not have been manufactured there. I tried to imagine how it could have got there. Too heavy to be carried in the waves, I supposed the pottery must have been brought to this island by people. In the absence of freshwater, people in the past could never have lived permanently on such islands; to do so today they import drinking water. Perhaps the pottery was brought as water containers for fishing parties.

We got back into the boat and set off for Moturiki. To get there from Leleuvia we first had to cross Davetalevu (great passage), a deepwater gap through the great barrier reef into which water from the deep ocean is driven by the southeast trade winds. As we reached the edge of Davetalevu, all of us in the boat were asked to remove our hats and refrain from talking until we reached the other side. The reasons for this are intriguing, as I found out when we landed finally on Moturiki.

Long ago, I was told, an island named Vuniivilevu had existed between Moturiki and Leleuvia: a chiefly island, one of immense cultural importance in the district. Then one day it abruptly disappeared; it slipped beneath the waves and only a few of its occupants survived. Sometimes today, my informants told me, when they cross Davetalevu, where the island once was, the shouts and screams of people can be heard coming up from the ocean depths. And that is why you must show respect by removing your hat and being quiet when you cross Davetalevu. For if you are loud, the cries from below cannot be heard and those uttering them will in frustration agitate the sea and capsize your boat. Only a few weeks earlier, I was told, a man from Moturiki (where alcohol is forbidden) went across to Leleuvia and had a few beers. On his way home, sailing alone through the dark, he sang at the top of his voice as he was crossing Davetalevu and his boat tipped over and he, quickly sobered by the cold water, swam ignominiously back to Moturiki. I suggested that the man in question may have been the architect of his own fate, but my informants merely smiled and told me that there had been many comparable incidents at Davetalevu that were not so easily explainable.

Since that time, and in more than twenty years of living and researching in the Pacific Islands, I have come across many tales of vanished islands that are often known only within very small communities. Yet the commonalities between these tales are striking, as are their similarities to the well-publicized story of Atlantis.

My career as a geoscientist was built on the study of islands in the middle of the oceans. A few years ago, the convergences between the stories of Atlantis and other vanished islands and what I understood about the life cycles of oceanic islands became too many and too compelling to ignore. For although Atlantis, like many allegedly vanished islands in the Pacific, never existed, many of the associated details derive from observations of a whole range of natural phenomena cemented together with human imagination.

There are mechanisms by which steep-sided oceanic islands can abruptly collapse, and the evidence that such megacollapses once occurred has been described from many oceanic islands. It was only a short jump from thinking of flank collapses that could remove part of an island to summit collapses, which, although they still affected mostly the flank of an island edifice, also carried away its top (above-sea) portion.

And then I acquired the 1984 book by oceanographer Henry Stommel entitled *Lost Islands*. It is a masterly review of all manner of such islands, from those incorrectly located, through those invented by unscrupulous navigators who wanted an island named after them, to those that really seem to have existed once . . . but now are no more. The latter group intrigued me most, especially because one of those Stommel described, Tuanaki, was within my immediate geographical sphere of interest. Discussions with Marjorie Tuaʻinekore Crocombe, a former colleague who had translated from Cook Island Maori the nineteenth-century narrative that gave most details about Tuanaki, convinced me that this island had been in existence in the 1830s but had somehow vanished by 1844 when missionaries went searching for it.

A period of immersion in Pacific Island origin myths led to numerous discussions with Paul Geraghty, a renowned Pacific Island linguist, who had traced the story of the supposedly sunken island Burotu back to near Matuku Island in Fiji. Burotu is a key component in the widespread myths about the homeland Hawaiki from which all Pacific Island people living east of Fiji, including those in Hawaiʻi and New Zealand, are reputed to have descended. Although Burotu no longer exists, according to the people of Matuku there are manifest signs that it is still present below the ocean.

As my interest in Tuanaki and Burotu grew, so information about other such vanished islands in the Pacific came to my attention, particularly through studies of Pacific Islander myths. I became interested in the hazard implications of island disappearance, particularly the possibility that islands may have vanished as a result of island flank collapses. If such collapses occurred in the past, then they will also occur in the future and cause massive disruption to humans in the Pacific, not just on the islands that collapse but also, through the impact of what are sometimes called megatsunamis, along other Pacific coasts.

It is easy to overstate details about vanished islands, particularly when dealing with myths and hazard potential. In this book, although relating the opinions of others, I have tried to ensure that no details are overstated, and that nothing that is unverifiable is presented as undeniable fact. That said, this is a field of enquiry that is in its infancy, and future research may well lead to the removal of some of the caution with which the accounts of particular vanished islands are treated in this book.

Myths are fictions that sometimes contain coded grains of historical truth. But anyone attempting to isolate these grains from the fiction must be cautious neither to identify them too uncritically nor to exaggerate them unduly. Yet conversely, it is self-defeating to argue, as some have done, that such myths can never be used as a source of historical information because the fictional overprint is just too great. Myths may be inherently difficult to interrogate, but, in the Pacific Islands, where most written history began only a couple of hundred years ago, they represent a massive archive of historical material that anyone interested in the region's long-term geological history would be foolish to dismiss.

It is also possible to exaggerate the nature of geohazards associated with island flank collapse and island disappearance. We have abundant evidence from the world's oceans of these events occurring throughout geological time, but the associated evidence of large-wave impacts is generally indistinct. This has led to the suggestion that flank collapses may not be the abrupt catastrophic events that they appear to be when viewed from a distance in time but slower, more gradual phenomena. The opposing view that such collapses have been catastrophic, albeit infrequent, events has been extended to modern hazard studies. Some scientists argue that large-scale collapses are threatening — in parts of Hawai'i and the Canary Islands, for example — and should be given appropriate attention by hazard planners.

This is a book about the entire Pacific, not just the islands and the ocean, but also the fringe of the continents that borders it. The Pacific Ocean is the world's largest, covering almost one-third of the earth's surface. By comparison, in the words of Robinson Jeffers, from his poem "The Eye,"

> The Atlantic is a stormy moat; and the Mediterranean
> The blue pool in the old garden,
> . . . but here the Pacific . . . the hill of water; it is half the planet:
> this dome, this half-globe, this bulging
> Eyeball of water, arched over to Asia,
> Australia and white Antarctica: those are the eyelids that never close;
> this is the staring unsleeping Eye of the earth . . .

It is remarkable how little was known for so long about the origin and nature of the Pacific Ocean, a fact that fuelled speculation and misinformation. Yet for all that, there were indeed islands that once existed in the Pacific and that have since vanished. And there are fragments of possible continents now hidden beneath the surface of this ocean and along its sides. What I intend in this book is to give a readable yet scientifically rigorous account of these phenomena in the hope that it will expose the falsehoods about this topic while also showing that there is substance to this intriguing subject.

A reviewer of one of my earlier books, *Environmental Change in the Pacific Basin,* rebuked me for having the temerity to recommend that readers interested in learning more about the ways in which native Australians interacted with their environment should read Bruce Chatwin's travelogue-novel *The Songlines.* I was told that such nonscientific texts should never be cited in a scientific book. I disagree. The current book has at its core a marriage between the arts and the sciences. Probably no other approach would work with such subject matter but that is not the only justification. The dangers that C. P. Snow foresaw more than fifty years ago from the continuing polarization of arts and sciences have been amplified since he wrote, not lessened as he hoped.[3]

Over the past twenty years, in addition to teaching at the international University of the South Pacific, I have had short stints at universities in Australia, Canada, Japan, New Zealand, and the United States. In all these places, students specialize either in science or arts. The arts students are generally terrified of science and are therefore ready prey for pseudoscience writers, and science students are often contemptuous of the arts. One subsidiary aim of this book is therefore to give those readers with a background in the arts some understanding of science and, more important, a critical notion of what constitutes pseudo-science. And for readers who have a science background, I hope that this book demonstrates the value of nonscientific information sources, such as myths, to scientific enquiry.

I have tried to write this book in a way that is accessible to every educated reader, whatever their particular interests. I have consigned formal referencing and technical detail to endnotes and have strived to avoid unnecessary jargon.

In chapter 2, the background to the dynamic condition of the Pacific Basin is outlined. I explain its evolution and describe the processes by which movements of the land (tectonic movements) have taken place.

In chapters 3 and 4, I discuss those islands and continents that may once have existed in the Pacific but that disappeared from view generally long before modern humans appeared on the scene. The principal purpose of these chapters is to show that, unquestionably, the history of the earth has been marked by the alternate appearance and disappearance of pieces of land in the ocean basins.

In chapter 5, by way of background to the accounts of vanished islands and hidden continents given by people from their own observations that occupy most of the remainder of the book, I explain when and how humans reached the Pacific Basin and spread across it.

Chapter 6 deals with vanished islands in the Pacific that are considered, on the basis of the available information, to be mythical but that may have some foundation in observations of the natural environment in the island groups concerned. In contrast, the accounts of mythical continents in the Pacific in chapter 7 have almost no basis in fact, being largely a result of incorrect inferences or the demonstrable products of people's imaginations.

In chapter 8, I describe the processes by which islands are disappearing, as observed by people. Then in chapter 9 I describe all those vanished islands from the Pacific that I regard as having once existed. Chapter 10 deals with the future, and when and why we might witness landmasses disappearing. Chapter 11 is a short concluding chapter.

2 The Earth's Dynamic Third

THE PACIFIC BASIN

Many of our ideas about the history of the planet Earth and the various processes that have shaped it are founded on observations made in northern Europe and eastern North America, places marked mostly by uncommon passivity of earth-shaping phenomena. These regions, for example, lack the climatic extremes of the tropics, they lack the proximity to the largest ocean on earth, and they are associated with an almost complete absence of deep ocean trenches where the earth's crust is being destroyed, a process associated with often-explosive volcanic eruptions and abrupt land movements. For many people, the Pacific Basin is the other side of the earth and consequently less well known and understood. The Pacific Basin occupies around one-third of our planet's surface yet remains marginalized in many textbooks of natural science. It stretches almost from pole to pole, contains the largest ocean on earth, and hosts most of the world's deep ocean trenches. In terms of understanding the history and the evolution of our planet, the Pacific is too important to be sidelined, but the fact that this has happened has made it fertile ground for the imaginings of people for centuries.

Much of what we know about the ancient history of the planet Earth has become known only quite recently, for example:

- the oldest rocks on earth formed around 4,000 million years ago
- the continents and ocean basins have changed their shapes and their positions relative to one another through time
- mountain ranges sometimes rise rapidly and are worn down slowly
- on a scale of several million years, many islands in the world's oceans are transient phenomena — going up, going down, appearing, disappearing
- just as vast new areas of the earth's crust (particularly on the ocean floor) have formed within the last million years or so, so others (including whole islands and bits of continents) have disappeared within the same period

It is not surprising that such ideas represent a radical departure from the ideas of most educated people a century or so ago. Today most earth scientists regard these old views as antiquated: worthy but irrelevant. Yet some of today's pseudoscience writers, either unaware of modern scientific thinking or dismissive of it, employ some of these old ideas to prop up their fanciful notions about human and planetary evolution. When you mention lost continents or vanished islands to some such people, their eyes light up with a neophyte's passion because they believe that knowledge of such ancient worlds holds the key to long-lost secrets about humans and their powers that will prove to be the panacea for many apparently insurmountable problems of modern human existence. Such assumptions are born more from wishful thinking than from rational scientific evaluation.

One hundred and fifty years and more ago, almost everyone in the Western world who pondered the history of the earth considered it a largely unchanging body whose principal features — mountains, valleys, ocean basins, and islands — had been fixed in both form and configuration since their creation. This view, styled fixist by subsequent detractors, was supported (and indeed prolonged well past the point of expected demise) by the Christian churches, which taught of a single act of divine creation unaffected by the subsequent behavior of the created (human or inanimate). Anything less could be taken as challenging God's omnipotence.

Our understanding of the nature of the earth and the ways in which it has evolved acquired over the past 150 years or so, particularly the last half century, contradicts this belief. Everywhere on the earth's surface there is change, and this is continuous. Nothing is truly fixed except as it sometimes appears to people who cannot extend their time horizons beyond that of a single human life span. Some change has been catastrophic, some slow. Some changes have gone unobserved, others were witnessed by thousands. Changes also took place before there were people to observe them, yet it is still possible to discover the nature of these changes through the study of various environmental proxies. Whatever way we look at it, the earth is a dynamic place.

The Pacific Basin

Most textbooks in geography and geology focus on Europe and North America, manifesting a bias toward regions of the world about which most is known and can be demonstrated, and where most potential buyers are located. The corollary is that such textbooks also manifest an ignorance, which is passed on to their readers, about many other parts of the world, including the massive Pacific Basin (Figure 2.1).

Perceptions of the Pacific by people who do not actually live there are generally uncomplimentary: people perceive it as empty space, occupied by only a handful of the world's population living on small islands. Yet to the people who have inhabited the Pacific Islands for generations, the perception is quite different. They regard the Pacific Ocean as an island-

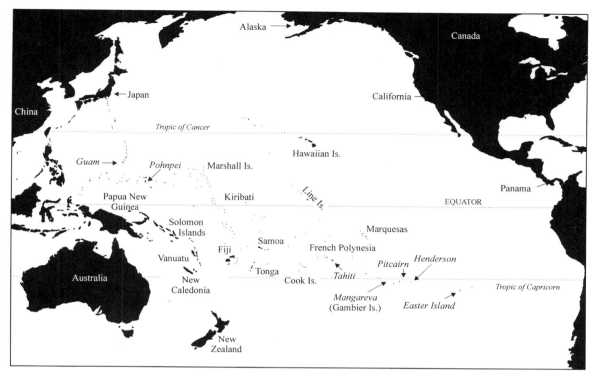

FIGURE 2.1. The Pacific Basin, showing the main places mentioned in this book.

peppered sea, huge pieces of which as easy to find your way around, using stars, clouds, and sea marks, as any state of the conterminous United States.[1]

The Pacific Basin is the subject of this book and comprises two distinct parts, the continental Pacific Rim and the oceanic Pacific Ocean and Islands. Rather than being a single giant basin, the modern Pacific Basin is actually a series of basins divided by ranges of mountains (mostly underwater). Despite their diversity, these ocean basins share a common oceanic origin, broadly unrelated to the origins of the continents, which are mostly much older.

The continental rim of the Pacific extends almost continuously around the Pacific Ocean, from the southern tip of South America to the western end of Alaska, and from easternmost Russia across some shallow ocean gaps to Tasmania off the south coast of Australia. The sizeable gaps that now exist between the Pacific coasts of Antarctica and the nearest parts of the rest of the Pacific Basin's continental rim formed comparatively recently, a mere thirty million years or so ago. Along most of its length, the continental edge of the Pacific Rim is shadowed by belts of fold mountains, including the Andes, the Rockies, and the Great Dividing Range of eastern Australia, that accentuate the geographical integrity of this vast region.

The Pacific Ocean is the world's largest, a residual of a much larger ocean (Panthalassa) that dominated the earth's surface around 220 million years ago when all the continents were amalgamated into a single one named Pangea. The islands of the Pacific[2] are most numerous in its southwest quadrant, where the processes responsible for forming islands are (and have been) most active. The archipelagoes of Solomon Islands, Vanuatu, Fiji, Tonga, and Samoa are examples from that area. The southeast quadrant, which includes the linear island groups of French Polynesia and isolated islands like Easter Island, and the northwest quadrant, which includes groups like the Marshall Islands and the Mariana Islands, are roughly equal in terms of their density of islands. The northeast quadrant is the least densely populated by islands, the Hawai'i group being the principal one there. It should also be noted that islands are generally more numerous in the Pacific in the tropics, within which ocean-surface temperatures are warm enough to allow coral reef to grow above a sinking volcanic island and continue to mark its spot long after it is completely submerged.

Some Definitions

Many geographers have caused themselves great problems by trying to define terms like islands and continents that probably originated as instinctive terms rather than ones intended to be clearly quantitatively definable.[3] After all, we all know what an island is when we see one, but do we always recognize a continent? One of the clearest expressions of this notion is that of Gillian Beer, who explained that "the concept 'island' implies a particular and intense relationship of land and water."[4]

The easiest kinds of islands to recognize are ones that we can clearly see to be surrounded by water. Most of us accept that there are other, larger islands that are also surrounded by water even though we cannot readily detect that. But then the distinction between islands and continents begins to blur, because Australia is a chunk of land surrounded by water, so should that not be called an island, just as the island of Hawai'i is? Well, no, say the encyclopedists and the people at Guinness World Records: Hawai'i is just an island, Greenland is the largest, and Australia is a continent. It is all arbitrary, unduly prescriptive, and certainly not instinctive. And as we shall see, the uses of the terms "island" and "continent" by many persons who write about the disappearance of islands and continents is even more arbitrary, often having more to do with making their arguments appear more compelling rather than any objective notion of relative size or geologic complexity.

Islands may be either continental — mere fragments of continent separated by ocean or the development of oceanic crust — or oceanic. It makes no sense to regard continental islands as geologically any different from their parent continents. Yet oceanic islands are quite different, the products of island-forming processes within the ocean basins. The rocks from which oceanic islands are made are different from those found in most parts of the continents, which are generally far older. So continents and continental islands can be distinguished from oceanic islands by their composition and their age.[5]

In this book, the terms "vanished" and "disappeared" are used interchangeably, generally in reference to islands. Some of the continents referred to are described as hidden to modern observers but never as lost, which implies an unintended degree of ignorance. Yet lost is an epithet that has been frequently applied to a variety of submerged, obscured, rumored, or fantastic lands and it is occasionally used in that sense in this book.

Geologic Time

Many people have only a slight acquaintance with the geologic time scale. These include many pseudoscience and new-age writers who often see catastrophe or at least uncommonly rapid change where a geologist would invoke no more than an average process regime to explain particular observations. Such problems typically arise when viewing the past from often great distances in time and failing to appreciate the long duration of the time periods within which particular changes took place. For such reasons, this section is devoted to a brief account of geological time as it is accepted by most scientists.

In at least two places on the earth's surface, rocks whose formation dates from around 4,000 million years ago have been found.[6] This provides a starting point for dividing the earth's subsequent history, which is the only part we can interrogate in detail, although it is clear that our planet must have become a solid body well before that point in time. Geologic time is divided into four eras (Figure 2.2). The earliest is the Precambrian, which ended 590 million years ago and was once thought to have been that period of time in which no life was present; it is now clear that organic life — simple organisms with low diversity — evolved during the Precambrian. Life exploded into complexity and diversity around the start of the Paleozoic era that ended around 250 million years ago. During the subsequent Mesozoic era, life continued to diversify, but much was brought to abrupt extinction at its end, about sixty-six million years ago, when an extraterrestrial body collided with the earth in Central America. Placental mammals, which include humans, first appeared during the most recent era, the Cenozoic, in which we are still living.[7]

The conventions used for age in this book are intentionally simple. Most ages acquired by scientific proxy methods, typically radiometric ones, are expressed as simply years ago but actually refer to calibrated years BP (Before Present) where Present is the year AD (Anno Domini) 1950. Recent radiometric ages have been converted to years BC (or BCE [Before the Common Era]) and AD (or CE [Common Era]), which are also used for directly recorded calendrical ages.

The Mobile Earth: Models and Ideas

As noted earlier, 150 years or so ago the prevailing thinking was that the earth's surface was in a fixed state. But since that time, our understanding of the earth's surface has changed radically, from one where everything is fixed to one where everything is dynamic. The first coherent expression of this school of mobile earth thinking was continental drift,

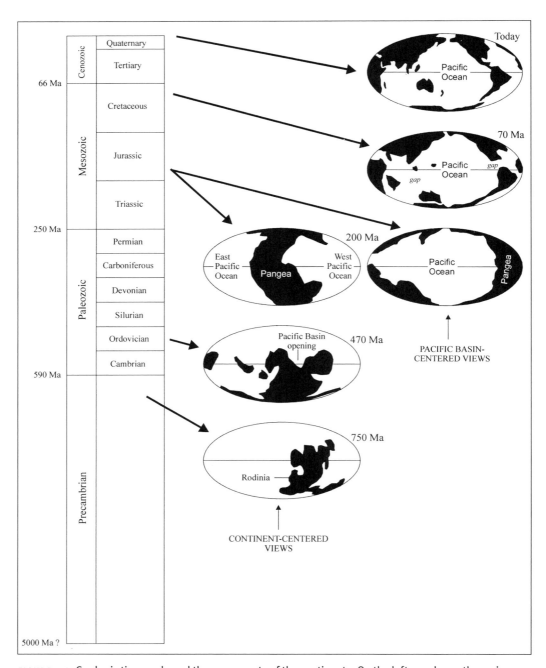

FIGURE 2.2. Geologic time scale and the movements of the continents. On the left are shown the main divisions of geological time (not to scale) with key dates (Ma means millions of years ago). The configuration of continents and oceans at various times is shown in the maps on the right. Note that two maps are shown for the time 200 Ma, one centered on Pangea, the other centered on the giant ancestor of the modern Pacific Ocean, also known as Panthalassa.

which evolved into seafloor spreading, and then into plate tectonics, the current ruling paradigm.

The idea of continental drift is hardly novel today, although half a century ago it was so radical that its adherents could be denied academic jobs for their beliefs. It was even more radical in 1915 when it was first suggested by Alfred Wegener as an alternative to the fixist orthodoxy of earth-surface evolution. From a comprehensive synthesis of the geometry (the fit) of continents and their geological, paleoclimatic and paleontological connections, Wegener argued that around 220 million years ago all the continents were amalgamated into a single supercontinent that he named Pangea. Surrounding Pangea was the massive forerunner of the modern Pacific Ocean named Panthalassa. Within the past 220 million years, Pangea has been breaking up, first into two large pieces, Laurasia in the north, Gondwana[8] in the south, separated by the equatorial Tethys Sea. The essence of Wegener's model of continental drift remains valid today.

The process by which Wegener envisaged the continents as moving came to be known as continental drift in English and involved fragments of continental crust moving independently of each other (and the ocean floor, about which hardly anything was known in Wegener's day) around the surface of the earth, cutting their way slowly like giant icebreakers through the lighter oceanic crust. This was one aspect of the continental drift model that has been proven incorrect — it is now known that it is the oceanic crust that moves, carrying the continents around like riders on horseback.

Chary of offending the establishment, doubting the all-embracing nature of Wegener's synthesis, and amplifying some of its minor errors, most scientists gave short shrift to the idea of continental drift when it was first announced. It took decades of patient scientific research, much of it focused on the Pacific Ocean floor, to demonstrate that the idea of continental drift was fundamentally correct. Much of the floor of the Pacific — like that of the other oceans — is more than 4 kilometers deep, and the technology and expertise to survey it systematically were not available until after World War II (1939–1945) when, fortuitously, the U.S. Navy among others found itself with trained personnel, the ships, and the time to undertake such surveys.

The results of these surveys produced a number of discoveries, particularly the observation that the age of the ocean floor increases uniformly as you move away from the volcanically young mid-ocean ridges, that supplanted the idea of continental drift and gave rise to a new theory known as seafloor spreading.[9] The key element of seafloor spreading was that new ocean floor is (and has been for millions of years) created along the mid-ocean ridges — the underwater mountain ranges that run for thousands of kilometers across the ocean floor — by eruption and intrusion of liquid rock (magma) from deep within the upper parts of the planet. The ocean floor on either side of the mid-ocean ridges has gradually been moved away from them as younger ocean floor (in the form of magma) wells up along their crestal parts.

It did not take scientists long to reconcile the facts of seafloor spreading with the ideas of Wegener regarding continental drift. But rather than have continents moving around the earth's surface by themselves, it was now possible to see them as riders on the moving ocean floor. In this way, it is clear that Pangea had indeed existed as Wegener had argued, and that the continents have a long history of motion. Indeed this history extended back in time beyond Pangea to times when other such supercontinents formed and then dispersed.[10] The earliest Pacific Ocean appeared at the time when the supercontinent named Rodinia aggregated, about 1,000 million years ago.

Oceanic islands were marginalized in much of the early thinking about continental drift and seafloor spreading, principally because so little was known about them compared with the continents. In the first half of the twentieth century, such islands were regarded by most earth scientists as generally unremarkable adjuncts to the continents, scarcely worthy of separate comment.[11] A shift in this view accompanied the development of the seafloor spreading model, because it became clear that islands, particularly those in the middle of oceans, held more information about earth-surface dynamics than many parts of the continents.

Continental drift and seafloor spreading were the first two mobilist models of earth-surface behavior. Their acceptance meant that none of the fixist models could be sustained. But there were still unanswered questions.

The key one was why, given that new oceanic crust is being created along a network of mid-ocean ridges worldwide, was the earth not expanding? There is actually a number of reputable scientists who have argued that this is precisely what the earth is doing,[12] but for the purposes of this discussion, it is convenient to go with majority modern opinion and assume that it is not, at least not simultaneously as a single body. If the earth is not expanding, then the new oceanic crust being created along mid-ocean ridges must be being destroyed in equal volumes elsewhere. And then in the late 1960s it became clear that the ocean trenches — the abnormally deep and little-known areas of the ocean floor — were the places where this destruction must indeed be happening. Most ocean trenches are found in the Pacific Ocean, adjoining and responsible for the Pacific Ring of Fire, the line of spectacularly active volcanoes that borders much of the Pacific Basin (Figure 2.3).[13]

With the realization that sections of the earth's crust were being simultaneously created and destroyed, a theory was developed to explain the process. The earth's crust, it appeared, could be divided into huge plates, each of which was effectively rigid and fitted together neatly with adjoining plates.[14] Thus when one plate moved, by virtue of being created and/or destroyed along one of its sides, so too did all the others. At last, there was a theory that was applicable globally and could successfully explain not only the past movements of continent and ocean floor but also, through the incidence of volcanic and seismic activity, the manifestations of their current movements. The theory was named plate tectonics.

It became realized that most of the volcanic activity in and around the Pacific Basin was a result of the process by which plates were destroyed along ocean trenches, a process named

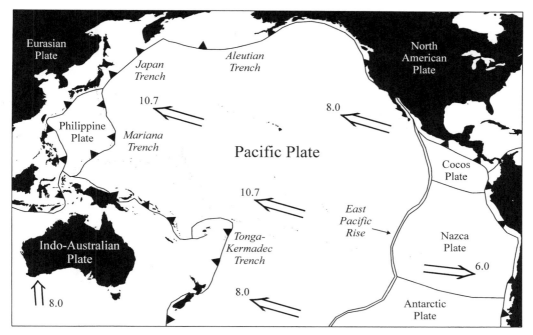

FIGURE 2.3. The Pacific Ring of Fire is the band of active volcanoes that runs along the borders of the Pacific Ocean. Its existence has often been used to justify the impression of the Pacific as a primeval place, the site of some great global cataclysm. This is incorrect. The Ring of Fire traces the places in the Pacific Basin where crustal plates are being destroyed along convergent plate boundaries. In this simplified map of plates in the Pacific Basin, the convergent plate boundaries (where one plate is being thrust under another) are shown as lines with filled triangles pointing in the direction of underthrusting. The arrows show the direction of plate (crustal) movement; numbers show the rates of this movement in centimeters per year.

subduction. Other active volcanoes, particularly those found underwater along the mid-ocean ridges, were linked to the creation of new ocean floor along mid-ocean ridges. The latter areas are generally not seismically active though, because earthquakes are associated with volcanism only where subduction is occurring.

In the Pacific, most of the volcanism and seismicity linked to subduction occurs along the Pacific Ring of Fire (see Figure 2.3). Some parts of the Ring of Fire are underwater, their existence signposted by active submarine volcanoes such as those in the Kermadec Islands and Tonga, north of New Zealand, or the Marianas-Izu island chain in the Northwest Pacific. In the western low-latitude Pacific, the Ring is actually a double line with the Marianas-Izu chain being shadowed by that running alongside the Philippines and the Ryukyu Islands farther west. North of this region, the Ring runs past Japan, a country where life is ever cognizant of the threats from volcanic eruptions and large-magnitude earthquakes, and thence through the Kuril Islands and into the Aleutian chain off the south coast of Alaska. Along the eastern borders of the Pacific, the best-known part of the Ring of Fire follows the coast of Central and South America, explaining the presence of some of the most spectacularly active volcanoes in the world.

If you picture a single rectangular crustal plate with one seafloor spreading boundary (where ocean-floor creation occurs) and opposite it one ocean-trench boundary (where subduction occurs), then the other two boundaries can be neither of these. They must be what are termed transform plate boundaries where one plate slides past another with no net convergence or divergence.[15] Such plate boundaries do not host volcanic activity — there are no conduits by which the magma can reach the ground surface — but they are seismically active because of the grinding of one giant plate past its neighbor. The best-studied transform plate boundary is the San Andreas Fault that runs along the western seaboard of the United States and along which have occurred some highly destructive earthquakes. The future threat from these to some of the world's largest population centers has made the San Andreas Fault the subject of intense study and monitoring.

There are a few places in the centers of plates that are volcanically active. These are where the earth's crust has thinned to a point where underlying magma has been able to punch its way through to the surface, leaking out and eventually building a volcano. Scientists term such a place a "hot spot." Where a hot spot has been in existence for a million years or more, often the combination of the lateral movement of the plate and the extrusion of magma at a fixed point gives rise to an almost-straight line of volcanoes, usually only one of which — that directly above the hot spot — is active. The best-studied example in the Pacific is the Hawai'i-Emperor island-seamount chain, which has been active for at least seventy-four million years. The largest active volcano in this chain, the so-called Big Island (Hawai'i Island), is coming to the end of its active life as it moves slowly but inexorably away from the unmoving hot spot to its southeast. Directly above the hot spot but with its head still about 1,000 meters underwater lies the active growing volcanic seamount named Lō'ihi.[16]

The plate-tectonics model can be employed to explain not only many modern natural hazards in the Pacific Basin but also much of its physical evolution. Some disingenuous new-age and pseudoscience writers capitalize on the uncertainties about plate tectonics and the other theories discussed by prudent scientists and pretend that there is such a thing as alternative geology. In this fictional paradigm, the Hawaiian Islands, for example, are not a line of hot-spot volcanoes but the mountain peaks of a sunken continent.[17] This, as explained in the next section, is impossible.

In the early part of the twentieth century and long before, a few scientists believed that the Pacific, because of its great size, was the scar left behind when the moon was ejected from the earth.[18] An associated idea was that the Pacific is the site of a hypothetical Pacific continent that subsequently disappeared.[19] Although today most geoscientists accept that large chunks of continental crust once existed in the Pacific, these never formed a single landmass that filled the Pacific Basin in the sense that many writers of supposedly "alternative" geology have averred.

A keystone of much new-age and pseudoscience writing about lost continents is that they sank in a gigantic cataclysm. The detail appears to be borrowed from Plato's account of fictional Atlantis. Yet continents cannot sink. They are made from lighter (less dense) rock than that composing the oceanic crust that surrounds them.[20] If something pushed (or pulled) a continent down into the surrounding oceanic crust, its tendency would be to rise again, just like a cork pushed down into a glass of water. Of course, such basic science hardly hinders writers bent on demonstrating otherwise,[21] but it does require elaboration near the start of a book such as this.

The History of the Pacific

The Pacific was named as such on November 28, 1520, by the Portuguese navigator Ferdinand Magellan in celebration of its passivity after an unrelentingly rough passage through the strait in southernmost South America that now bears his name.[22] Far earlier the Pacific was given the name Moana by some of its inhabitants. But it existed as a geographically definable entity long before, first becoming recognizable around 1,000 million years ago.

Around that time, most of the continental crust of the earth had aggregated into the supercontinent Rodinia. The residual ocean was not the ancestor of the Pacific — this appeared only as Rodinia began to break up, the earliest Pacific being marked by rifting between what is now the southwestern United States and eastern Antarctica.[23] The edge of the nascent Pacific Ocean is likely to have been a good place to live because, as Rodinia broke up, in perhaps a mere 180 million years, the accompanying increase in continental coastline produced an unparalleled diversification of climates. This led to an increase in erosion of the land areas and the accumulation of sediments offshore that led in turn to a rise in the proportion of oxygen in the earth's atmosphere from some 3 percent about 1,000 million years ago to around 20 percent some 800 million years ago.[24] The disaggregation of Rodinia created a huge diversity of environments for occupation by specially adapted life-forms, and the increase in atmospheric oxygen provided the important impetus for their evolution. The explosion of biodiversity during the early Paleozoic era has been attributed to these factors.[25]

Evidence that plate tectonics was active during the early history of the Pacific in much the same way as it is today comes from a variety of sources. Perhaps the most compelling is the accretion of terranes. These are fragments of anomalously thick crust — sometimes fragments of continents, sometimes island arcs formed parallel to ocean trenches by prolonged subduction — carried around the earth's surface on moving plates. Sometimes a terrane might be carried to a subduction zone where, unable to be subducted beneath the upper plate, it may become accreted onto its edge. One example is the Ontong Java Plateau, whose continuing pushing against the northern edge of the Solomons Arc in the western Pacific may explain the disappearance of a number of islands. Yet in the case of subduction that happened perhaps 500–600 million years ago in places where all trace of plate convergence vanished long ago, often the only way in which we can infer that subduction ever happened is by looking for the presence of terranes.

Along the eastern borders of the Pacific Basin, particularly along the western seaboard of North America between California and Alaska, there are countless exotic terranes.[26] These are fragments of Pacific Ocean floor, occasionally continental but more commonly of oceanic origin, that moved east across the ancient Pacific to collide with the North American craton (continental core). Collision led to accretion, and often to uplift, as the terrane continued to press against the craton. Some parts of the western United States are formed from ancient Pacific islands now elevated thousands of meters above their original position. These terranes are termed "exotic" because they are now found in places far from where they originated.

The fragments of what had once been Rodinia eventually aggregated once again, albeit in a different configuration, around 220 million years ago as the supercontinent Pangea. The residual ocean (Panthalassa) at that time was an ancient yet clearly identifiable ancestor of the modern Pacific. It was a giant ancestor too, more than twice as wide as the modern Pacific and extending from the North Pole to the South Pole. As Pangea broke up, continents moved into Panthalassa from all sides, progressively reducing the size of the ocean, a process that continues today.

The vertical movements that have periodically affected parts of the Pacific Rim during the past seventy to eighty million years are similar to those that can explain the disappearance of some of the landmasses that once existed within the Pacific Ocean itself. A good example is the land connection between North and South America that was established for only a few million years around the end of the Mesozoic era. It allowed a variety of animals, including hadrosaurs,[27] from North America to invade South America. The land bridge subsided soon after but eventually reemerged around three and one-half million years ago, cutting off ocean water exchange between the Pacific and the Atlantic thereafter, and allowing an interchange of biota between North and South America that devastated many long-established ecosystems.[28]

Dinosaurs were just one group of animals that rose in numbers, diversity, and habitat adaptability during the Mesozoic era. In the Pacific, we also see the numbers of teleost fish genera increasing dramatically from around 20 to 400,[29] principally as a result of the sea-level rise that occurred almost continuously throughout that era. Other groups of marine organisms also showed huge increases in diversity during the Mesozoic, and an important factor in their spread (a preliminary to diversification) across the then-larger Pacific is thought to have been the presence of smaller landmasses that served as stepping-stones across the ocean. Among these landmasses, most of which are now thought to have disappeared, are atolls and mid-ocean terranes. Where they have gone is no real mystery. Most of these atolls have been drowned, a result of the cooling of the ocean surface waters during the Cenozoic era (that followed the Mesozoic) that caused the region within which corals could grow (to maintain atoll reefs) to shrink. And most such terranes have become accreted onto the continental margins of the Pacific — there are far fewer terranes (per unit area) in the modern Pacific than there were in the Mesozoic Pacific.

But before we leave the subject of Mesozoic-era terranes, we must briefly consider what is arguably the most famous of these: Pacifica. This has been regarded as a continent-sized piece of Gondwana, the southern part of the disaggregating Pangea, that separated from the rest about 180 million years ago and drifted into the center of what is now the Pacific Ocean Basin.[30] Recent work contradicts this conclusion: the likely remnants of Pacifica that remain in the Pacific (the Ontong Java Plateau and the Manihiki Plateau) are more likely to be the remnants of Large Igneous Provinces than continental fragments.

Like everywhere else on the earth's surface, the nature of life-forms occupying the Pacific was profoundly and abruptly changed at the end of the Mesozoic era, about sixty-six million years ago, by the effects of a sizable meteorite crashing to earth at Chicxulub in Yucatán.[31] In terms of the focus of this book, this so-called K-T Boundary Event brought few changes, although it did render the dinosaurs extinct (along with uncountable others)[32], thereby clearing the way for the extraordinary rise of mammals, particularly placental mammals, during the succeeding Cenozoic era. The most successful group of placental mammals has been our own — the human family.

Although the outline of the geography of the Pacific Basin was laid down as Pangea began disaggregating around 220 million years ago, the foundations of the modern biogeography of the Pacific were laid down by the devastating K-T Boundary Event, the like of which has not been experienced since,[33] together with the cooling and sea-level fall that marked the earliest part of the Cenozoic.

Plate tectonics continued to exert great influence on the development of the Pacific Basin during the Cenozoic. As the Americas continued to move westward, a consequence of the still-continuing breakup of Pangea, numerous terranes were accreted on to their advancing edge, and several plate boundaries were swallowed up. The end result was that the Pacific Basin became less complex geologically than it had been for much of the Mesozoic, being dominated by the vast Pacific Plate.

The single most important event in the Cenozoic history of the Pacific, indeed of the entire planet, was the isolation of the continent of Antarctica some thirty to twenty-five million years ago in a location centered on the South Pole. Until that time, as part of Gondwana, Antarctica had been connected to both South America and Australasia, and its climate was kept warm (and ice-free) because warm air masses and warm ocean currents both had access to it. The subsidence of the Tasman Rise between Antarctica and Australasia began in the Cretaceous but accelerated about thirty million years ago, around the same time as Drake Passage, between Antarctica and South America, began opening. The combination of these two events led to establishment of the Circum-Antarctic Current that effectively insulated Antarctica from the influence of both warm air and warm water masses, leading to rapid changes in ocean circulation and ice buildup on Antarctica itself.[34] Subsequently, but particularly within the last ten million years or so, terrestrial ice sheets have shown themselves to be very sensitive to long-period fluctuations in solar radiation. Their regular waxing and waning helps explain the oscillations in sea level that have been experienced within the last few million years.

The most recent period within the Cenozoic is known as the Quaternary (last 1.8 million years) during which regular climate and sea-level changes have had well-understood effects on environments and their living occupants. In terms of islands that have vanished, sea-level rise at the end of the last ice age (glaciation) was a significant factor.

Profound changes have been caused to the oceanography and climate of the Pacific through time as a result of the alternate opening and closing of gaps along the Pacific Rim, particularly resulting from the subsidence of the Tasman Rise and the opening of Drake Passage discussed earlier. But there are other examples. These are instructive because they demonstrate the important role played by emergence and submergence in the history of the Pacific Basin and its inhabitants, human and other.

Consider that today there is an ocean gap in the northernmost Pacific — the Bering Strait — through which water moves between the Pacific and the Arctic Ocean. In winter, the cold Oyashio Current flows south through the Bering Strait injecting icy water into the northwestern part of the Pacific, especially the northern islands of Japan, where winters are extremely cold. But the Bering Strait has not always existed. Perhaps its most significant time of closure was when it dried up as a result of low sea level during the last ice age about 30,000 to 17,000 years ago, allowing humans to step on the soil of the Americas for probably the first time.

The gap (now closed) between North America and South America was open for most of the last few hundred million years. The main oceanographic effects of this were to allow the exchange of ocean water between what is now the Atlantic and the Pacific and, for much of the history of the Pacific Basin, the equatorial circulation of water around the earth. The fossil record of Pacific-Atlantic interchange through the gap between North America and South America shows its role in the exchange of ocean biotas.[35] Studies of the Hawaiian monk seal, for example, suggest that it originated in the Atlantic, passing through the gap between North and South America sometime before fourteen million years ago.[36] The most recent closure of the gap between North and South America was caused by the uplift of the Panama isthmus. Beginning around twelve million years ago with the uplift of a sill, closure took almost nine million years to complete.[37]

The final gap in the Pacific Rim is that (still narrowly open in a few places) between Asia and Australia. For most of the history of the Pacific it was much wider and much deeper. It began closing during the Mesozoic, becoming effectively closed (meaning that deep water could no longer pass along it between the Indian Ocean and the Pacific) by about 7.5 million years ago.[38] The cause of closure was the progressive northward movement of the Australian continent, a fragment of Gondwana, that simultaneously led to the opening of the gap (marked by the Tasman Rise) between Australia and Antarctica to the south.[39]

Some gaps in the Pacific Rim closed because of land uplift (and opened because of land subsidence), and others closed because of lateral plate movements. Over long time periods, vertical movements may not be as important in terms of earth-surface evolution as lateral movements, yet, in an account of islands and continents that have appeared and disappeared, vertical movements are of great importance. For this reason, they are discussed in more detail in the following sections. But first some definitions.

Uplift and subsidence are both movements of the land relative to a fixed point. The only fixed point available is the earth's center (more than 6,370 kilometers below sea level) so, in practice, uplift and subsidence are terms used relative to sea level. The danger of this is that over long periods of time, sea level has been going up and down; thus, what might appear to be uplift (or subsidence) might in fact be a result of sea-level fall (or sea-level rise). So when there is any doubt about whether the land has really gone up (rather than the apparent uplift being a result of sea-level fall), the term emergence is used. The equivalent term for relative subsidence (possibly sea-level rise) is submergence.

Some vertical movements are rapid and some slow. Often the rapid ones, both uplift and subsidence, are associated with earthquakes and are thus called coseismic. Vertical movements that are not associated with earthquakes are termed aseismic.

Uplift in the Pacific

Plate tectonics is an idea that at a global scale is concerned mainly with the lateral movements of the earth's crust. Major changes in the history of the Pacific can be explained by such long-term movements. Yet vertical movements are also important, at least locally, in the model of plate tectonics. When continents collide, for example, brought together by subduction, huge ranges of fold mountains may develop. The greatest range of fold mountains in the world, the Himalayas, is believed to have formed in precisely this manner, when the plate carrying the continent of India collided with the Asian continent some ninety million years ago. The Himalayas continue to rise today, because India is still being pushed (largely in vain) down the trench that once existed along its northern borders.

A smaller-scale version of the same situation is to be found in the western Pacific. A thickened piece of ocean floor on the Pacific Plate, the Ontong Java Plateau, was carried southwest toward the trench along the northern side of the Solomons Arc where the plate was being subducted. About eleven million years ago when the Ontong Java Plateau reached the trench, it proved too thick be readily subducted, so the strain was eventually accommodated by a shift in subduction — a subduction polarity reversal. This was manifested by the formation of a new trench along the southern side of the Solomons Arc. Indications that the Ontong Java Plateau is still being forced against the Solomons Arc come from the extraordinary evidence for recent and continuing uplift along the north coasts of some of the islands in the Solomons Group: on Choiseul, for example, fossil coral reefs are found as much as 800 meters above the modern reef, testimony (as with the Himalayas) to the vertical expression of horizontal stress.

The present-day composition of many islands in the Pacific attests to their long history of uplift. Much of the elongate continental island of New Caledonia (actually La Grande Terre) in the Southwest Pacific is composed of an assemblage of rocks known as ophiolites that represent part of the deep-ocean floor thrust up onto the land.[40] Ophiolites are generally rich in metallic minerals like nickel (mined extensively on New Caledonia) and produce soils that are highly toxic to most plants, something that explains the scrubby vegetation in much of the island (Figure 2.4A).

In other cases, it is the appearance of a particular coastline that shows it has probably been uplifted. The characteristic stepped appearance of island coastal profiles of high limestone islands often shows that these parts are a result of continuous (or repeated bursts of) uplift that has caused the living coral reefs that fringe the island's coast to be uplifted.[41] A similar landscape is found along those parts of the continental rim of the Pacific that have been uplifted (Figure 2.4B).

Most parts of the Pacific Basin that have been uplifted are close to places where two plates are converging. In such places, the edge of the upper plate is often being thrust upward as it attempts to move across the lower plate that is being thrust down and subducted. All these processes happen along most convergent plate boundaries in the western Pacific and typically produce on the upper (overriding) plate a line of nonvolcanic (commonly reef-limestone) islands running parallel to the axis of the associated ocean trench; good examples are provided by the Mariana Islands between Saipan and Guam.

Sometimes the surface of the lower (downthrust) plate is irregular — perhaps it has a ridge running along it — and when such significant irregularities are subducted, those parts of the upper plate immediately above are also forced upward at a faster rate than nearby parts. For example, the oblique subduction of the Louisville Ridge along the Tonga-Kermadec Trench in the South Pacific led to the anomalous uplift of the central part of the nonvolcanic island arc, which includes Tongatapu Island. A similar situation is happening in Vanuatu where the attempted subduction of the D'Entrecasteaux Ridge is causing the conspicuous uplift of Malakula and Espiritu Santo islands.

All these examples come from the overriding plate at convergent plate boundaries, but the lower (downthrust) plate also flexes upward before it is able to go down beneath the upper plate, causing any seamounts or islands that pass across the flexure (bulge) to be uplifted. The Loyalty Islands of New Caledonia provide good examples, as does isolated Niue Island in the central South Pacific. Around 600,000 years ago, Niue was an atoll, barely rising above the ocean surface. Then, as it was carried westward on the moving Pacific Plate toward the Tonga-Kermadec Trench, it began ascending the eastern side of the associated bulge. It has now been uplifted about 70 meters and is about halfway up the bulge. After the island has passed over the crest, it will begin to subside, eventually being drowned and perhaps ending up wedged against the western wall of the Tonga-Kermadec Trench just as Capricorn Seamount is today (Figure 2.4c).

Islands may also experience uplift in the middle of plates, far from plate boundaries, but such uplift is usually slower and less enduring (and therefore of lesser magnitude) than that which occurs near plate boundaries. Most such intraplate uplift occurs when an island is carried across a crustal swell. A good example comes from the Tuamotu Islands chain in French Polynesia that straddles a conspicuous swell; some islands are rising up it, others lie atop it and exhibit signs of slight emergence, and others have moved away down its western flanks.[42]

Uplift also occurs as a result of more localized processes. For example, when the massive volcano that now forms the island of Rarotonga in the southern Cook Islands began forming around 1.1 million years ago,[43] the ocean floor around is thought to have been undeformed — flat like most ocean floor is today. But as Rarotonga grew larger and higher, its weight caused the ocean floor on which it rested to bow down, creating a moat around the island itself and an arch (or crustal bulge) slightly farther away. At the time Rarotonga began forming, a number of seamounts surrounded the island. As it grew and the surrounding crust bulged, those seamounts that found themselves on the arch were uplifted. Today, although Rarotonga has ceased growing (its volcano is extinct) and is in fact becoming smaller as denudation takes its toll, it remains surrounded by a corona of islands, including Mangaia, Mauke, and Atiu, that show signs of significant amounts of uplift. We might expect, as Rarotonga becomes smaller and more lightweight, and therefore less of a burden on the underlying ocean floor, that the bulge will slowly collapse, leading to the subsidence of the islands thereon.[44]

Much uplift (and most subsidence) is a slow process, yet there is some debate about how it happens and how fast it happens. Most accounts of how uplift happens are generalized, and rates of uplift are invariably given as long-term averages. Most aseismic uplift appears to operate at variable rates in both space and time, to accommodate stresses created by plate movements. Rates of aseismic uplift of islands in the Tuamotus average 0.76 millimeters/year. In contrast, parts of Taiwan, an island overlying an active convergent plate boundary, have risen aseismically at rates of around 6 millimeters/year during the last few thousand years.[45]

Coseismic uplift occurs typically along the coasts of nonvolcanic islands close to ocean trenches. Shallow earthquakes that have their foci along the subsurface boundary between the plates often produce rapid vertical uplift (sometimes subsidence — see following section) on these islands. Numerous examples have occurred within the past 100 years. The February 3, 1931, Hawke's Bay earthquake in New Zealand caused some 13 square kilometers of the

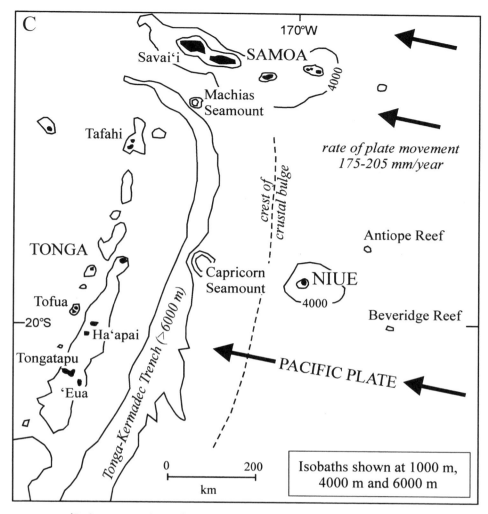

C

170°W

Savai'i   SAMOA

Machias
Seamount

4000

Tafahi

rate of plate movement
175-205 mm/year

crest of
crustal bulge

Antiope Reef

TONGA

Capricorn
Seamount

NIUE

4000

Tofua

Beveridge Reef

−20°S

Ha'apai

Tonga-Kermadec Trench (>6000 m)

PACIFIC PLATE

Tongatapu

'Eua

0   200

km

Isobaths shown at 1000 m,
4000 m and 6000 m

FIGURE 2.4. (*Facing page and above*) Evidence of uplift in the Pacific Basin. **A.** The scrub vegetation that covers much of La Grande Terre (New Caledonia) is a product of the weathering of rocks known as peridotites, part of a slice of ancient ocean floor (an ophiolite) that has been thrust up onto the land. **B.** The stepped coast of southern Erromango Island, Vanuatu, is an example of an uplifted coastline. In this case, the steps in the landscape are the product of the successive uplift of fringing coral reefs. (Photo by Shane Cronin, used with permission.) **C.** Map of part of the Southwest Pacific showing plate boundaries and movements in the Niue-Samoa-Tonga area. The island of Niue is being carried westward on the Pacific Plate at a rate of some 175–205 millimeters/year and, for the past 600,000 years has been carried up the bulge in the ocean floor that is associated with the downthrusting of this plate beneath that to the west along the Tonga-Kermadec Trench. Note that Niue has almost reached the crest of the bulge, Capricorn Seamount has crossed it and has gone down the other side, and the seamounts represented by Antiope and Beveridge reefs are just beginning their ascent. (Rate of plate movement from Bevis et al. [1995]; details of the geological history of Niue from Nunn and Britton [2004].)

Ahuriri area, comprising a village and harbor, to be rapidly uplifted by 2.4 meters. In parts of Vanuatu and Solomon Islands, coseismic uplift occurs regularly; a 1965 event on Malakula Island in Vanuatu involved 1.2 meters of abrupt uplift.[46]

The calculation of long-term uplift rates is complicated by the evidence for variability in uplift rates in space and time. Along a 60-kilometer stretch of the coast of the Huon Peninsula on the island of New Guinea, uplift rates vary from 1 to 4 millimeters/year. In southern Taiwan, the Honchun Peninsula and Tainan Tableland are rising at 6.1 millimeters/year, yet the adjoining Tainan lowland is subsiding at 1 millimeter/year. For both the Huon Peninsula and southern Taiwan, it has been suggested that uplift rates have slowed during the last 10,000 years.[47] Some areas exhibit uplift that is occasionally aseismic, occasionally coseismic. Sometimes bursts of uplift alternate with subsidence. This explains the Quaternary history of certain islands in the Lau Group of eastern Fiji.[48]

Subsidence in the Pacific

Most Pacific islands are sinking (subsiding). It is the usual fate of islands formed in oceanic settings that, when they come to the end of their active (volcanic) life, they cool down and are moved into deeper ocean waters. This involves subsidence. Yet because sea level is not constant it is not always straightforward to determine whether an island is subsiding. Basically, when an island shows signs of long-term submergence, and this is clearly not a result of sea-level rise, then it can be regarded as subsiding.

It was the nineteenth-century American geologist James Dwight Dana who first pondered the evidence for coastal subsidence (actually submergence), concluding that shoreline embayment was the main clue.[49] Today the idea seems self-evident. If, for example, a volcanic island having many river valleys subsides, then the mouths of those valleys will be drowned and the island coastline will become embayed (Figure 2.5).

One hundred and twenty years ago, when Dana and his group of natural scientists were becoming increasingly intrigued by earth movements, almost the only evidence for subsidence that they could ponder was above water. The floor of the oceans, which cover 73 percent of the earth's surface, was largely a closed book to them. But increasingly during the subsequent period, the contents of that book became revealed, and today most of the evidence that we have for the subsidence of landmasses comes from beneath the ocean surface. Such evidence comes from ancient coasts that have been displaced downward, far lower than the level at which they could possibly have formed. Excellent examples are known from Hawai'i, where numerous drowned coral reefs calibrate the long history of island subsidence.[50]

In the warmer parts of the world's oceans, it is coral reefs that also provide, by growing upward from subsiding volcanic foundations, some of the clearest evidence for long-term island subsidence. It was Charles Darwin who, in 1835 during the voyage of HMS *Beagle,* first outlined a connection between island subsidence and coral-reef upgrowth. On Febru-

FIGURE 2.5. Embayments along the south coast of Viti Levu Island, Fiji. Such embayments were once thought to be indicative of subsidence, formed when the mouths of rivers were drowned as islands sank. Today it is more prudent to refer to such coasts as submergent, given that they might be stable, affected only by sea-level rise. Note also the fringing reefs that exist around headlands along this coast. These reefs cannot usually grow up to the ocean surface within the embayments because of the freshwater from the rivers that empty into them.

ary 20 that year, Darwin experienced an earthquake in Valdivia (Chile) and later traveled to Concepción, where the associated devastation made a lasting impression. Here he also noticed beds of freshly killed mussels above high tide, evidence of nearly 3 meters of coseismic uplift. Later, while traveling up the Andes, Darwin found fossil seashells in the rocks, clear evidence of a long history of uplift of the Pacific coast of South America. From this deduction it was only a short step to supposing that, if uplift of such magnitude (along continental margins) could occur, so too might subsidence (within ocean basins) occur. And thus, before he ever saw an atoll, Darwin formulated what became his Subsidence Theory of Atoll Formation.[51]

Today Darwin's idea also seems deceptively simple. Yet in the nineteenth century, when Darwin and Dana were working, the community of scientists was small; the majority believed that their thinking should be irrevocably tied to the teachings of Christianity. These teachings sat most comfortably with fixism so, as a consequence, ideas involving the downward movement of islands were regarded as somewhat suspect.

To understand the Subsidence Theory of Atoll Formation, it is necessary to understand that a coral reef that appears to fringe an island (such as that in Figure 2.5) rises vertically from the underwater flanks of that island. Most of such reefs are dead; only the uppermost parts comprise living corals and other organisms. The living part of a coral reef lies mostly within the photic zone, that zone of the uppermost part of the ocean into which sufficient sunlight penetrates for the photosynthetic algae living within the corals to produce food. This food feeds the corals, thereby powering the whole coral-reef ecosystem. When the foundations of an island begin to subside (or when sea level begins to rise), the veneer of living coral within the photic zone suddenly finds itself being moved downward out of the photic zone. Darkness and death threaten, so corals respond by growing upward, building on the skeletal remains of dead corals (and associated material). Often at such times, branching (staghorn) corals will replace those species less well adapted to growing upward.[52]

The starting point of the Subsidence Theory of Atoll Formation involves a volcanic island with a fringing coral reef (Figure 2.6A). As the island sinks, its coast becomes embayed (Figure 2.6B). The coral reef responds to the subsidence of its foundations, and the simultaneous submergence of the living coral-reef surface, by building its surface upward so that it remains within the photic zone. In this way it stays alive even though the island from which it grew is subsiding. In the process of building vertically upward on sinking foundations, the fringing reef becomes gradually transformed into a barrier reef, separated from the island's coast by a lagoon. In many Pacific island contexts, this stage is represented by the development of an almost-atoll, in which a large barrier reef encloses a few, comparatively small islands that are subsiding; a good example is the Chuuk (Truk) Group in Micronesia.

The final stage of the Subsidence Theory of Atoll Formation (Figure 2.6C) involves the disappearance beneath the ocean surface of the landmasses from the flanks of which the foundations of the surrounding coral reefs rise. These reefs will continue to grow upward until they become the only surface expression of a sunken volcanic island. Such reefs are called a ring reef or an atoll reef. Islands of reef rubble or detritus that form on these reefs are commonly called *motu* (atoll islands).[53]

Not all islands that sink beneath the ocean surface have their positions forever marked by atolls. When islands outside the warm coral seas subside, they disappear without above-sea trace. But as their uppermost parts sink slowly below sea level, they are flattened by marine erosion to create a distinctive type of drowned island called a guyot. Guyots are found in all parts of the world's oceans and are distinguished from other undersea island-type edifices by their flat tops; islands that have never poked their heads above the ocean surface typically have a more conical form and are called seamounts.[54]

It is also worth mentioning at this point that sometimes even within the coral seas, islands do sink without trace. Much clearly depends on the rate at which the subsidence occurs; rapid subsidence, perhaps coseismic, may drop an island coast below the photic zone in a matter of seconds. A good example comes from studies of the Kohala Volcano in the

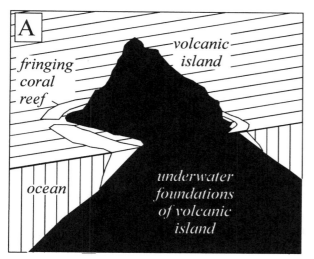

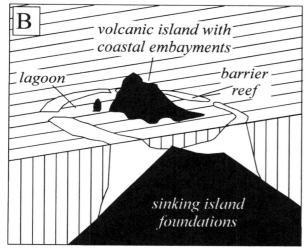

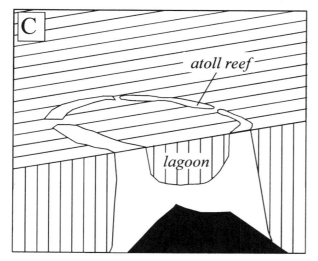

FIGURE 2.6. The Subsidence Theory of Atoll Formation. **A.** The process of atoll formation begins when a volcanic island in the coral seas, surrounded by a fringing reef growing upward from shallow foundations, begins to sink. **B.** Subsidence begins to drown the volcanic island, causing its shoreline to become embayed or scalloped. To stay alive, the coral reef will build itself upward, which results in a fringing reef being transformed into a barrier reef, separated from the island's shoreline by a lagoon. **C.** The process of atoll formation is complete when the volcanic island sinks completely below sea level. The coral reef has continued to build itself upward so that a ring reef (or atoll reef) is formed, enclosing a lagoon.

northwestern part of Hawai'i Island. Coral reefs grew around the edge of Kohala on at least two occasions during its history but are now dead and submerged well below the ocean surface in the area: a result of the combination of recent sea-level rise and comparatively rapid subsidence of the volcano.[55] Rapid sea-level rise may have the same effect. Indeed, within the Pacific, there are areas of drowned atolls that are thought to have been submerged without trace by rapid bursts of postglacial sea-level rise.[56]

Subsidence is endemic within the ocean basins, particularly in areas far from plate boundaries. Yet subsidence also occurs along those plate boundaries where plates either diverge (mid-ocean ridges) or where they converge (ocean trenches).

Generally few islands are associated with mid-ocean ridges, largely because most of these are deep underwater and have steep flanks, down which broken-off parts of the ridge move. Notwithstanding such constraints on island formation at those locations, some islands that formed along mid-ocean ridges have subsequently subsided; good examples are those in the Southeast Pacific on which unique biotas, now found only on Easter Island, were preserved.

Convergent plate boundaries are generally more complex places in terms of uplift and subsidence, but examples of the latter process are also found there. Many of the examples in chapter 9 of islands that have probably vanished in the Pacific within the period of the region's human occupation appear to have done so because of subsidence along convergent plate boundaries.

Finally, when we think of the places in the ocean basins where subsidence occurs, we should acknowledge the existence of localized subsidence. This might be because of areally restricted crustal loading, as in the case of Rarotonga discussed earlier, or because of localized landsliding resulting from oversteepened slopes. Good examples of the latter come from the Hawaiian Islands, which rise from the Hawaiian Ridge, one of the most steep-sided natural structures on earth, where island-flank landslides are common.[57]

Parts of the continental rim of the Pacific are also subject to subsidence, for some of the same reasons that it occurs in the ocean basin. One exception is when subsidence occurs around the mouths of large rivers where the huge bodies of sediment they deposit around their mouths weigh down the underlying crust in much the same way as large islands like Rarotonga can do. The Pacific Rim is cut by extraordinarily few large rivers and probably only those in China, the Changjiang (Yangtze) and the Huanghe (Yellow) rivers, qualify as among those in the world that cause subsidence of the continental rim. Part of the ground subsidence that is experienced in Chinese cities such as Shanghai is attributable to delta subsidence resulting from alluvial sediment loading.[58]

One of the most dramatic forms of subsidence is that associated with large-magnitude earthquakes: coseismic subsidence. This is less common than coseismic uplift but is still significant, occurring primarily at convergent plate boundaries. A representative example occurred during the 1964 Prince William Sound earthquake in the northernmost Pacific, where coseismic uplift and subsidence occurred simultaneously; parts of Kodiak Island subsided 2.3 meters within a few seconds.[59]

Sea-Level Changes in the Pacific

Before we leave the whole issue of the geoscientific framework within which island uplift and subsidence occurs, it is necessary to acknowledge that islands can emerge or be submerged without any movement of the land itself. This process involves sea-level changes; rising sea level causes drowning and falling sea level causes emergence of a coastline.[60]

When contemplating the long-term history of the Pacific Basin, we can link long-term changes in biotic evolution and even supercontinent breakup to sea-level changes.[61] Long-term sea-level rise during the later half of the Mesozoic, for example, led to increasing environmental diversity and an associated increase in biodiversity.

Most swings of sea level over the past few million years have occurred in approximate synchrony with swings in temperature. These swings have been approximately regular within this time period, a result of regular changes in the earth's orbit. Thus the earth's climate has alternated recently between times of comparative coolness (and dryness in the tropics) termed "ice ages" or "glacials," and times of intervening warmness (and wetness in the tropics) termed "interglacials," a process that began only after Antarctica became covered with ice. The slow prolonged cooling at the start of an ice age was associated with the growth of ice sheets on the land, a process that was fed by the evaporation of ocean water leading to sea-level fall. Conversely, at the end of an ice age, land ice began melting in response to rapid warming. Sea level rose rapidly in response.

Toward the end of the last ice age, about 15,000 years ago, global sea level began to rise. This period of postglacial sea-level change accounts for many of the shorter-term disappearances and appearances of islands. The comparison of maps of the Southwest Pacific today and at the maximum (coldest time) of the last ice age (approximately 18,000 years ago), when sea level was around 120 meters lower than today, illustrates the point (Figure 2.7). Many islands were drowned subsequently as sea level rose. The reverse process happens when sea level falls at the beginning of an ice age.

Despite being interrupted by short-lived intervals of sea-level fall, sea-level rise continued in the Pacific until about 4,000 years ago. At that time, sea level reached a level about 1.5–2.1 meters above its current level.[62] Although minor compared with the massive swings of sea level during the earlier Quaternary period, the ensuing sea-level fall proved critically important in creating environments that attracted people to various parts of the region. For example, many Pacific Islands, particularly in the central eastern Pacific, became inhabitable only when their coastal plains emerged sufficiently for them not to be inundated at high tide,

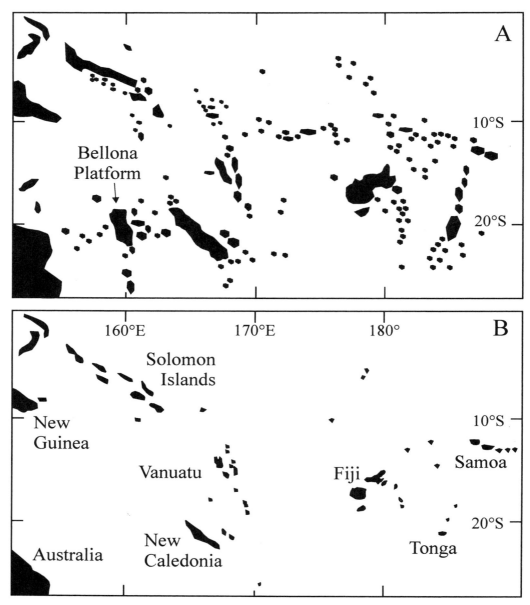

FIGURE 2.7. The geography of the Southwest Pacific. **A.** About 20,000 years ago during the lowest sea level (-120 meters) of the last ice age (Last Glacial Maximum), much more land was exposed in this region. Note particularly the large island between New Caledonia and Australia, marked today by the Bellona Platform from which a few isolated reefs rise. **B.** The modern geography of the region for comparison.

something called the crossover point. The attainment of the crossover point is coincident with the earliest-known date for human settlement of many such islands.[63]

Many atoll reefs became habitable only within the past 2,000 years or so as sea level fell. In most such places, it appears that coral reefs were growing at the ocean surface (low-tide level) when the sea level was higher than at present about 4,000 years ago. When sea level began falling subsequently, the tops of these reefs emerged, becoming loci for sediment accumulation. In this way the islands (*motu*) that exist today on these reefs were formed.[64] Such islands are transient by nature; they formed as a result of recent sea-level fall and may disappear in the future as a result of sea-level rise.

Uncertainty: The Achilles Heel of Modern Global Geoscience

Theories (or models) that seek to explain historical processes of earth-surface evolution, particularly globally or even at the scale of the Pacific Basin, can never be verified in the way that other scientific theories (particularly in the physical sciences) can. The reasons are obvious. They deal with things that happened in the past but that are not happening now and will never happen again in precisely the same way; thus they cannot be predicted, only retrodicted. They deal with things that happened very slowly across vast areas, and un-imaginable quantities of data would be needed to try to verify them. So models or theories like plate tectonics are destined to forever remain as theories — coherent collections of ideas that appear to explain most relevant observations.

Observation and interpretation are carried out largely by humans or by human-guided machines. And, as noted earlier in this chapter, the experiences of some humans have convinced them that plate tectonics is wrong and that the earth is actually expanding. Thus there is a degree of uncertainty, and this provides grounds for many nonspecialists to be skeptical of entire issues: for example, to doubt the theory of plate tectonics or the existence of earth mobility per se. But this type of skepticism often misses the point; it does not acknowledge the function of scientific uncertainty. Science progresses when erroneous ideas are recognized as such, and attention then turns to alternative ideas that are not so erroneous. No sane earth scientist today doubts the evidence for earth-surface dynamism that is manifested by contemporary seismicity and volcanism, and that is largely responsible for the modern configuration of continents and oceans.

Earth scientists interested in long-term earth evolution can rarely verify the big ideas. Rather they evaluate probabilities, seeing which explanation best accounts for most observations. The process is littered with discarded explanations and is rife with uncertainty, but it has still moved a long way forward since Wegener's mold-breaking ideas about continental drift were first published in 1915.

The scientific method is sometimes its own worst enemy. In its efforts to be rigorous, to implement all necessary cross-checks, duplicate results, and agree on interpretations, it exposes a soft underbelly into which persons with nonscientific agendas can sometimes make some headway by prodding. Such pseudoscience writers often attract large audiences of nonspecialists who may delight in learning that the mysterious and supposedly infallible beast called science is apparently otherwise. But this is incorrect. For all its failures, all its well-publicized false starts and dead ends, science provides the only plausible explanations of our dynamic planet's history, its current condition, and its future.

3 Islands That Vanished Long Ago

The periodic disappearance of islands and continents is part of the natural evolution of the earth's surface. It happened before people were present to witness it and, as will be seen later in the book, it happened after they arrived. There is no need to invest the process with undue importance simply because it was witnessed by the first species that would come to recognize it for what it was, then record it and ponder its meaning.

Questions concerning the presence or absence of large continental landmasses in the ocean basins at particular times in the past are understandably easier to prove or refute than questions concerning smaller landmasses, like islands. There are innumerable islands whose highest parts currently lie beneath the ocean surface but that rose above it in the past. The change may have occurred because the islands either subsided or were drowned by sea-level rise (or both). Ascertaining the time(s) when these sunken islands were emergent is often problematical, yet critical, for example, to the reconstruction of transocean biogeographic tracks and for linking the existence of such islands to oral traditions concerning their disappearance.

This chapter gives an account of those islands in the Pacific that were formerly emergent but are now submerged. It focuses exclusively on those islands that were submerged before humans settled the region in large numbers. The purpose of this exercise is to show that island submergence, for a variety of reasons, has been a relatively common phenomenon during the history of the Pacific. This supports the contention that incidences of island disappearance recalled by humans through oral traditions (discussed in chapters 6 and 9) are not necessarily fanciful or even wildly exaggerated versions of something that actually happened.

At the outset, we need to distinguish islands from continents. Unfortunately, this has never really been done successfully although continents are generally considered to be larger, older, and geologically more complex than islands. As a working definition, this has to suffice.[1]

Islands go up and down for many reasons, and sea-level changes also cause some islands to become either emergent or submergent. The purpose of this chapter is to show, using a number of well-documented examples, how such processes led particular islands to vanish.

Islands That Have Disappeared into Ocean Trenches

Ocean trenches, those asymmetrical steep-sided chasms etched into the ocean floor, are often portrayed as smooth-sided: places where one plate slides effortlessly beneath the other, part of the great engine that is plate tectonics. The reality is very different. Oceanic crust is everywhere marred by irregularities — islands, ridges, swells — that are carried into ocean trenches and distort or sometimes disrupt the process of subduction.

Some of the best examples in the Pacific come from lines of islands that have been carried along the moving ocean floor, pulled down into ocean trenches, and finally thrust underneath the overlying plates. The Louisville Ridge is an underwater ridge in the South Pacific with several high spots that were once probably islands. It collided obliquely with the Tonga-Kermadec Trench around six million years ago and has since been dragged down beneath the Tonga Island Arc (see Figure 2.4C). The section of the Tonga Island Arc beneath which the northern part of the Louisville Ridge has disappeared is noticeably wider than what is regarded as normal for this island arc. Specifically in this section there is a 60-kilometer gap between the nonvolcanic arc, which includes islands like Tongatapu and those in the Ha'apai Group, and the active volcanic arc, including islands like Tofua, that is attributed to the subduction of the Louisville Ridge.[2]

Farther west, plate convergence is occurring along the Vanuatu (New Hebrides) Trench west of the islands of Vanuatu (shown in Figure 9.5A). In the central part of this trench, the D'Entrecasteaux Ridge is being dragged down into the trench and subducted in much the same way as the Louisville Ridge. But something quite different is happening in Vanuatu. The D'Entrecasteaux Ridge is not slipping down beneath the overlying plate quite as readily as we imagine the Louisville Ridge to have done. Perhaps because the D'Entrecasteaux Ridge is thicker, perhaps because the slope down into the trench is steeper, it is clear that the D'Entrecasteaux Ridge is bulldozing into the central Vanuatu Arc, pushing it eastward at around 50 millimeters/year. In the process, the islands immediately east of the place where the D'Entrecasteaux Ridge meets the arc (Malakula and Espiritu Santo islands) are being deformed and fragmented.[3]

It is easier to envisage how individual islands can vanish along convergent plate boundaries, and there is no shortage of examples along the Tonga-Kermadec Trench (shown in Figure 2.4C). One of the most impressive is the 5,000-meter-tall Capricorn Seamount, a circular guyot at least 450 meters below sea level that began moving down into the 9,000-meter-deep Tonga-Kermadec Trench within the past million years, tilting markedly trenchward in the process. Capricorn Seamount is poised at the top of the outer wall of the Tonga-Kermadec Trench, but, because it is so large (it has a basal diameter of 100 kilometers), its lower western flank is already at the bottom of the trench being subducted. If convergence continues here at the same rate as it is today, the summit of Capricorn Seamount will reach the bottom of the trench in about 500,000 years. Data from seismic profiling and analysis of samples dredged from the surface of Capricorn Seamount suggest that it was once a volcanic island, the summit of which, like all proper guyots, was planed flat by wave erosion as it sank slowly beneath the waves. Continued sinking was followed by the development of a coral capping that undoubtedly took the form of an atoll. This in turn was followed by more rapid subsidence that drowned the atoll and carried it downward toward its date with destiny at the bottom of the Tonga-Kermadec Trench.[4]

The other sizable sunken island along this (northern) part of the Tonga-Kermadec Trench is Machias Seamount, located 140 kilometers south of Savai'i Island in Samoa, near the unusual hook at the northern end of the trench (see Figure 2.4C). The hook is likely to be a place where the Pacific Plate, which is moving westward into and past the trench, is being ripped apart, something that Machias Seamount shows well. Machias is a classic guyot, with a flat, planed top some 700 meters below the ocean surface, perched precariously on the upper part of the slope leading down to the bottom of the trench some 7,000 meters below. The Pacific Plate on which Machias was carried to its present position is moving approximately westward, yet the guyot is now being pulled to the southwest into the trench. These dual forces have caused the original form of Machias Seamount to become elongated along a northwest-southeast axis, causing the breakup — what has been called "tectonic dismemberment" — of the guyot structure.[5]

Subsided Islands in the Middle of Plates

Like most other processes that affect the earth's crust, including its islands, in the middle of plates, subsidence in such places is commonly slower than along plate boundaries. Some atolls in the tropical Pacific have been subsiding at apparently the same rate for very long periods of time; Enewetak in the Marshall Islands, for example, has been subsiding at a rate of 0.03 millimeters/year for the past forty-five million years.[6] But Enewetak is an atoll, still very much a part of the landscape of the Northwest Pacific, so it cannot be strictly regarded as a sunken island.

Most sunken islands in midplate locations in the Pacific disappeared either because they sank beneath the ocean surface in areas outside the coral seas (and therefore could not develop into an atoll) or because, having developed into an atoll, they were moved out of the coral seas. For all their resilience, shown by their astonishing persistence in the fossil record,[7] the corals that are essential to building coral reefs have become very fussy organisms, basically unable to survive more than a few weeks if ocean-surface temperatures fall below 20°C or exceed 30°C. So when an atoll is carried slowly across the 20°C isotherm into cooler water, corals begin dying and, as it continues to sink, so too will its coral cap.

This situation is well illustrated by the long Hawaii-Emperor island-seamount chain, a classic hot-spot island chain. The modern hot spot is today located below the underwater volcano named Lō'ihi at the southeastern end of the chain.[8] Moving northwest, the island in the chain farthest away from Lō'ihi is Meiji Seamount, 5,860 kilometers distant. Although today Meiji is teetering on the brink of the Kuril Trench, soon to be consumed beneath the margin of the North American Plate, approximately eighty-five million years ago Meiji was over the hot spot (where Lō'ihi is today).[9] The Meiji Volcano probably sank below sea level when it still lay within the coral seas, so it would be expected to have developed a capping of coral reef that would itself have been submerged as the island later moved outside the coral seas. Since that time its surface has accumulated a great thickness of deep-sea ooze, chalk, and clay, coring of which has allowed much to be understood about the changes through time of the chemical character of the ocean within which Meiji has existed.[10]

Yet many islands have sunk in the center of plates without forming atolls, at least not atolls that exist today. In 1956, the Geological Society of America published a book by Edwin Lee Hamilton called *Sunken Islands of the Mid-Pacific Mountains,* an account of what we now know to be hundreds of submerged reef-capped guyots in the central Pacific between Hawai'i and the Mariana Islands. These guyots were all emergent during the Cretaceous period, about 100 million years ago, and Hamilton commissioned a painting by Chesley Bonestell to show exactly how the mountains would have appeared.[11] The reason for the submerged nature of the Mid-Pacific Mountains has to do with their great age; they became submergent at a time when regular sea-level changes were not occurring, so subsidence eventually carried them well below the photic zone. Today the submerged guyots are covered with deep-water sediments.[12]

Islands That Drowned When Sea Level Rose

During the last few million years, the most widespread cause of island disappearance in the world's oceans occurred at the ends of ice ages when the surface of the ocean rose rapidly as water from melted land ice poured into it. Outside the coral seas, as shown in Figure 2.7, many islands were drowned by the most recent period of postglacial sea-level rise, between about 15,000 and 4,000 years ago when sea level rose some 120 meters. But within the coral seas, islands at least had a fighting chance of remaining visible at the ocean surface.

In the tropical Pacific, some coral reefs were evidently able to grow upward at the same rate as postglacial sea level rose ("keep-up" reefs), or at least "catch up" with rising sea level if they were temporarily left behind. There are coral reefs living today at the ocean surface above the Bellona Platform in the Southwest Pacific (shown in Figure 2.7A), so it is clear that these reefs were able to keep up or at least catch up with the rising postglacial sea level.

But some reefs evidently were not so fortunate, and today their largely lifeless surfaces lie well below the photic zone of the tropical ocean, leading to their classification as give-up reefs.[13] In 1960, an article published in the journal *Deep-Sea Research* described Alexa Bank, a drowned atoll on the Melanesian Border Plateau, an area of the central tropical Pacific labeled "most curious" by the article's authors.[14] Through echo-sounding surveys, Alexa was found to be one of some forty flat-topped underwater banks with raised rims typically 20 meters below sea level and lying 16–20 meters above the general floor of the bank. In some of the shallower areas of Alexa Bank, isolated colonies of living coral were found (Figure 3.1), although most of its underwater surface was formed from "dead coral boulders encrusted with algae and covered with a white (foraminiferal) mud."[15] Later research in that area found that banks like Alexa Bank had formerly been atolls that had been drowned during the most recent period of postglacial sea-level rise, largely during bursts of rapid sea-level rise.[16] In other words, they are typical "give-up" reefs.

It is wrong to instantly classify all shallow-water drowned reefs and atolls in the modern Pacific (and elsewhere) as give-up reefs because it is possible that subsidence also played a role in their drowning. Alexa Bank is probably a give-up reef because it occurs in an area of the Pacific where there are many other similar submerged islands. The most obvious explanation for an entire region of atolls that failed to survive during the most recent period of sea-level rise is that oceanographic conditions across that region at the time did not allow optimal rates of upward reef growth.[17] The ocean current that sweeps across this region, the South Equatorial Current, comes from a largely landless area and may therefore have been carrying few if any coral planulae (by which a new reef could have been seeded).

Another area in the tropical Pacific where today there are no coral reefs living at the ocean surface is that around the Marquesas Islands of French Polynesia. The absence of reefs here is puzzling; most authorities attribute it to the steep underwater slopes of these islands, on which coral reefs could not find suitable footholds.[18] But the South Equatorial Current also skirts the area to the north of the Marquesas and may also explain why, as with the Alexa Bank area, no reef is growing there today as it does in most other parts of the tropical Pacific.

Before we leave the subject of islands that drowned as a result of postglacial sea-level rise, it is worth examining the nature of this process briefly. As is seen in chapter 5, certain aspects of the human history of the Pacific during the period of postglacial sea-level rise are explainable by the effects of catastrophic sea-level rise events (CREs). These CREs resulted from the rupture of natural dams that held back huge lakes of meltwater on the continents. The water poured down valleys and into the ocean, raising its level comparatively rapidly. For example, CRE-1, which occurred about 14,200 years ago, involved 13.5 meters of sea-level

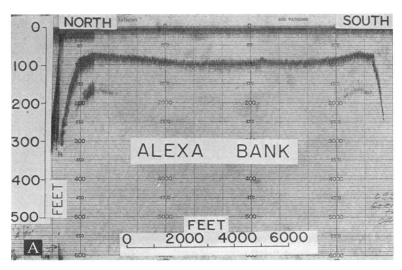

FIGURE 3.1. **A.** Echo-sounding trace by RV *Horizon* across the top of Alexa Bank, one of the first submerged islands to be mapped and examined. Note the rims of the bank at 18–21 meters (60–70 feet) and the saucer-shaped depression at 27.5 meters (90 feet), suggesting that Alexa Bank was once an atoll. (Reprinted from Fairbridge and Stewart, copyright 1960, *Deep-Sea Research* 7:100–116, with permission from Elsevier.) **B.** Walter Munk examining a giant coral *(Echinopora lamellosa)* on the surface of Alexa Bank during Operation Capricorn in 1952. Although some coral species can grow at such depths, most hermatypic (reef-building) corals grow in much shallower water. (With permission of the Scripps Institution of Oceanography archives at the University of California, San Diego.)

rise in less than 290 years.[19] Although obviously this could have proved more than merely inconvenient for humans, it may also have provided the final blow to coral reefs struggling to reestablish themselves in certain parts of the world's tropical oceans.

If past history can be seen as a good guide to the future, drowned atolls and other tropical islands may not remain that way forever. For in the past, some such islands have been alternately exposed as sea level fell at the start of an ice age and then drowned as sea level rose at its termination. This has happened many times. One of the clearest examples comes from Midway Island (an atoll in the Hawai'i Group), where drilling through the reef limestones encountered the underlying volcanic basement rocks at a depth of 381 meters.[20] The limestones comprise a number of ancient reefs of contrasting ages lying one on top of the other with sometimes abrupt transitions termed "unconformities" between them. Unconformities represent times when rock formation stopped or where the intervening rocks are missing, perhaps removed by erosion. The latter explanation was used for the Midway limestone sequence, the idea being that when sea level falls around an atoll (at the start of an ice age), the limestone of which the atoll is formed is exposed and begins to be weathered and eroded by rain and waves. This results in the lowering of the atoll surface, which involves the removal of a large amount of reef limestone. Then when at the end of an ice age (usually at least 70,000 years later than its start) the sea level rises and drowns the ancient atoll surface, new (young) coral reef will begin to grow on this much-older surface (Figure 3.2). In the case of the Midway drill hole, an unconformity found at a depth of 169 meters was interpreted as having formed during a low stand of sea level, probably associated with an ice age about six million years ago.

Atolls: Drowned or Not?

It may appear paradoxical to question whether a particular type of island is drowned or not. Surely, it is patently obvious. Yes and no. Yes, if you cannot see the island above the ocean surface, but you know by various means that it was once emergent. No, if the top of the original island is definitely below the ocean surface but its presence there (and perhaps its outline) is marked by subtidal ecosystems dominated by sedentary animals living on tall precipitous columns of biogenic material (coral reefs) built upward from the deeply submerged underwater flanks of that island. As seen in Figure 2.6C, this is essentially what atolls are.

Atolls or ring reefs are some of the most remarkable landforms on earth: vertical, tall, rigid, enduring, and yet organic in origin. Clearly they are buffered by the ocean water — it prevents them from collapsing as they might were they exposed. But they are also buffeted by the ocean waves, and, to many a casual observer, it seems nothing less than fortuitous that they have not all been toppled by such processes over the millennia.

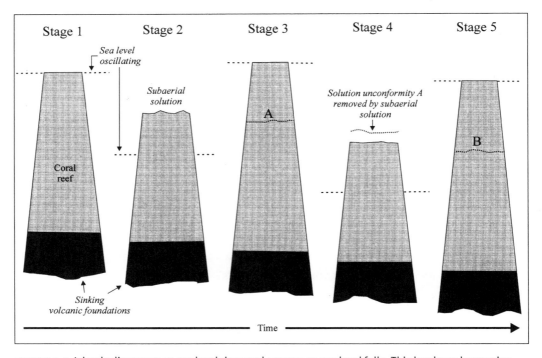

FIGURE 3.2. Islands disappear as sea level rises and emerge as sea level falls. This has been happening every 70,000 years or so over the past few million years; sea level typically fluctuated more than 120 meters within each such period. Traces of vanished islands are preserved as former land surfaces (known as solution unconformities) within the foundations of atoll reefs. This diagram shows a five-stage illustration of the process. Stage 1, An atoll exists, comprising a stack of coral reef (mostly dead) above sinking volcanic foundations. Sea level is relatively high. Stage 2, Sea level has dropped, exposing the reef surface: a high limestone island has formed. Exposed to rain and sea spray, the surface of the island is gradually lowered through erosion. Stage 3, The sea level rises again (to form an atoll), drowning the island that existed in Stage 2. Younger coral reef grows upward from the foundations of that island as they are slowly inundated. Where the younger reef overlies the land surface of the former island (made from older reef), there is a solution unconformity (A). Stage 4, The sea level falls again, exposing the island surface, which is gradually lowered by erosion as it was in Stage 2. Solution unconformity A is removed during this period of erosion. Stage 5, Sea level rises again, forming an atoll once more. Someone drilling down through the coral reef might eventually hit solution unconformity B, where younger reef overlies older reef. This solution unconformity is evidence of an island that "vanished" (that which existed in Stage 4 but was drowned by subsequent sea-level rise and overgrowth by younger reef). Of the older "vanished island," which existed in Stage 2 and was drowned in Stage 3, there is now no trace. (Adapted and redrawn from Nunn [1994].)

By definition, all atolls mark places where larger higher islands were once emergent; a few possible exceptions are those where submarine platforms have been pushed up into the photic zone from below and so developed a coral-reef cover.[21] Those high islands would have been able to support a far greater diversity of life than they do (as atolls) today. Thus these former islands are often used in models of biotic dispersal at particular times in the past; an example given later in this chapter explains the origins of endodontoid land snails in Hawai'i and Palau as having derived from high islands marked today by atolls.

Ancient Island-Flank Collapses

Early suggestions that the steep-sided flanks of oceanic islands were prone to megascale landsliding were not regarded as seriously by many scientists as they are today because the technology needed to acquire corroborative data from deep beneath the ocean surface was insufficiently developed.[22] Once a controversial view, it now seems clear that an important reason for the gradual long-term disappearance of islands below the ocean surface is collapse of their flanks, which sometimes removes their crests (see Figure 3.6).

When University of Hawai'i marine geologist Barbara Keating first saw Johnston Atoll, central Pacific, she was struck by the island's asymmetry. Much of the southern flank, where, being the windward coast, you would normally expect the broadest reef development, appeared to be missing. Keating was on Johnston to conduct the first marine survey of the island's geology.[23] Given that the island was used, after the end of the cold war, for the storage and destruction of chemical weapons, Keating's survey was timely.[24] For it appeared that the southern flank of Johnston has slipped downslope underwater in a series of huge landslides. Keating mapped the headwall of the giant landslide(s) that truncated the southern margin of the island edifice. At the foot of this headwall was a pile of collapse debris 700 meters thick. So great is its weight that it has deformed the underlying ocean floor, causing the entire island to tilt to the southeast.[25]

Then consider the islands of the Marquesas in French Polynesia, a group of high volcanic islands, a hot-spot volcanic chain rising from a steep-sided ridge. For all their celebrated boldness in profile,[26] the islands of the Marquesas are falling to pieces. They are surrounded by an apron of sediment derived from the erosion of the islands themselves; the volume of material in that apron is 240,000 cubic kilometers, and the volume of the volcanic islands themselves (from 4 kilometers depth) is a mere 50,000 cubic kilometers (see Figure 6.1). Evidently far more of the islands of the Marquesas lies as debris scattered around their bases than remains in the islands themselves. The corollary to this startling observation is that to produce the volume of material in the modern debris apron, the islands of the Marquesas must have had to build themselves up, then waste away, then build themselves up again, then waste away again, and so on.[27]

Large-scale flank collapse through landsliding is a normal and expected part of the history of an oceanic island.[28] It is well established that younger, particularly volcanic, islands have steeper sides that are more prone to landsliding than older islands, whose flanks have become gentler over time as a result of landsliding and other processes of mass wasting.[29] Yet age is not the only guide to the relative vulnerability of islands to flank collapse. Many islands, despite being comparatively old, still maintain steep sides that are as liable to collapse as those of younger islands; such islands are found along much of the Hawaiian Ridge, which has the steepest underwater slopes of any emerged oceanic structure on earth,[30] and elsewhere, particularly where uplift (rather than subsidence) is occurring.

One of the largest landslides to have occurred in a single event during the history of the Hawaiian Ridge is the Nuʻuanu Slide, which occurred about 2.1 million years ago off the northeastern coast of Oʻahu Island and involved the movement of an estimated 5,000 cubic kilometers of material. It is inferred that this slide took place at high speed (perhaps 80 meters per second) because some large blocks were moved down the ridge flanks and then 350 meters up the other side of the submarine moat that surrounds the Hawaiian Ridge. The Nuʻuanu Slide produced waves 50–60 meters high that washed across Oʻahu and nearby Molokaʻi Island, and sent a tsunami across the Pacific that was about 20 meters high when it reached the land in the northeastern part of its continental rim (Alaska to California).[31]

The youngest landslide off the west side of Hawaiʻi Island (the Big Island) may have happened during the coldest time of the last ice age around 22,000 to 18,000 years ago but more likely around 100,000 years ago.[32] Known as the ʻĀlika Slide, it involved movement of as much as 2,000 cubic kilometers of material downslope in two major phases. Some of the major submarine landslides around the southeastern islands of Hawaiʻi are shown in Figure 3.3A.

There are innumerable volcanic islands in the Pacific that, when viewed without their debris aprons, are unmistakably asymmetrical as a result of large-scale failure. The island of Oʻahu in Hawaiʻi formed from the coalescence of two large volcanoes, Koʻolau and Waiʻanae, of which, respectively, the eastern and western flanks are missing, presumed collapsed.[33] The northeastern part of the Koʻolau Volcano appears to have been removed more than two million years ago by the Nuʻuanu Slide (see description earlier in this section), leaving an impressive headwall on the island (Figure 3.4), and the southwestern part of the Waiʻanae Volcano was removed slightly earlier by the Waiʻanae Slump.[34]

As noted earlier, the high volcanic islands of the Marquesas Group are falling to pieces. The island of Guam in the Northwest Pacific is likewise incomplete: one-half has collapsed.[35] A "giant landslide" removed half of Volcán Ecuador in the Galápagos Islands.[36] Shown in Figure 3.3B is the extent of the giant (1,150 cubic kilometers) landslide that occurred on the flanks of Tahiti Island 650,000–850,000 years ago;[37] numerous giant flank collapses have occurred around volcanic islands elsewhere in French Polynesia.[38] And even Lōʻihi Volcano, the youngest in the Hawaiʻi chain, whose summit is still 1 kilometer or so below the ocean surface, has had more than half its surface modified by landsliding.[39]

Some volcanic islands have cappings of coral reef that render their upper parts even more steep-sided and therefore more liable to collapse. These collapses may affect low-level atolls, such as Johnston (see earlier in this section), or emerged atolls such as Niue. Niue is an isolated limestone island in the central South Pacific, a classic emerged atoll. The coast of Niue is conspicuously embayed, each giant scallop likely to represent a collapse of the island's flanks. One of the most recent landslides is that centered on Avatele Bay, and it appears to have occurred about 700,000 years ago.[40]

A similar picture appears with many Pacific atolls.[41] It is somewhat worrisome that the

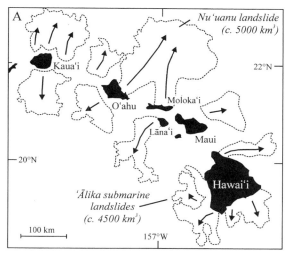
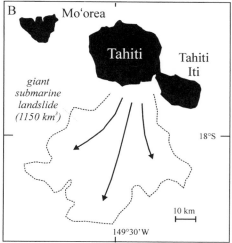

FIGURE 3.3. Examples of submarine landslides resulting from island-flank collapses in the Pacific. Note that the two maps have different scales. **A.** The main submarine landslides around the southeastern Hawaiian Islands. One of the most recent of these landslides, the two-phase 'Ālika Slide, is believed to have occurred about 100,000 years ago and produced a giant wave that washed over nearby islands Lāna'i and Moloka'i. (After Moore et al. [1989] and McMurtry et al. [1999].) **B.** The giant submarine landslide that occurred 650,000–850,000 years ago off the south coast of Tahiti Island, French Polynesia. The form of the landslide is not readily seen on Tahiti itself because the concavity at its head became filled with lavas from eruptions of the Tahiti volcanoes after the slide took place. (After Clouard et al. [2001].)

two atolls in French Polynesia where nuclear testing was commonplace during the 1980s and 1990s, Moruroa and Fangataufa, seem to be excellent examples of unstable atolls, nothing to do with the weapons testing. When Moruroa was an emergent volcanic island, nearly twelve million years ago, it experienced a period of landsliding that created concavities on its flanks that are still visible on the underwater slopes of the atoll and have been the foci for subsequent submarine landslides and fissuring. Nuclear testing has itself caused landslides and tsunamis.[42]

Rapid large-scale flank collapse of an island not only involves displacement of a large amount of solid material but also the effectively instantaneous displacement of a large amount of ocean water, something that will usually produce an unusually high wave (series) commonly called a tsunami. Sometimes, when we peer back into the distant past of the ocean basins, the evidence for rapid large-scale island flank collapse appears nebulous, often obscured by subsequent collapses, yet the evidence of the associated tsunamis is sometimes clearer.

A particularly interesting, and latterly somewhat controversial, example of possible megatsunami deposits comes from the Hawaiian islands of Lāna'i and Moloka'i.[43] The earliest views were that in situ (unmoved) deposits that included pieces of coral reached as much as 326 meters above sea level on Lāna'i and must have been deposited by a wave at least that high. The obvious candidate for the generation of this wave is the giant 'Ālika debris slide that occurred perhaps 100,000 years ago on the western flank of Hawai'i Island (shown in

FIGURE 3.4. Headwall (*pali*) of the giant Nuʻuanu Slide above Kāneʻohe Bay on the northeastern side of Oʻahu Island in Hawaiʻi. This slide took place around 2.7 million years ago, probably in a single event, sending waves 20 meters high across the coasts of the western seaboard of North America. (Photo by Jennifer Paduan, used with permission.)

Figure 3.3A). Some doubt has since been cast on this interpretation; a closer examination of the deposits suggested that the higher ones are not in situ and may not even be tsunamigenic.[44] Notwithstanding this, it is likely that huge waves did indeed wash across the Hawaiian Islands following some of the giant collapses that they have experienced.

Parts of the New South Wales coast of eastern Australia are bleak and rugged, formed from bare rock platforms against which the waves crash with seemingly unremitting fury. And well we might understand their anger, if that is what it is, for some of the water in these waves has traveled tens of thousands of kilometers without encountering any obstruction until this point. Yet looking carefully at the geological structure of this coast, one can see evidence of what must have been even larger, more violent, waves that once hit this coast (Figure 3.5). Many rock structures are truncated by wave-cut platforms, some extending hundreds of meters inland, littered with giant boulders ripped off the underwater continental shelf and tossed onto the shore as though they were pebbles. Some people are skeptical about this interpretation, which is championed by Ted Bryant, professor at the nearby University of Wollongong. Bryant believes that these features, that extend for 600 kilometers along the New South Wales coast, are consistent with the impact of at least two huge tsunamis during the past few thousand years.

FIGURE 3.5. Deposit of huge boulders in a narrow reentrant at Jervis Bay on the coastline of New South Wales, East Australia. The rocks are not cliff collapse because there is no evacuation zone in the cliffs above, and the deposit rises to the top of the cliffs. Further, the boulders are imbricated, meaning that their long axes line up within the deposit, showing that they were deposited here by a giant wave. Note the circled figure for scale. (Photo by Edward Bryant, used with permission.)

The presence of megatsunami deposits in various places in the Pacific[45] requires an explanation. One that involves large-wave generation by large-scale flank collapse appears most plausible, and an emerging area of research is trying to link particular tsunami events to particular collapses. One possible example involves the AD 1800 flank collapse of Fatu Huku Island in the Marquesas (French Polynesia), which has been suggested to have generated a tsunami that strewed huge boulders across the reef of Rangiroa Island, almost 1,200 kilometers southwest.[46] Studies of the impacts of the 1946 Aleutian tsunami in the Hawaiian Islands and the Marquesas have also proved illuminating.[47]

The absence of megatsunami deposits from many islands where they would be expected, given the numbers of flank collapses they appear to have experienced, also demands an explanation. In one view, compelling to anyone who has stood atop a windswept island cliff being drenched by sea spray, this absence may be "a function of scanty preservation."[48]

Large-scale island-flank collapse seems a plausible mechanism for some of the likely observed incidences of island disappearance in the Pacific described in chapter 9, but it is important to review the simple theoretical model for such collapses (Figure 3.6). The casual observer of an island naturally attaches greater important to its readily visible above-sea parts without always thinking that, in every case, the greatest part of the island's bulk is below the ocean surface. When the side of a smaller island experiences a large landslide, it might matter very little in terms of the overall history of the island edifice whether its top is carried away in the landslide or not. But it matters hugely to the occupants of the island's top, particularly if it is the only dry land associated with the island edifice. By way of example, the changes in Ritter Island following its 1888 eruption and collapse are shown in Figure 8.2; although this event did not cause the island to disappear completely, most of it did, demonstrating the possibility that whole islands have done so in a similar situation.

The hypothetical model in Figure 3.6 shows how the top of an island could be removed in its entirety by two flank collapses. Of course, it could be removed in a single event or, probably more common, in a series of smaller flank collapses such as appear to be affecting the island of Hawai'i (see earlier in this section), perhaps occasionally punctuated by a larger-scale collapse.

Islands That Blew Themselves to Pieces

A long time before humans came to the Pacific, there were islands within it that apparently blew themselves up. Such enormous island-obliterating eruptions have occurred at least twice in the Southeast Asia–Pacific region within the past millennium: at Kuwae in Vanuatu in AD 1453 (see Figure 9.4) and at Krakatau (Krakatoa) in AD 1883,[49] implying that they have occurred on innumerable occasions in earlier times. Unfortunately, and somewhat paradoxically, the evidence is hard to find.

The most common landform left behind when the center of a volcano explodes is a caldera, basically a vast hollow within which eruption detritus accumulates and which may be surrounded by the remains of the original volcano's flanks. Calderas formed at both Kuwae and Krakatau following their eruptions and are widespread throughout the Pacific Basin. It remains uncertain, often practically undemonstrable, whether every such caldera formed only after an entire island disappeared. The most likely candidates are in those parts of the Pacific that are susceptible today to explosive volcanism, and these are mostly in the volcanic island arcs running alongside the convergent plate boundaries in the western part of the Pacific Basin.

Claimed as the largest volcanic eruption on Earth within the past 10,000 years, the eruption that led to the formation of the Kikai Caldera in the Ryukyu Islands of southern Japan occurred some 6,300 years ago.[50] Although the Kikai eruption is not strictly within the pe-

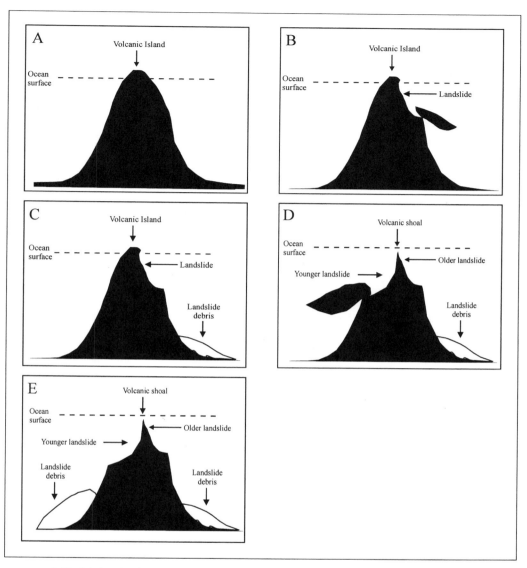

FIGURE 3.6. Model showing how a volcanic island can "disappear" as a result of successive flank collapses. **A**. A typical volcanic island; note how the underwater slopes are steeper than those above the ocean surface. **B**. A large landslide (flank collapse) undermines one side of the island, perhaps creating a large wave that backwashes over the island. **C**. The debris from that landslide settles on the ocean floor. **D**. A second, larger landslide removes part of the opposing flank of the island and the remainder of the island top (which includes all the above-sea parts). **E**. The island is now a volcanic shoal surrounded at its base by an apron of landslide debris.

riod before humans were in the Pacific, the population was obviously far less than in recent millennia, and no memory of this distant event is known. The modern Kikai Caldera, some 19 kilometers in diameter, is today largely underwater. As at Kuwae and Krakatoa, recent eruptions have led to the formation of small islands along the caldera rim (Figure 3.7A). The most tangible legacy of the Kikai eruptions is the 150 cubic kilometers of ash that was deposited across a huge area of southern Japan[51] (Figure 3.7B). An eruption of comparable magnitude occurred farther north on the Kamchatka Peninsula about 7,600 years ago and also produced a caldera that is now largely underwater.[52]

Some good examples of volcanoes that blew themselves to pieces are found in the Aleutian volcanic arc in the northernmost Pacific. This corner of the world is one of the most seismically and volcanically active; it was the site of the great 1964 Prince William Sound earthquake, mentioned in chapter 2. One of the Aleutian Islands that evidently blew itself to pieces is the ancient Mount Kanaton, which is marked today by a vast underwater caldera. Since its formation, a new volcano, Kanaga, has grown within it and has erupted at least seven times since 1763.[53]

Islands that blow themselves to pieces sometimes leave a rich legacy behind them. For when the heart of a volcano is blown out, many of the fragments later come back to rest in the hole that was formed. Of course, these fragments are loose, what geologists call unconsolidated, with many spaces between them. And sometimes beneath these calderas, superheated liquids rich in precious minerals leached from these broken rocks seep upward, cooling and precipitating out in the gaps to form (hydrothermal) deposits. Some modern Pacific Island calderas are the sites of mineral riches that are worked today. They include Vatukoula in northern Viti Levu Island in the Fiji Group, and Lihir (Niolam) Island in Papua New Guinea, where a vast gold deposit was emplaced after the formation of the Luise Caldera.[54]

Only in the last decade or so has the mineral potential of underwater (submerged) calderas become properly understood. It now seems that a submerged caldera is better than an emerged one as a location for a rich deposit of precious hydrothermal minerals. Such deposits, called Kuroko-type deposits, are formed when superheated hydrothermal liquids rising up through the fragmented fills of underwater calderas suddenly encounter cold ocean water. The response is for the minerals to abruptly precipitate out in concentrations that are typically forty times what they are in a comparable above-sea situation. At Myojin Knoll, some 400 kilometers south of Tokyo, one rich hydrothermal deposit of gold and silver forms a mass 400 meters in diameter and at least 30 meters thick. And of course, such deposits are forming continuously as Myojin's hydrothermal system keeps depositing fresh minerals in its caldera.[55]

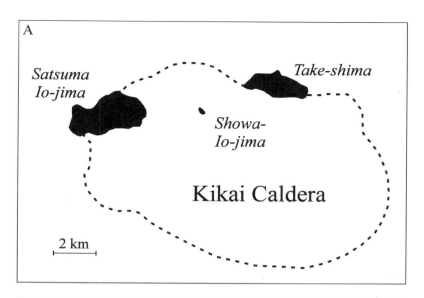

FIGURE 3.7. **A.** Map of the Kikai Caldera. Showa-Io-jima is a lava dome rising 26 meters above sea level that formed during the 1934 underwater eruption. (Adapted and redrawn from Kuno [1962].) **B.** Deposits of ash (tephra) and pyroclastic materials from huge caldera-forming volcanic eruptions has built much of southern Kyushu Island in Japan. At the base of this roadcut inland from Ibusuki is a local tephra of uncertain provenance. Overlying this tephra are pumice and pyroclastics from the Aira Caldera eruption about 22,000 years ago, which are in turn overlain by tephra from the Kikai Caldera eruption dating from 6,300 years ago.

Vanished Islands Inferred from Biogeography

Long before people came to occupy the Pacific Basin, land- or shallow-water–dwelling plants and animals moved around and across it, something that can be readily inferred from the current distribution of their fossils. How they accomplished this has been the subject of considerable speculation. In the days before the acceptance of earth-surface mobility, it was thought that improbably long linear land bridges (now vanished) once existed, along which these plants and animals duly moved from one landmass to another. No direct evidence of such land bridges was ever found and, with the advent of mobile-earth theories such as plate tectonics, the concept of land bridges has been rendered unnecessary. Yet many current ideas about transpacific biotic dispersal involve the belief that stepping-stones of land once existed, enabling the crossing of what now appears to be a long ocean distance in a series of shorter hops. Some cross-Pacific biogeographic tracks may have involved continent-sized stepping-stones in the central Pacific at various times, but most appear to have involved dispersal between islands, some of which have since been submerged.

The idea that such island stepping-stones could help explain the current distribution of particular organisms in the Pacific Islands has been around for a long time, ever since scientists began to contemplate the origins of the endemic biotas of remote islands. For biogeographers concerned with these issues, considerations of the abilities of particular organisms to disperse across the ocean combined with observations of where they are now found have led to inferences about the presence of now-submerged islands and island chains in particular locations.

Among the most remote islands in the Pacific Ocean are those of its southeast quadrant (Figure 3.8). A good example of the role of now-submerged island stepping-stones in biotic dispersal comes from the Pitcairn Group, which comprises high volcanic Pitcairn, uplifted limestone Henderson (a former atoll), Ducie, and Oeno, both atolls. Together with neighboring islands in French Polynesia (but excluding Easter Island), the biotas of these islands form a distinct biogeographic province, within which biotic exchange between islands has occurred regularly for millions of years.[56] Explaining the biotic affinities between the islands of the Pitcairn Group is challenging, not only because of the great distances involved, but also because high (interglacial) Quaternary sea levels would periodically have drowned Ducie and Oeno (and Henderson before it was uplifted), obliterating their terrestrial biotas. It has been suggested that successive biotic colonizations of the Pitcairn Group from eastern French Polynesia (probably the Gambier Islands) was facilitated by the exposure during the ice ages of now-submerged islands, some of which are shown in Figure 3.8. The absence of any shallow submerged islands between the Pitcairn Group and Easter Island, a distance of some 2,200 kilometers, explains why there has apparently been no natural biotic exchange between the two in recent millennia.[57]

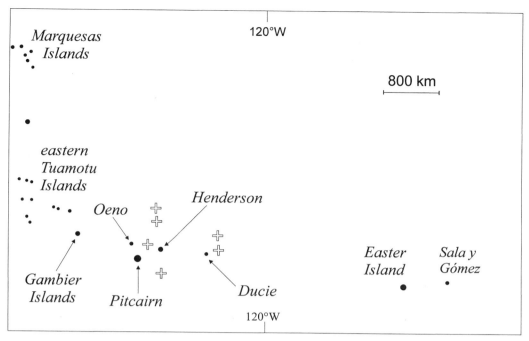

FIGURE 3.8. Some of the islands of the east-central and southeastern parts of the Pacific Ocean. Note that island shapes and sizes are not accurate. The biota of high Pitcairn Island shows affinities with those of the Gambier and eastern Tuamotu island groups. Henderson Island was uplifted during the late Quaternary (about 200,000 years ago) but was periodically submerged in earlier times. Ducie and Oeno atolls have regularly been submerged during high sea-level stages of the Quaternary; their terrestrial biota is derived from Pitcairn yet consequently much less diverse. The existence of shallow-water (< 130 meters) submerged islands (marked by crosses on the map) that emerged during the low sea-level stages of the Quaternary (most recently between 22,000 and 16,000 years ago) has undoubtedly aided the successive recolonization of Ducie and Oeno. The lack of such submerged islands between the Pitcairn Group and Easter Island–Sala y Gómez accounts for the marked lack of biotic similarity between these island groups. (Locations of submerged islands from Kingston et al. [2003].)

The following five sections provide examples of the ways in which the existence of now-submerged islands has been inferred from biogeography.

Of Hawaiian Flies, Marshallese Snails, and the Shore Fauna of Easter Island and Sala y Gómez

Certain animals that inhabited oceanic islands were evidently able to successfully colonize new islands as their home islands were being gradually submerged. Several examples come from the Hawaiian Islands, where both fruit flies (drosophilids) and honeycreepers dispersed from submerging islands to younger higher ones.[58]

Hawaiian fruit flies comprise more than 400 species, more than 15 percent of the world total, yet the islands on which they are found represent less than 0.01 percent of the world's land area. This has been interpreted to mean that these fruit flies have existed in Hawaiʻi for a very long time, far longer in fact than they could have been present on the island of Kauaʻi, which, having formed five to six million years ago, is the oldest on which they are found today. Studies of the DNA of Hawaiian fruit flies suggest rather that they originated in these islands thirty-three to fifty-three million years ago, first occupying Koko Island, what is now submerged Koko Seamount, 3,000 kilometers northwest of the modern hot spot. It is envisaged that, as Koko Island began to be submerged, the flies colonized the emergent island immediately to the southeast. As that island began to be submerged, the flies colonized the next emergent island . . . and so on and so on. In this example, even if we did not know that submerged islands existed northwest of the modern emergent Hawaiian Islands, we could infer this from the biogeography of drosophilid flies.

Another example is provided by the endodontoid land snails of the Pacific Islands. More than two million years ago, these moisture-dependent organisms lived on three islands (Bikini and Enewetak in the Marshall Islands, Midway in Hawaiʻi) at a time when they were high islands, but as they began to be submerged (today they are all atolls) the habitats for these snails vanished and they succeeded in colonizing other islands.[59] One genus, *Aaadonta,* lives today only in the high islands of Palau, and *Cookeconcha* appears to have colonized the high wet Hawaiian Islands southeast of Midway as it subsided and became too dry for these snails to survive.[60]

But for both these examples, it is legitimate to ask how a rather small fly and a land snail actually managed to colonize a new island. Even small flies, it appears, are capable of flying, or at least being carried in the wind and surviving, between islands tens of kilometers apart; in this way the ancestors of the modern Hawaiian fruit flies colonized most modern emergent islands in the group.[61] The dispersal of the land snail is more difficult to explain but probably required that eggs or small individuals became attached to birds that flew between islands.[62]

A more compelling example comes from the Southeast Pacific. Although only 415 kilometers apart, Easter Island and (the uninhabited rocks named) Sala y Gómez are extraordinarily remote. The Pitcairn Group, itself quite remote, is the nearest land to Easter Island, some 2,200 kilometers away. The Galápagos Islands are nearly 3,500 kilometers northeast, tiny Juan Fernández some 3,000 kilometers east-southeast, and the nearest point on the continental Pacific Rim (the Peru coast of South America) about 3,800 kilometers due east. It might be expected that this degree of remoteness would confer a low biotic diversity and even lower endemism, but the biota of Easter Island and Sala y Gómez exhibit quite the opposite.[63] The shore fauna of these islands exemplifies the case. For example, more than 42 percent of the shallow-water marine mollusks of Easter Island and Sala y Gómez are endemic[64] — they occur nowhere else in the world. The puzzle for biogeographers is that such a high

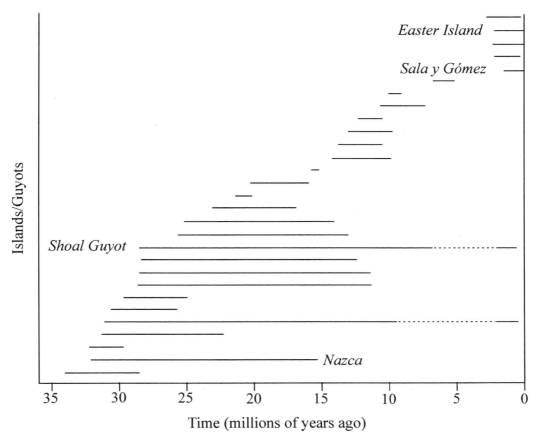

Islands/Guyots

Easter Island

Sala y Gómez

Shoal Guyot

Nazca

35   30   25   20   15   10   5   0

Time (millions of years ago)

FIGURE 3.9. The ancient character of the shallow-water marine fauna of Easter Island and Sala y Gómez Islands in the Southeast Pacific Ocean requires that land in the area has been emergent far longer than the two and a half million years since these islands formed. One explanation is that islands, now mostly drowned, have indeed been emergent in the area for almost thirty-five million years and that by the time one island was submerged, another had emerged sufficiently for it to be colonized by biota from the other. This graph shows 29 islands and guyots in the area that are believed to have been emergent at certain times (solid lines), temporarily submerged at others (broken lines), and nonexistent or submerged at others (no lines). (Figure modified from Newman and Foster [1983].)

degree of endemism generally takes a long time to develop, considerably longer than the two and a half million years that Easter Island and Sala y Gómez have been dry land.[65]

An unusual explanation has been developed for the origin of the shore fauna of Easter Island and Sala y Gómez. It involves emergent islands in the area having been colonized as much as thirty-five million years ago and, before sinking into deep water, successfully passing on their shallow-water marine fauna to a nearby, more recently emerged island.[66] And so on and so on until the ancient shore fauna was passed on to Easter Island and Sala y Gómez, the youngest representatives of a line of islands that first emerged thirty-five million years ago and have since mostly subsided (Figure 3.9).

Ancient Island Arks

As discussed in chapter 2, terranes — chunks of oceanic or continental crust that have moved across the ocean basins to places where they clearly did not originate — are known to have crossed the Pacific during its early history. The role of large (continental?) terranes in the speciation and dispersal of Pacific plants and animals is briefly considered in chapter 4. Smaller terranes — emergent fragments of continents or (oceanic) volcanic islands — are also thought to have played a role in biotic dispersal in transporting particular groups of organisms within and across the Pacific, enabling them to colonize places they could not otherwise have reached.

Two Pacific "island arks," perhaps important in introducing Gondwanan shallow-water marine genera to the Hawaiian Islands, are Wentworth Seamount and Necker Island. Now both incorporated into the Hawai'i-Emperor island-seamount chain, neither is considered to have formed over the associated hot spot.[67] Both Wentworth and Necker formed south of the equator, possibly along the East Pacific Rise, and then moved slowly northward, meeting the Hawaiian chain, respectively, thirty million and ten million years ago. Among the Gondwanan genera they carried, presumably from some continental terrane such as Wrangellia or from an older island, and introduced to the oceanic islands of Hawai'i were many shallow-water marine biota. These include most extant genera of Hawaiian mollusks together with fish such as *Hyporhamphus acutus* and *Crystallodytes*, both of which are represented globally by just two subspecies, one in Hawai'i and the other elsewhere in the Pacific Islands.[68]

There appear to be even clearer affinities between the terrestrial biota of the Hawaiian Islands and that of other islands on the Pacific Plate,[69] suggesting that many of these plant and animal species reached the Hawaiian Islands not during occasional long-distance landings from the nearest parts of the continental Pacific Rim but on drifting islands that became incorporated into the Hawaiian hot-spot chain.[70]

Island Stepping-Stones in the Tasman Sea

New Zealand and Australia in the Southwest Pacific Rim were both once part of Gondwana. Today they are separated by the 1,800-kilometer-wide Tasman Sea, which began opening eighty to ninety million years ago as that part of Gondwana began to break up.[71] For a long time, scientists thought that the ancestors of the modern native New Zealand (terrestrial) flora existed there *before* the breakup of this part of Gondwana. It now seems probable that most elements of the New Zealand flora arrived significantly *after* this breakup,[72] a journey that would have entailed a wide ocean crossing. Although airborne dispersal is demonstrably possible across the modern Tasman Sea, the principal method of dispersal appears to be the West Wind Drift, a west-to-east–moving ocean current that has operated across the Tasman Sea ever since it began opening.[73]

This all seems very neat until you examine the dominant directions of trans-Tasman biotic dispersals. If you assume that the West Wind Drift is indeed the main agent of dispersal, that would mean that dispersal would be significantly more frequent eastward from Australia to New Zealand rather than the other way round. Unfortunately this is not the case. For although most (not all) plant dispersals are indeed eastward, most animal dispersals across the Tasman Sea have been westward.[74] One solution to this puzzling observation is to suppose that the Tasman Sea, in contrast to its current geography, was once dotted with islands that facilitated biotic dispersal in both directions. The most obvious locations for such stepping-stone islands are the Lord Howe Rise and Norfolk Ridge, the latter of which may have had a significant above-sea presence as late as the late Miocene, eight to ten million years ago.[75]

There is another mystery of sorts concerning the origins of the respective biotas of New Zealand and New Caledonia. Both are fragments of Gondwana, connected by the Norfolk Ridge that allowed a sporadic land connection, possibly with short ocean gaps, until about thirty million years ago. One would therefore expect their biotas to be similar, but in fact they are remarkably different. New Zealand's biota is grouped closely with those of Australia and South America; that of New Caledonia has marked affinities with those of New Guinea and the Southwest Pacific islands.[76] Given that both New Zealand and New Caledonia were probably submerged completely during the early Tertiary (about fifty million years ago) with the result that their (Gondwanan) terrestrial biotas were wiped out, their modern biotas must be a result of subsequent long-range cross-ocean dispersal from particular source areas. The reason why this dispersal apparently did not occur between New Zealand and New Caledonia may have to do with a lack of island stepping-stones in that area after the time when the two island groups last emerged. In contrast, such stepping-stones are inferred to have existed between Australia and New Zealand, allowing trans-Tasman biotic interchange, and between New Caledonia and New Guinea, permitting similar movements, but not between New Caledonia and New Zealand.[77]

Long-Distance Dispersal into the Central Pacific: The Long-Jawed Spiders of the Society Islands and Other Biogeographic Enigmas

Some stories of biotic dispersal across the Pacific are so remarkable that they seem to demand that landmasses once existed as stepping-stones along their dispersal paths. Yet this is not so, and it is worth examining precisely why with two well-documented examples.

One of the most puzzling discoveries of recent years concerning the native plants of the

Hawaiian Islands (Northeast Pacific) is that, contrary to earlier views, most of these plants originated in the south and west of the Pacific Rim rather than from the much-nearer western seaboard of North America.[78] Yet until recently it was unclear precisely how these plant colonizers reached Hawai'i, particularly given that it was necessary for them to have crossed the Pacific equatorial zone, the climate of which is not conducive to the survival of such subtropical and temperate species.

One recent study of the genus *Metrosideros,* which occurs in both Hawai'i and New Zealand and many other Pacific Islands,[79] showed that it may have originated in New Zealand and moved out into the Southwest Pacific island groups (Vanuatu, Fiji, and Samoa) by about five million years ago, most likely through wind dispersal at times of lower sea level when islands were larger and closer together. In this model, a second dispersal occurred during the Quaternary period and led to the colonization by *Metrosideros* of islands in the central tropical Pacific (southern Cook Islands, French Polynesia). The third major phase of dispersal to Hawai'i has attracted special attention because it presumably involved wind dispersal against the direction of the northeast trade winds.[80] It has been suggested that initial dispersal to Hawai'i took place during the late Quaternary, at least 350,000 years ago, from the Marquesas Islands in northern French Polynesia. The most plausible explanation is that this happened at a time when the intertropical convergence zone (ITCZ) formed south of the Marquesas, as it has occasionally been known to do in historical times. In such a situation, seeds from plants like *Metrosideros* could have been removed from these islands and carried northward within the Northern-Hemisphere Hadley Cell, descending to ground level at about 25° N, where the islands of Hawai'i lie.

The interisland dispersal of *Metrosideros* appears similar to those of many organisms in the Pacific, including birds and marine taxa.[81] But it is not always that simple. Consider the biogeography of the long-jawed orb-weaving spider genus *Tetragnatha,* which is found in three Pacific Island groups, Hawai'i, the Marquesas, and the Society Islands, the last two being part of French Polynesia. One might expect that, like *Metrosideros, Tetragnatha* first gained a foothold in one of these archipelagoes, then reached a second, and finally the third. But studies of DNA from populations of *Tetragnatha* in these three island groups show instead that they were each colonized separately from a continental source, not from each other.[82] The *Tetragnatha* of the Pacific Rim are excellent long-distance colonizers, typically living over water and being able to balloon across water. Yet, once they had successfully colonized these remote island groups, they formed lineages that were best adapted to terrestrial living and not to long-range cross-ocean dispersal. The similarities in form and lifestyle that the *Tetragnatha* of Hawai'i, the Marquesas, and the Society Islands share are explained by parallel evolution on similar islands, not by the derivation of one island population from another.

Pumice Mats and Dead Whales:
Novel Forms of Biotic Dispersal

In recent decades biogeographers have become increasingly aware that vanished islands, of the solid varieties described in this chapter, may not always be necessary to explain biotic dispersal across long, otherwise seemingly uncrossable, ocean gaps. For awareness has grown of the incidence of more transient, more ephemeral islands—those typically made from mats of vegetation or pumice—that can remain intact for months tossing about in the ocean, sometimes carrying a whole range of organisms.[83]

In a few islands in Fiji and Samoa (Southwest Pacific) we find species of iguanas that are descended from those in the Galápagos Islands (easternmost Pacific). The journey of at least 11,500 kilometers involved might have been achieved had an underwater volcano in the Galápagos produced a raft of pumice on which a pregnant female iguana became stranded. It is possible that the hatchlings survived to reach Fiji and Samoa.[84] Pumice mats may be the most enduring form of floating island, but others made from soil and vegetative debris can also survive several weeks in the open ocean.

When we think of biotic dispersal by long-range organisms, such as the probable dispersal of endodontoid snails by birds and large insects (see earlier in this chapter), we generally think of dispersal through the air. Yet it has recently become apparent that many types of marine biota can be dispersed by large ocean-dwelling organisms such as whales. The discovery of clams associated with thirty-five-million-year-old fossils of baleen whales led to suggestions that these, alive or dead, might have been important agents of clam dispersal across the Pacific,[85] an idea subsequently extended to include the large-bodied ocean-dwelling predecessors of whales.[86] Of course whales are not islands. Yet manifestly they can fill the same role in biotic dispersal as true islands.

Serendipity and the Human Exploration of the Pacific

Many people know the word serendipity from the movie of the same name that had the tagline "can once in a lifetime happen twice?"[87] The original word was coined in 1754 by Horace Walpole after he had read a fable about Sri Lanka (once rendered as Serendip) entitled *The Three Princes of Serendip*. Walpole's definition of serendipity is somewhat at odds with the fable.[88] He viewed serendipity as the predisposition of the three princes to make discoveries, by accident and sagacity, of things they were not in search of, a definition that appears today in most standard dictionaries. In this sense, serendipity describes very well the way in which some writers have explained the earliest (pre-European) explorations of the Pacific Ocean by people.

The idea that the first inhabitants of the Pacific Islands were incapable of sailing the vast distances between these islands to colonize them is a judgment that has its roots in the

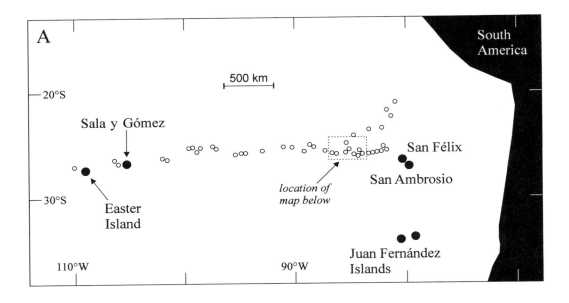

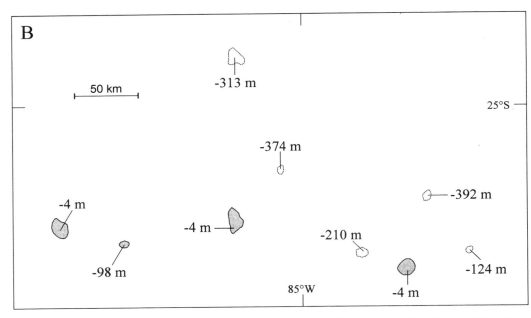

FIGURE 3.10. Vanished islands in the eastern South Pacific that could have been utilized by island-hopping biota to cross the Pacific from west to east. **A.** Map of the eastern South Pacific showing existing islands (filled circles) and guyots-seamounts (open circles) that may have been emergent during Quaternary ice ages when sea level was as much as 120 meters lower. Note that island-guyot size is not proportional to circle size. **B.** Map of the inset from Map A showing guyot minimum depths (meters below mean sea level), the 200-meter isobath where appropriate, and the form of guyot crests. (Adapted and redrawn from Wyatt [2004].)

prejudices of some seventeenth- and eighteenth-century European commentators. At that time, one solution to the apparent dilemma was to propose that these islands had once been the mountaintops of a vast central Pacific continent that had sunk, leaving the survivors huge distances apart yet sharing cultural traits that unmistakably showed that they came from the same ancestral stock.

Contrary to this, most scientists in the past fifty years or so have had no difficulty in accepting that Pacific Islanders succeeded not only in colonizing all the larger island groups but also in crossing the entire Pacific Ocean from west to east — a straight-line distance of around 14,500 kilometers — before Europeans even knew this ocean existed.[89] Yet perhaps Pacific Islanders would not have needed to cross the longest ocean distances involved without landfall, as they would have to do today. Perhaps islands existed, located serendipitously, where none does today. On the face of it, to suppose that such islands once existed, enabling people to make shorter voyages than they need to today in the drowned Pacific, appears plausible, given that the same mechanism is invoked to explain the dispersal of certain nonhuman biota.

But the most recent expression of this view, although founded on some solid scientific data, must be regarded as an extreme one. It is the view of Steve Wyatt that humans crossed the Pacific from west to east far, far earlier than is generally thought, utilizing island stepping-stones that were subsequently drowned by postglacial sea-level rise. Wyatt's principal reason for proposing this idea is that there remain huge questions about how and when the first humans reached the Americas, and who they were. Many islands were undoubtedly emergent during the last ice age in the now almost islandless expanses of the eastern tropical Pacific (Figure 3.10) and could have provided staging posts for any organisms crossing the Pacific.[90] But this would have required people to have occupied Pacific Islands at least 10,000 years earlier than they are known to have done, and no one has been able to demonstrate that there were people anywhere in the Pacific Islands outside western Melanesia (Papua New Guinea and Solomon Islands) at that time. Of course, because of subsequent sea-level rise, it is conceivable that the coastal settlements of early inhabitants of island groups east of Solomon Islands have not been found — these settlements perhaps now are under tens of meters of water and overgrown by coral reef.[91] But nevertheless this scenario seems highly unlikely, not least because there are no environmental indicators (such as evidence for burning) from the still-emergent parts of these islands of a human presence before about 3,000 years ago, at the most.

The role of serendipitously located islands, now vanished, in the movement of people through the Pacific Islands cannot be emphatically denied, but it is certainly far from being regarded as plausible.

4 Ancient Continents Hidden by Time

Many continents have been claimed as having once existed in the Pacific before disappearing subsequently, but hardly any of these claims are true. Accounts of some of the undoubtedly mythical continents claimed to have existed in the Pacific are discussed in chapter 7; in this chapter I discuss two continents that have been suggested to have once existed in the Pacific. One has indeed become hidden; fragments of the other, if it ever really existed, may lie hidden beneath the ocean surface.

The Paleozoic Continent in the Southeast Pacific

The structure and composition of some of the oldest rocks found along the west coast of South America, particularly in northern Chile, Peru, and Bolivia, have long puzzled geologists. Some of these rocks, ancient sediments called turbidites, form on the steep slopes along the edge of a continent, where particles of terrestrial rocks accumulate in huge sediment aprons (or fans). No obvious surprise then that there should be turbidites along the high steep western margins of the South American continent. But what is surprising when you come to look closely at these turbidites is that they become thicker and coarser to the west, the complete opposite of what you would expect had they been washed off the continent of South America.[1]

It is not only the reversed turbidites that caught the attention of early geologists working along the west coast of South America but also the folded rocks in belts trending parallel to the coast and the truncation of what are today northwest-trending structures.[2] The only mechanism that appeared credible to early geologists working in this area was that a huge landmass — *un ancien continente Pacifique* — had once existed off the west coast of the South American continent.[3] It must have been large to produce enough sediments to create the thick turbidites off its east coast, and it must have occasionally collided with the modern continent of South America, causing some of the rocks there to be folded parallel to the coast. In addition, it must have moved sideways along the western border of the South American continent, causing some of the preexisting structures to be partly scraped off.

Handicapped by working with fixist models of earth-surface behavior, early geologists could do little more than simply infer the existence of a Southeast Pacific continent. It is possible that they were encouraged in this radical interpretation by the widespread belief in the late nineteenth century that the Pacific was a site of a sunken continent.[4]

Reconstructing what the earth looked like during the Paleozoic era (590–250 million years ago) is understandably a major challenge because of the massive overprint from later processes and overlay of younger formations. Yet considerable advances in understanding the Paleozoic world have been made from examining those areas of the continents that have remained comparatively undisturbed since that time.

The model of plate tectonics has helped this process immeasurably, by providing various options for reconfiguring the positions of existing continental fragments at various times in the past. One of its most successful applications has been the SouthWest US–East Antarctica (or SWEAT) hypothesis, championed by Ian Dalziel at the University of Texas, which involved the juxtaposition of the Pacific margin of North America (Laurentia) with East Antarctica–Australia during late Precambrian times (about 1,000–800 million years ago) when the supercontinent Rodinia existed (see Figure 2.2). At the end of the Precambrian period, after the opening of the first Pacific Ocean between these continents, Laurentia moved a short distance away from (what is now) the west coast of South America, and the turbidites were produced. But Laurentia came back again, colliding several times with South America, causing the folding noted earlier, and also producing, along the collision suture, a line of fold mountains that today comprise the Andes of South America and the Appalachians of eastern North America.[5] The truncation of South American structures was caused by the northward movement of Laurentia along the west coast of South America.

The identity of the Paleozoic Southeast Pacific continent and its current whereabouts have not been known for long. Mainly through the increasing recognition that the Andes and Appalachian mountain ranges formed at the same time as a result of the same processes, it now seems beyond doubt that this continent (Laurentia) is simply an ancestor of the modern continent of North America. From northern Canada to Georgia in the south, and from Greenland in the east to Wyoming in the west is hidden the ancient Paleozoic continent of the Pacific. The continent vanished from the Pacific Ocean and became, overlain and hidden by younger rocks, part of the continental frame of the modern Pacific Basin.[6]

The next example comprises elements of a possible former continent that are today both hidden and sunken within the Pacific, and is altogether more contentious.

Pacifica: Fact or Fiction?

When you look at a map of the Pacific Basin you can see that, unlike the other ocean basins, it is surrounded almost entirely by lines of fold mountains that trend parallel to modern coasts. Along the western seaboard of North America, you have the Rocky Mountains; in South America, the Andes. In Antarctica, the Transantarctic Mountains that separate that continent's east and west parts run parallel to the Pacific coast. Similarly, there is the Great Dividing Range of East Australia and mountain ranges forming the spines of the largest islands in New Zealand and New Guinea. Numerous ranges of mountains run parallel to the coast of East Asia, from the Wuyi Mountains of southern China to the Sikhote Mountains and, farther north, the Koryak-Kolyma Ranges of far eastern Russia. This is manifestly an unusual situation — instinctively you might expect the edges of large landmasses to be lower than their interior parts — and one that has exercised the minds of geologists for decades.

An influential explanation for the origin of all these mountain ranges came in 1977 in an article published in the prestigious journal *Nature* entitled "Lost Pacifica Continent" by Amos Nur and Zvi Ben-Avraham. They argued that these mountain ranges owed their existence to collisions between continents at some time in the past.[7] The crux of their innovative idea was that these fold-mountain belts had all been created along convergent plate boundaries. All went smoothly as long as it was only oceanic crust that was being subducted along the trenches adjoining the continents. But many of the oceanic plates also carried terranes of uncommonly thick crust, and, when these reached the trench, they could not be subducted. So they pushed against the continental rim (a process that geologists, accustomed to thinking very long term, call collision), causing it to buckle, fold, and rise, creating lines of fold mountains at right angles to the direction of collision. The idea is compelling and had a major well-documented precedent in the collision between India and Asia that led to the formation of the Himalayas ninety million years ago.

Much of the evidence for the idea presented by Nur and Ben-Avraham came from the undeniable evidence of terranes having been accreted onto the edges of the continents around the Pacific. Terranes are thick, fault-bounded chunks of earth's crust, of either continental or oceanic origin, that are sometimes today found in places far away from where they formed. Hence they are labeled exotic, in the sense of foreign or alien.[8] Nur and Ben-Avraham proposed that most terranes found accreted onto the margins of the modern Pacific were part of an underwater continent that once existed in its center — a continent they named Pacifica (Figure 4.1).

Pacifica was originally part of Gondwana, the southern part of Pangea, that had separated from the northern part (Laurasia) by about 200 million years ago. Gondwana itself then began breaking up, with the Pacifica section moving away into the Pacific from the New Guinea–New Zealand border of Gondwana about 180 million years ago. Later, Pacifica itself began disaggregating, with various fragments being carried northwest where they

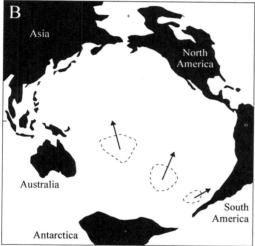

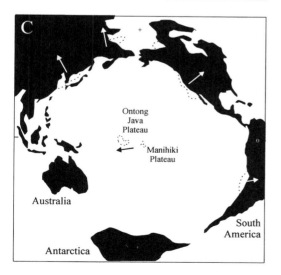

FIGURE 4.1. Simplified maps showing the existence and breakup of the Pacifica continental fragment of Gondwana, as suggested originally. A few elements of this model have been retained, but several parts of what were once thought to be Pacifica are now known to be of oceanic, not continental, origin. Note that in all three maps only the modern Pacific Rim is shown, but this was significantly different at each of the times shown. **A.** 225 million years ago. At this time, Australia was farther south abutting Pacifica. This was when the supercontinent Pangea existed and the Pacific Ocean (named Panthalassa) was consequently far larger than today. **B.** 180 million years ago. Pangea has split into Laurasia in the north and Gondwana in the south. A sea-floor spreading ridge (divergent plate boundary) has opened up between the Australia-Antarctica part of Gondwana, causing Pacifica to split from it. Pacifica itself has been broken up by the development of sea-floor spreading ridges, dividing at this stage into three parts. **C.** 65 million years ago. Most parts of Pacifica have now become accreted onto the continental rim of the Pacific, both because of their own movement and also because of the shrinking of the Pacific Basin associated with the breakup of Pangea. The Ontong Java and Manihiki plateaus in the central Pacific (also shown in Figure 4.2) were originally suggested as having been part of Pacifica but are now known to be of oceanic, not continental (Gondwanan), origin. (Adapted from various sources, principally Nur and Ben-Avraham [1977].)

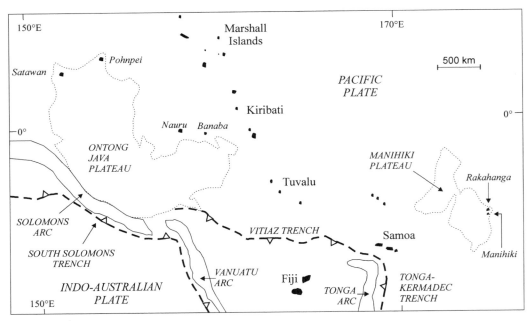

FIGURE 4.2. Map of part of the western tropical Pacific showing the Ontong Java Plateau and the Manihiki Plateau, two uncommonly thick pieces of ocean crust, once suggested to be fragments of continental Pacifica (see Figure 4.1). The southern edge of the Pacific Plate on which both these oceanic plateaus lie was formerly being thrust downward along the Vitiaz Trench. Yet when the Ontong Java Plateau reached the trench, it could not be readily subducted because of its thickness. As a result, about eleven million years ago, a new ocean trench formed south of the Solomons Arc and west of the Vanuatu Arc, along which subduction of the Indo-Australian Plate began occurring. Islands alleged to have vanished in the eastern part of the Solomons Arc, discussed in chapter 9, may have done so because of the crustal instability resulting from the continued pushing of the Ontong Java Plateau toward the southwest. (Adapted and redrawn from Nunn [1994].)

collided with the East Asian continental margin, to the east where they collided with the western margin of South America, and to the northeast where they are found today along the western seaboard of North America, from Alaska to California.[9]

According to Nur and Ben-Avraham's original model, not all the fragments of ancient Pacifica are currently stuck onto the continents along the Pacific Rim. In the Northwest Pacific, at the latitude of Japan, are two exotic terranes—the Hess and Shatsky rises that were believed to be fragments of Pacifica. Their movement northward took them through low latitudes during the late Mesozoic, and their progress has been precisely logged by studies of the deep-water limestones that accumulated on their surfaces during that time.[10]

Other fragments of Pacifica were not thought to have moved so far. These were proposed as including both the Manihiki Plateau in the central Pacific and the Ontong Java Plateau in the Southwest Pacific (Figure 4.2). But this is where the argument of Nur and Ben-Avraham begins to falter. Today most geologists regard the Manihiki Plateau not as a fragment of Pacifica but as a Large Igneous Province formed 120 million years ago at a junction between three plates.[11] It subsequently broke up, with pieces moving south and east.[12] The origin of the Ontong Java Plateau is similar. With a volume of lava of some 60 million cubic kilometers, it is the largest Large Igneous Province in the ocean basins;[13] there is no evidence that any continental crust underlies it.[14] The Hess and Shatsky rises are oceanic plateaus that formed in the same way.

As if all that was not enough, the Pacifica model has come in for further criticism, with many of its supposed fragments actually shown to be of oceanic origin and as being accreted onto the continents fringing the Pacific well before the disaggregation of eastern Gondwana.[15] Clearly, the Pacifica model needs reexamination to see whether there is anything in it bearing on the origins of the Pacific Basin that is worth salvaging. It cannot now be said that the continent named Pacifica ever existed within the Pacific.

The story of the rise and fall of Pacifica is instructive because it illustrates how scientists, particularly those having only limited information about the nature of particular landmasses at their disposal, sometimes erect theories that have wide appeal, largely because they are tidy and explain a large number of observations by a single process. It is normal in the course of scientific advancement for some such theories to be toppled as more information is gathered, and there is no shame in having been wrong. But a danger appears when wrong theories become accepted outside the boundaries of science and begin to inform popular prejudices and misconceptions. The normal processes of scientific verification, modification, and repeated testing of models are almost impossible to apply once an idea has entered the popular realm, especially when that idea is convincingly and energetically promoted. And this, although it does not apply specifically to the Pacifica model, is the main reason why so much nonsense has been written about hidden continents and vanished islands in the Pacific.

Hidden Continents Inferred from Biogeography

It is perhaps a little unfair to suggest that Pacifica was wholly an invention of geologists, because a similar idea was proposed at least a decade earlier by biogeographers seeking to explain the ancient transpacific dispersal of particular organisms.[16] As explained in chapter 3, island biotas have long been an object of fascination to biologists and others, awed by the apparent ability of particular animals and plants to reach and multiply on specks of land in the middle of the oceans.[17]

Some scientists have questioned the fact of long-range cross-ocean dispersal, leading them to propose that large landmasses (now disappeared) once existed in ocean basins like the Pacific. One such influential person was J. W. Gregory, who, writing in 1930, supposed that a continent-sized landmass must once have been present in the central Pacific, extending from Rapa Island in the south to the Marquesas in the north, and from the Society Islands of French Polynesia to Pitcairn Island in the east. Gregory's evidence was that certain endemic animal species exist on these islands, with manifestly close affinities to each other. He assumed that such animals, particularly the Lepidoptera (butterflies and moths), would simply have been unable to colonize islands separated by thousands of kilometers of ocean and so concluded that these islands had once been part of a single continent, now disappeared, and that its ancient endemic fauna was represented today only by the remnants on these islands.[18]

Yet Gregory and many others underestimated the ability — one might say the ingenuity — of particular organisms to cross enormous ocean gaps. Obviously those organisms that live in the air or the ocean have an advantage, but even this does not necessarily render their feats of dispersal less awesome.[19] For example, during the Triassic period (250–210 million years ago), when the supercontinent Pangea was assembled, the residual ocean (Panthalassa) was much larger than the modern Pacific Ocean (see Figure 2.2). Yet on either side of this ocean, the same species of shallow-water-dwelling marine organisms lived. The most compelling explanation is that larvae of these organisms were moved across this ancient ocean by an equatorial ocean current, an extreme example of long-distance dispersal.[20]

On occasions, biogeographers have invoked the existence of continent-sized landmasses, now hidden, to explain cross-ocean dispersals. The idea of Nur and Ben-Avraham, reviewed in the preceding section, that such a landmass (Pacifica) once existed in the central Pacific was welcomed by many biogeographers who found in it a ready explanation for transpacific dispersal, particularly between Northeast Asia and western North America.[21]

Notwithstanding the demise of the model of Pacifica per se, it remains clear that large terranes have played an important role in the dispersal of Pacific biota. The singular biota of the Galápagos Islands in the eastern tropical Pacific provides a good example. Thought for a long time to have originated on the South American continent, just 900 kilometers east, the Galápagos biota is now known to have come from the west, most plausibly from the Wrangellia Terrane that lay in that area about 140 million years ago.[22] Wrangellia is one of the terranes that is now part of western North America, with fragments dispersed from Alaska in the north to Idaho in the south.[23]

To illustrate some of the varied issues in understanding why some biogeographers are particularly keen to have had continental landmasses moving around the Pacific aiding biotic dispersal, an example from the South Pacific is discussed in more detail in the following section.

Why Only One Tongan Island Has a Gondwanan Tree Species

Most islands in the Kingdom of Tonga in the central South Pacific are arranged in two parallel lines: a line of (mostly active) volcanic islands and a line of high limestone islands, on which most people live. The southernmost limestone island is 'Eua, which is home to a most

FIGURE 4.3. Detail of the podocarp (*Podocarpus foliolatus*) found only on 'Eua Island in southern Tonga. (Photo by Art Whistler, used with permission.)

unexpected tree, the gymnosperm *Podocarpus foliolatus*.[24] This tree is in an unusual place because podocarps are endemic to Gondwana, the southern part of Pangea, and 'Eua is an oceanic island that owes its origin and continuing uplift largely to processes along the adjacent plate boundary, marked by the Tonga-Kermadec Trench (see Figure 2.4c for location). The nearest part of old Gondwana is New Zealand, nearly 2,000 kilometers from 'Eua.

The Tongan podocarp (Figure 4.3) may have reached 'Eua with frugivorous birds or bats, flying between islands that formed the Vitiaz volcanic island arc in this region during the early Tertiary period twenty-five to forty million years ago.[25] In support of this argument is the observation that 'Eua was part of the Vitiaz Arc; volcanic rocks that peep out from beneath the thick pile of uplifted reef limestone at the foot of the cliffs along the island's east coast are the same age and have the same origin as equivalent arc volcanic rocks in Fiji, Vanuatu, and Solomon Islands.[26] In this scenario, 'Eua at the time of Vitiaz Arc activity would have been closer than today to islands like Viti Levu in Fiji and Guadalcanal in Solomon Islands. Birds and bats would have moved along the chain of islands in the Vitiaz Arc.

Yet there are reasons for supposing this explanation to be too simple and that another one, altogether more controversial, should be entertained. This involves a small fragment of Gondwana breaking off the edge of the ancient continent in the vicinity of the Norfolk Ridge about forty-one million years ago and being carried eastward across the South Fiji Basin, eventually crashing into the Vitiaz Arc, where it became incorporated into the island of ʻEua.[27] This would explain why Gondwanan podocarps are found within the Tongan Islands only on ʻEua Island — they were actually carried there on an emergent piece of the original continent that is now almost wholly submerged.

There is supporting evidence for this proposition from the Fiji Islands, the flora of which contains some unmistakably Gondwanan elements, much to the puzzlement of geologists,[28] which has led biogeographers to propose for decades that Fiji must have once been part of Gondwana.[29] The evidence is impressive. Despite the same podocarp that occurs on ʻEua being found on Viti Levu, Fiji's largest island, the archipelago contains 476 genera of flowering plants, 90 percent of which are also found in New Guinea, 75 percent in New Caledonia, and 65 percent in Australia, all parts of Gondwana. Countering the inference that at least Viti Levu must have once been part of Gondwana is the geological evidence. Some early geologists to work in Fiji thought the islands continental, mainly because the size of the shallow platform from which they rose was considered too large to be oceanic.[30] As geological interest in Fiji grew, and as its rain forests were increasingly cleared to reveal the underlying rocks, it became clear that all its islands had been formed in an ocean basin and that there was no shred of geological evidence that any had once been part of Gondwana.[31]

So how can the apparent contradiction between biogeography and geology be resolved? The most plausible answer lies with the fragment of Gondwana that slewed off the Norfolk Ridge forty-one million years ago and moved eastward to ʻEua (see earlier in this section). In its transit across the South Fiji Basin, this fragment of Gondwana may have collided with the Fiji Platform, allowing Gondwanan biota to hop off onto some of these oceanic islands, before continuing its eastward journey to ʻEua. Although more data are needed to test it properly, this idea[32] is a clever solution to a problem that was once thought almost intractable.

5 The Coming of Humans to the Pacific

Our story begins in the distant past, about 40,000 years ago and perceived today only very dimly through the haze of history, when modern humans[1] first encountered the Pacific Ocean. This event took place somewhere in East Asia, possibly Southeast Asia: humans had been living for perhaps the preceding 80,000 years in the wide, fertile valleys of rivers like the Changjiang (Yangtze) and Huanghe (Yellow).[2] Today we might be tempted to think that these humans were traveling down those valleys to get to their mouth, but they almost certainly had no such notion. On any particular day they sought sustenance from the environment that surrounded them, sometimes following herds of animals but not persisting in following a particular direction for any less-pragmatic purpose. Also bearing in mind the comparatively low numbers of humans living in that region at the time, we can thereby understand why it apparently took them 60,000 years or so to move about 5,000 kilometers to reach the Pacific Ocean.[3]

Reaching the Pacific Coast

When humans first encountered the Pacific Ocean, the available indications are that they did not think much of it, perceiving it as a barrier rather than an opportunity.[4] We can infer that it took perhaps 10,000 years of intermittent contact with the ocean for the relationship to change significantly and for the first coastal-dwelling humans in the Pacific to come to depend on marine foods. This change may have taken place initially when humans from southern China moved south into Southeast Asia.[5]

In his landmark 1993 book, *Self-Made Man*, Jonathan Kingdon argued that (for archipelagic Southeast Asia) this lifestyle change was an important contributor to the diversification of human skin color. According to Kingdon, when humans depended only on foods available on the land, they would choose to acquire that food at cool times of the day, thereby limiting their exposure to the sun, resulting in their skin becoming, over tens of thousands of years, light brown. But when humans began gathering food from the coast, much of that was accomplished from shore-reef flats at low tide. Such locations were exposed to the sun, and people had no choice about what time of the day they visited them. As a result, such people gradually acquired darker skins than those who consumed only land foods.

As people began to make more use of ocean foods, they naturally became more adept at interacting with the ocean. They devised innovative techniques for collecting marine foods, developing the technology and vessels to enable them to venture increasingly farther offshore in search of foods from deeper waters. Perhaps for many people along the coasts of the Pacific Ocean (and those of Southeast Asia) 30,000 years or so ago, a switch to largely marine food subsistence was embraced because it lacked the more manifest dangers of life on land. In the rain forests of Southeast Asia, humans, who had no special advantage at the time, needed to compete for the same foods with fierce predators like tigers; farther north in what is today eastern China these were also present, along with bears, leopards, and a now-extinct lion-sized hyena that appears to have routinely attacked humans.[6]

There is some evidence that the scenario just sketched is somewhat conservative in terms of its chronology. The principal evidence suggests that humans reached the Australia–New Guinea landmass at least 40,000 years ago, perhaps significantly earlier. Stone tools on emerged coral reefs at the Huon Peninsula in Papua New Guinea have been dated to at least 40,000 years ago.[7] And in Australia, although most evidence points to human arrival around the same time, some scientists have interpreted rapid sediment accumulations in lakes and nearshore areas as evidence for much earlier human impact on natural landscape processes.[8]

To cross from Southeast Asia (Sunda) to Australia–New Guinea (Sahul) would have required, even at the time of lowest sea level during the last ice age, an ocean crossing of more than 70 kilometers.[9] Although there have been many explanations as to why humans crossed from Sunda to Sahul, the most parsimonious is that people living along the coast of Sunda had become so adept at interacting with the ocean 40,000 years ago that several groups at different times and from different places intentionally crossed to Sahul.[10] The other main possibility — that groups of people fishing off the Sunda coast were accidentally swept farther offshore and carried to Sahul — appears to some to be more plausible given our knowledge of contemporary human seafaring ability.[11] Yet it appears as something of a deus ex machina and really underlines the point that our knowledge of Pleistocene human settlement, lifestyles, and movements in Sunda-Sahul is currently inadequate.

The earliest people in Australia–New Guinea are characterized today as Papuan language speakers, to distinguish them from the Austronesian language speakers who arrived later in prehistory, mostly in the outer islands of Papua New Guinea. The first people in Australia encountered a naive biota that had evolved in the absence of large predators and had no defenses against humans. Life was far easier for them than it had been in Sunda, and they developed a nomadic lifestyle punctuated by short periods of sedentism, typically at times of environmental stress. The situation in Papua New Guinea was somewhat different — the first inhabitants found little to sustain them in the islands' tropical rain forests. These people responded by beginning to manipulate forest structure and composition from at least 35,000 years ago,[12] a world record, and had started irrigated agriculture at least 9,000 years ago.[13] It is almost certain that this development represented a local innovation, not one that reached Papua New Guinea from elsewhere.

There is evidence that, although most of the Papuan inhabitants of Papua New Guinea stayed within the main islands, some did venture farther out across the ocean into the archipelago and probably along the chain of Solomon Islands to its east. One of the main reasons for this may have been access to valued lithic resources for the manufacture of stone tools. A good example is provided by obsidian, a volcanic glass that, when broken, forms razor-sharp flakes (see Figure 5.3A). The obsidian from the island of Lou, just south of Manus, has been mined for much of the past 10,000 years, and it was transferred, perhaps as a trade item, throughout the New Guinea mainland, as well as many other places in this region.[14]

The Peopling of the Pacific Rim

Although people occupied islands in what today we call Papua New Guinea and the western Solomon Islands during the Pleistocene, this represented the farthest that people ventured into the Pacific Ocean at that time.[15] The initial settlement of the islands east of the western (and central?) Solomons was accomplished much later, by groups of Austronesian language speakers. To understand where they came from and why they came, we need to go back to about 30,000 years ago. At that time, along the Pacific Rim to the north, there were concentrations of people around the mouths and in the lower valleys of the Changjiang, Huanghe, and other major rivers in East and Southeast Asia. Owing to the lower sea level during the Last Glacial Maximum, mainland Asia was contiguous with or much nearer to what are today the islands of Japan and Taiwan, among others.

Nomadic people accustomed to including marine foods in their diet would have tended to wander along the coastal fringe of East Asia, rather than going far up the valleys down which their ancestors had come. In this way, people may have spread along the entire western rim of the Pacific, reaching what is now easternmost Russia by at least 20,000 years ago.[16] Around 15,000 years ago they had crossed what is today the Bering Strait, entering the Americas for the first time (Figure 5.1).

Some authors have tentatively suggested that the first Americans may actually have entered the continents through South America, where the earliest traces of human settlement are, rather than through North America, the northernmost part of which may have been covered by an impassable ice sheet for much of the latest Pleistocene.[17] There is certainly solid scientific evidence that the entire Pacific was traversed by humans (from Papua New

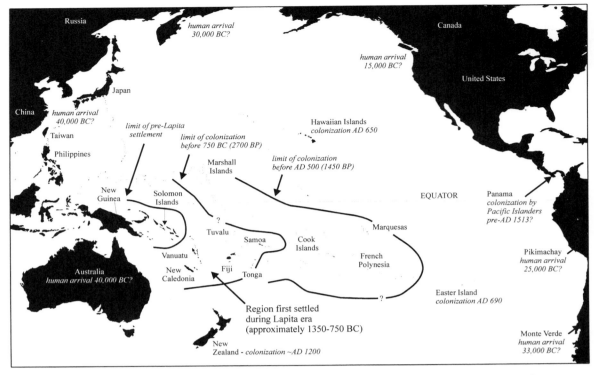

FIGURE 5.1. Map of the Pacific showing the pattern of human spread. The term "human arrival" is used to show the presence of humans in the area, the term "colonization" to mean permanent settlement.

Guinea to perhaps Panama or somewhere in South America) probably by around AD 1400, certainly before the first European saw the ocean in AD 1513 but no clear evidence that this was (or could have been) accomplished significantly earlier. Apparently the jury is still out on the origins of the first Americans.

The First Pacific Islanders

There have been various, largely unverifiable, ideas about the colonization of the more-remote Pacific islands before about 1330 BC that require but brief mention here. One idea is that people crossed the entire Pacific from west to east during (or just after) the last ice age when sea level was lower utilizing as stepping-stones landmasses that are now underwater (see chapter 3). Another idea, much quoted by diffusionists,[18] is that which involved a fishing party from Japan about 3000 BC being swept accidentally to Ecuador where the Jomon culture (from Japan) became fused with that in South America to give rise to the distinctive Valdivia culture.[19] The ceramics that define this culture allegedly contain elements such as castellated rims that have been argued to be Jomon introductions although, as research on South American ceramic traditions has increased, there have been many refutations of this idea.[20]

The actual story of the first Pacific Islanders is less serendipitous, far more firmly rooted in fact, but no less a source of pride to their descendants. It seems likely that the ancestors of the first people in the tropical western Pacific Islands came from coastal East Asia, particularly from the area that we now call southern China and Taiwan, where the diversity of Austronesian languages, of which most Pacific Island languages are part, is greatest.[21] There has been some suggestion recently, particularly from DNA tracing, that the earliest settlers of Pacific Islands east of Solomon Islands originated in coastal Southeast Asia,[22] but for the purposes of the following discussion the difference is largely immaterial.

Let us say that the first modern humans in East Asia reached the Pacific coast perhaps 40,000 years ago and began systematic marine-food exploitation there 30,000 years ago or less.[23] The numbers of people living in coastal East Asia over the next 10,000–15,000 years may have remained low, because the earth passed through the coldest time of the past 150,000 years — the Last Glacial Maximum.[24] When the Last Glacial Maximum came to an end, perhaps 18,000 years ago, the climate of the earth began warming comparatively rapidly, and humans, like many other groups of living things, flourished as existing food supplies became greater and expanded their range as new possibilities for subsistence opened up. In coastal East Asia, this time is likely to have seen population densities begin to increase, with small groups wandering the valley floors and coastal lowlands, especially those around the mouth of large rivers. The sea-level rise that accompanied postglacial warming may have been largely unnoticed by these humans because they were largely nomadic.

But then between about 11,200 and 10,200 years ago came the Younger Dryas event.[25] The earth was plunged abruptly into extreme cold, comparable to that which characterized the Last Glacial Maximum, and ecosystems everywhere on the planet's surface faced a crisis. For humans, the crisis was mostly about finding sufficient food to survive in an environment where conditions had become far colder and drier than they had been in recent memory, and where forest and woodland ecosystems were disappearing together with the food sources they habitually contained.[26] It has been suggested that humans in East Asia (and elsewhere in the world during the Younger Dryas) responded to this food shortage by taking over food production: in other words, by beginning agriculture.[27] Involving both horticulture and animal domestication, lowland East Asia is one of the centers where agriculture appears first in the history of humans, but apparently only around 8,400 years ago, much later than the end of the Younger Dryas.[28]

Of course, a connection between Younger Dryas cooling and the start of agriculture might still be valid in this region. For by imagining that the population dwindled during the Younger Dryas to such a low level that it could subsist on the available wild foods, and that it was not until the Younger Dryas ended and populations increased to a density that could not be sustained solely by wild foods that the imperative for agriculture arose. Realizing that wild foods were in short supply, we can imagine that people began deliberately planting the seeds of wild foods to supplement what could be gathered. In the same vein, domesticating animals obviated the need to expend increasing amounts of energy locating and killing perhaps dwindling numbers of wild animals.[29]

The important adjunct to the development of agriculture was the development of sedentism, or a sedentary mode of life, in contrast to the nomadic way of life that people in East Asia followed before the Younger Dryas. Sedentism implies a tie between people and a place, and obviously when you start to plant crops (and to a certain extent when you domesticate animals) you begin to be tied by necessity to a particular place. This leads you to perhaps set up permanent dwellings and associated infrastructure, to invest energy in establishing ties with a particular piece of land. If this all sounds like a positive development though, then consider that it also automatically increases the vulnerability of your lifestyle to ruin by natural forces. And so it may have proved for at least some of the early agriculturalists of coastal and lowland East Asia.

If we jump forward in time in coastal East Asia from the Younger Dryas to about 8,000 years ago, we might see the lowlands around the mouths of the Changjiang and Huanghe rivers dotted with small settlements of agriculturalists, mostly planting rice or foxtail millet, and keeping a range of domestic animals. A plausible scenario is shown in Figure 5.2. Sea level was rising slowly: the earth was still warming following the Last Glacial Maximum and land ice was still melting. But then about 7,600 years ago, sea level rose rapidly, as much as 6.5 meters in 140 years. This was one of several catastrophic rise events (CREs) that took place during the rise of postglacial sea level.[30]

It is likely that several islands in the Pacific (and elsewhere) drowned rapidly during such CREs. If you look at Figure 2.7 and see all the islands between the Queensland coast of East Australia and New Caledonia that drowned as a result of postglacial sea-level rise, it is easy to imagine that some may have finally vanished within a couple of generations 7,600 years ago as a consequence of CRE-3.

It was not just in those areas that islands must have vanished at that time. In particular, islands must have drowned throughout the Southeast Asia region, as the Sunda landmass became a vast archipelago, gradually reduced in land area, under the influence of postglacial sea-level rise. The rapidity of CRE-3 might clearly have left an enduring impression on the area's human inhabitants, probably sufficient to become the subject of oral tradition and myth, including the common land-raiser myth-motif.[31]

But CRE-3 did not just drown islands. In places along the low, flat deltaic coasts of East Asia it may have caused the shoreline to move hundreds of kilometers landward, displacing in the process countless communities of agriculturalists and destroying their means of subsistence, as shown in Figure 5.2D. Of course it did not happen overnight, but it probably happened so rapidly that many groups were unable to adapt, except in the short term. A longer-term adaptation option for a significant number of such people may have been to take to the sea.[32] Then perhaps, finding drowned coastlines and displaced peoples everywhere they landed, they went further than merely crossing the sea in search of new lands

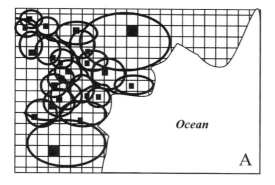
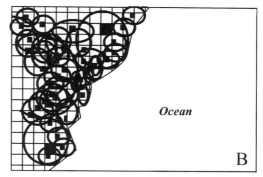
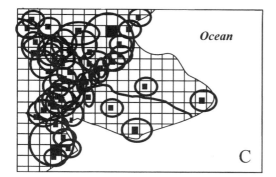
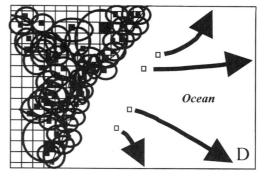

FIGURE 5.2. A scenario for the occupation and abandonment of the East Asia coastal margin under the influence of sea-level rise, particularly the "catastrophic rise event" (CRE-3) about 7,600 years ago. **A.** 9,000 years ago: Human communities (represented by filled squares) gathered and hunted wild foods from within elliptical areas. Where these overlapped, this signifies competition for the same resources; some communities consequently found it difficult to obtain sufficient food. **B.** 8,400 years ago: Postglacial sea-level rise led to shoreline recession, increasing markedly the degree of overlap between the areas on which adjoining communities depended. This situation forced the start of plant and animal domestication (agriculture) and an increased dependency on coastal (including offshore) wild foods. **C.** 8,000 years ago: Slower relative sea-level rise, particularly around the mouth of large rivers, created deltas that became occupied by agriculturalists. Elsewhere, as represented by the ellipses extending into the ocean, people became increasingly dependent on marine foods. **D.** 7,600 years ago: Then came CRE-3, involving perhaps 6.5 meters of sea-level rise in less than 140 years, which caused rapid shoreline recession and increased dependence on marine foods for many people. For a significant number of people, the best option was to sail in search of new lands to settle.

to settle. They decided that the sea was a better choice than the land, and became the first sea nomads.[33]

At least some groups displaced by CRE-3 from this area of East Asia who took to the sea moved south past the Philippines and into part of island Southeast Asia, eventually forming the remarkable Lapita culture in the offshore islands of Papua New Guinea.[34] The antecedents of Lapita pottery design have been traced back to Halmahera Island in eastern Indonesia,[35] and some of the rock art associated with the Lapita people in Papua New Guinea may have been derived from that in Borneo.[36]

The Lapita People and Their Origins

The original Papuan speakers who settled Papua New Guinea and the main part of Solomon Islands 30,000–40,000 years ago were joined around 1650–1350 BC by a distinctive group of people. These Austronesian language speakers, popularly referred to as the Lapita people, differed from the earlier peoples occupying this area in many ways, including their pottery-making abilities, their livelihoods based on a mixture of land-based horticulture and marine foods, and their extraordinary seafaring abilities (Figure 5.3). Many writers have commented on the latter, all agreeing that they were the greatest sailors of their age, able to successfully cross ocean distances many times greater than their counterparts anywhere else in the world at the time. The 1923 comment of ethnographer Elsdon Best says it all. For him, the first Pacific Islander

> was the champion explorer of unknown seas of neolithic times. For, look you, for long centuries the Asiatic tethered his ships to his continent ere he gained courage to take advantage of the six months' steady wind across the Indian Ocean; the Carthaginian crept cautiously down the West African coasts, tying his vessel to a tree each night lest he should go to sleep and lose her; your European got nervous when the coast-line became dim, and Columbus felt his way across the Western Ocean while his half-crazed crew whined to their gods to keep them from falling over the edge of the world.[37]

A clue as to why the Lapita people were such advanced seafarers may lie in their origins, described earlier, particularly their ancestors' enforced interaction with the sea resulting from postglacial sea-level rise and, ultimately, their being compelled by a rapid burst of sea-level rise (CRE-3) to begin the series of voyages that brought some of them to the Pacific Islands (see Figure 5.2).

From island Southeast Asia and the Philippines, the ancestors of the Lapita people began moving southeast, reaching the outer islands of Papua New Guinea, particularly those in the Bismarck Archipelago, before 1350 BC (see Figure 5.1). And it was in the Bismarcks around that time that we see the efflorescence of the phenomenal culture known as Lapita.[38] Lapita was a culture to which the ocean and its resources were central. Many of the earliest Lapita settlements in Papua New Guinea, such as that at Talepakemalai off Eloaua Island, involved dwelling houses built on stilts across the reef flat, with a connection to the land only for access to food gardens of taro and yam (Figure 5.4A). There is evidence that stilt-house occupations, which probably signify a preference for the ocean rather than the land, were built throughout the Lapita realm, through the eastern Solomon Islands, even to Fiji (Figure 5.4B).

The Lapita people colonized the entire western tropical Pacific, including parts of the island groups of Papua New Guinea and Solomon Islands, as well as Vanuatu, New Caledonia, Fiji, Tonga, and Samoa. Their voyages of colonization, which are judged to have been inten-

FIGURE 5.3. The Lapita people, the first Pacific Islanders, traveled thousands of kilometers across the ocean to colonize distant lands. **A**. The piece of obsidian, a type of volcanic glass, was found during excavations I directed at the early Lapita settlement site at Bourewa in Fiji in July 2005. It was found in layers that date back to around 1100 BC, marking the earliest-known human settlement of this island group. Obsidian does not occur naturally in the Fiji Archipelago. This piece has been traced to the Kutau-Bao area of New Britain Island in Papua New Guinea, some 3,300 kilometers distant in a straight line. Two hundred years ago, this discovery might have been taken to confirm that a continent once connected Fiji and Papua New Guinea but had since sunk. This is not true. Today this piece of obsidian is proof that more than 3,000 years ago the Lapita people sailed thousands of kilometers across the South Pacific, long before people elsewhere in the world could sail such distances, with the intention of settling far-flung island groups. **B**. The reconstructed head of Mana, a woman from the Lapita era whose skeleton was disinterred in 2002 at the Naitabale site on Moturiki Island in central Fiji during excavations I directed. This was the first time that a Lapita skull had been found sufficiently well preserved to allow the head to be reconstructed.

tional from the plants and animals they carried with them, involved open-ocean crossings of more than 1,000 kilometers (see Figure 5.1) against the prevailing wind.[39]

Questions about what motivated the Lapita people to set out on deliberate voyages of colonization from Papua New Guinea and the western Solomon Islands (what archaeologists call Near Oceania) over the eastern horizon and out into the open Pacific (what archaeologists refer to as Remote Oceania) are largely beyond the scope of this book. Yet what is relevant is to note that such voyages (and those of the ancestors that may have been enshrined in Lapita oral traditions) involved close observation of the ocean, and the islands within it.

After Lapita: The Peopling of the Remaining Pacific Islands

The Lapita diaspora, beginning around 1330 BC and accomplished incredibly within a couple of hundred years, reached Tonga and Samoa, both a little more than 4,000 kilometers of mostly ocean from the Lapita homeland in the Bismarck Archipelago. When we look at a

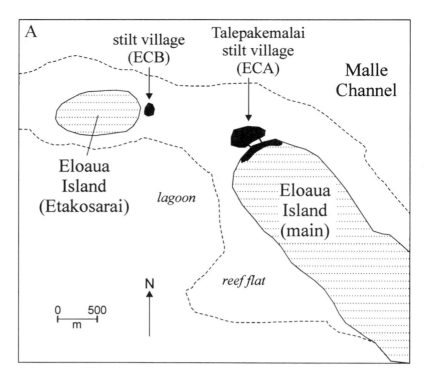

FIGURE 5.4. Evidence of the preference of the earlier Lapita people for building villages of stilt houses rather than living on the land. The Lapita settlement in A is about 3,700 kilometers in a straight line from that in B. **A**. Stilt villages at Etakosarai and Talepakemalai on Eloaua Island, Mussau Islands, Papua New Guinea, during the period of Lapita occupation (approximately 1350–850 BC). (Adapted and redrawn from Kirch [2001].) **B**. Photo of postholes that once possibly supported stilt houses, from excavations I directed at the early Lapita settlement (1150–550 BC) at Bourewa, Viti Levu Island, Fiji.

map of the entire Pacific today, that achievement may not strike us as truly incredible, but for a Neolithic people 3,000 years ago, likely traveling on bamboo rafts (possibly canoes), with probably no certain knowledge that they could encounter land, it ranks as one of the world's greatest-ever colonization movements.[40]

A few hundred years after the Lapita colonization of Remote Oceania, the Lapita culture, as indicated by elaborately decorated pottery, disappears from the archaeological record and is replaced in situ by cultures marked by greater intraregional diversity and perhaps less complexity. The people who lived after Lapita times were undoubtedly their direct descendants but did not decorate their pottery nor, does it seem, did they have the same drive to search for new lands across the eastern horizon. Such a judgment may be somewhat unfair, but it accords with the facts known currently and explains why a millennium or so apparently elapsed before islands east of Tonga and Samoa were eventually colonized.

This colonization process began with the tropical islands of French Polynesia, the Cook Islands, and Niue. On all these islands, the colonists seem to have been less well prepared than their Lapita ancestors, suggesting to some commentators that these islands were colonized unintentionally rather than deliberately.[41] Most brought no pottery with them, at least none that has been found, and apparently no one who could fashion pots from the raw materials available in these island environments.[42] There is little evidence that the first people on these islands, unlike their Lapita forebears, subsisted from any crops that they had brought with them, instead plundering the wild foods, including the birds (and their eggs) that commonly inhabited these islands in large numbers, before moving on to a neighboring island, where the process was repeated.[43]

Eventually, humans in the eastern tropical Pacific islands reached an accommodation with their environments, and the latest phase of long-distance voyaging and island colonization is likely to have been accomplished within a more viable social-economic framework. This phase included the colonization of the Hawaiian Islands about AD 650 and Easter Island about AD 690[44] and then, finally, somewhere in South or Central America perhaps around AD 1000 and New Zealand about AD 1200 (see Figure 5.1). The latter two journeys involved ocean crossings, probably without landfall, of at least 3,800 and 2,900 kilometers, respectively.[45] These incredible journeys serve to underline the close affinities between Pacific Islanders and the Pacific Ocean, affinities that have sometimes been underrated in assessments of Pacific Islander achievements.[46]

Environmental Crisis about AD 1300

The peopling of the entire Pacific (with a very few exceptions) before Europeans first arrived in this part of the world was conducted largely in peace. Obviously there would have been occasional flashpoints, occasional crises, but for the most part the achievements of the first Pacific Islanders were untarnished by conflict. This begs the question, why then are many of the earliest written accounts of Pacific Islanders so focused on warfare, on violence that is portrayed as endemic, and on practices like cannibalism and head-hunting that both enliven museum displays and shame the modern generation of (mostly) God-fearing islanders? Ultimately the answer lies, in my opinion, in climate change.[47] This is not an explanation that appeals to many archaeologists, human-focused by vocation, who prefer explanations involving human societal changes.[48]

Most Pacific Island societies, and many of those along the Pacific Rim, went through a profound crisis just after AD 1300. The crisis was marked not only by the outbreak of conflict on many islands, but also by the abandonment of (unprotected) coastal and lowland settlements in favor of ones on hilltops, in caves, or on offshore islands that could be defended. Cross-ocean contacts between islands ceased abruptly.[49]

The principal cause of all this was an abrupt decrease in the amount of food available to coastal dwellers, both from nearshore areas (coral reefs and lagoons) and in lowland areas; these people had become accustomed to subsisting off marine foods supplemented by root crops and vegetables grown on land. The clearest regionwide cause for this drop in the amount of food available, perhaps by as much as 80 percent in places, was sea-level fall. Around AD 1300, sea level across the Pacific fell by at least 70–80 centimeters, exposing the most productive parts of offshore coral reefs and lowering inland water tables, making it more difficult to grow root crops. And, to complete the story, the sea-level fall was driven by temperature cooling.[50]

The relevance of understanding the effects of this so-called AD 1300 Event in a book about vanished islands and lost continents is twofold. First, it demonstrates that, contrary to some recent ideas about human-environment (society-nature) interactions, humans in the Pacific Islands during the last millennium remained as close to the natural environment, and therefore as capable of creating and maintaining oral traditions recalling its vagaries, as they had ever been. Second, it makes the point that the framework for Pacific Islanders observing and interpreting the natural environment may have changed around AD 1300, with more reflection subsequently on the changes it exhibited. Many of the earliest written accounts of Pacific Islander beliefs (that all postdate the AD 1300 Event) tell of their fascination with evil, their apprehension of natural phenomena, their mistrust of other groups' actions and motives, and it seems plausible to suppose that all this was linked, ultimately, to the climate change of the AD 1300 Event.

The Latest Migrants to the Pacific Islands: The Period of Post-European Contact

Sustained European contact with most Pacific Island groups began in the first half of the nineteenth century and was followed by the colonialization of many a few decades later. Contact brought about massive changes in Pacific Island environments and societies, the latter changes exacerbated by the drastic reductions in populations on many islands from the introductions of alien diseases against which they had no natural defenses.[51]

Some of the first Europeans to live in the Pacific Islands were energetic recorders of Islander traditions and beliefs that had been passed down orally through many generations. These records are especially valuable because their only real alternative, modern oral and written traditions, have often been markedly changed from the ways that they were recounted 150 years earlier. Particularly influential as a source of recent change in Pacific Islander oral traditions has been Christianity, which taught many island people to abjure their traditional beliefs and lifestyles and to embrace their new beliefs to the point of almost complete denial of the old ones.

Care must therefore be taken in accepting modern oral and published versions of Pacific Island oral traditions to ensure that they record authentic rather than modernized or sanitized versions of particular stories. For example, an element that occurs in some modern oral traditions tells of a historical movement of people from one island to another being facilitated by the ocean parting to allow them through. This is likely to be an embellishment of a traditional tale derived from the parting of the waters of the Red Sea to allow the Israelites across in the Old Testament of the Christian Bible. Another example comes from Hawaiian mythology, where an account of a great flood involves a heroic survivor named Nuʻu, latterly identified as a derivative of Noah.

There are also oral traditions in some Pacific Island societies that were invented only after European contact. An outstanding example comes from Fiji, where today most indigenous people believe that their ancestors came from a place called Tanganyika in East Africa, in a canoe named the Kaunitoni, led by a warrior chief called Lutunasobasoba, who settled in Fiji first at Vuda on the west coast of the largest island, Viti Levu. The story is a total fabrication. It was a result of a newspaper competition in 1892–1894 in the Fijian-language *Na Mata* that invited people to submit an origin story for the Fijian people. This extraordinary competition, won by the Kaunitoni myth, was prompted by the frustration felt by missionaries who, upon repeatedly asking native Fijians where they came from, were told that they did not come from anywhere; they had "always been here."[52]

As part of the revival of indigenous knowledge and an appreciation of its key importance to both Pacific Islanders' cultural identity and the solution of many issues of environmental and resource management, there have been efforts to restore authentic oral traditions in many islands. Among these, perhaps those oral traditions of native Hawaiians and New Zealand Maori have been most fully researched and published; those of other island countries have generally been published with a view to informing schoolchildren rather than adults. There are very real concerns that globalization, in one or more of its many manifestations, will see an end to the unique traditions of many Pacific Island peoples.

But not if the new-age and pseudoscience writers have their way. For many of them, keen to create a sense of mystery where there is none, and to shamelessly capitalize on the ignorance of people who could not be expected to know otherwise, the pointers to a hidden land and a hidden past are everywhere. They are found especially in the character of the modern people of the Pacific Islands, a discussion of which forms the last section in this chapter.

Melanesians, Micronesians, and Polynesians: Real or Invented Categories?

For most new-age writers, the most cherished of all Pacific Island "races," because of their fairer skins and (in consequence) their supposed descent from the inhabitants of a lost continent (such as Lemuria or Mu), are the Polynesians.[53] Yet "Polynesian" was a term originally conceived and still widely used today by scientists to refer only to a person native to the geographical region named Polynesia. Its use as the name of a distinct ethnic category is far more tenuous because of problems of definition. As long as a Polynesian is merely any inhabitant born within the geographical subregion known as Polynesia, the definition is straightforward, but as soon as physical and cultural traits are assigned to the Polynesian "race," then problems arise. Some of these arise because many other Pacific people, who were not born in Polynesia, nevertheless share particular physical and cultural traits with those who were. The Fiji Islands, for example, are popularly regarded as the crossroads of Polynesia and Melanesia, and, although distinct physical traits and languages, for example, can be found in particular parts of the archipelago, it is manifestly naive to talk about a simple twofold ethnic distinction. And the same is true of the rest of the Pacific Islands.

It has long been the practice of geographers to divide the Pacific Islands into smaller, more readily describable geographical subregions named Melanesia (dark islands), Micronesia (smaller islands), and Polynesia (many islands). Yet considerable misunderstanding has accompanied the subsequent identification of Melanesians, Micronesians, and Polynesians as supposedly distinctive cultural groups inhabiting these (and other) subregions. In particular, misunderstandings have arisen because these groups are often assumed to be exclusive rather than part of the single cultural continuum that numerous scientific studies show them to be.[54]

Pseudoscience writers have long refashioned the origins of Pacific Island peoples for their own ends. Consider this nonsensical passage from a book published in 2001: "It almost seems as if Polynesians came out of nowhere . . . or from a place no longer on our maps, at least not on those of the ordinary variety."[55]

Many twentieth-century new-age writers go into even greater detail about aspects of contemporary Pacific Island culture and that of the supposed lost continents of Lemuria and Mu, about which you will soon read more. Some such writers envisage Polynesians as the surviving descendants of the "root race" that inhabited these continents. Other such writers, however, regard Polynesians as somewhat less advanced than their ancestors. At this point, it is worth noting that such writers not only invariably make massive and blatantly wrong assumptions about the history of humans in the Pacific Basin but compound this by regarding Polynesians, Melanesians, and Micronesians as distinct ethnic categories. The corollary to this error is that some explanation needs to be found for the presence of Polynesians on islands in the middle of the Pacific, surrounded by people of supposedly different races.

The truth is of course that the inhabitants of Polynesia and Melanesia today represent the result of a series of eastward migrations, perhaps initially by people who looked like Mana (see Figure 5.3B). The racial characteristics of these early settlers have become altered through time by intermarrying with people arriving in later migrations from a variety of places. It is as simple as that.

The fixation of such writers with Polynesians seems to stem from the naive perception that they are a fairer-skinned race than other Pacific Island peoples and therefore have to be the descendants of the root race of the advanced people who supposedly inhabited ancient Mu or Lemuria. Such rubbish may appear palatable, even implicit, because most new-age and many pseudoscience writers are probably writing for a fairer-skinned audience. Its racist implications, part, it must be said, of a long tradition,[56] have no scientific evidence in their support and appear abhorrent to right-thinking people in the twenty-first century.

6 Mythical Islands in Pacific Islander Traditions

Many Pacific peoples have a comprehensive body of oral traditions. Of course today, because the traditional way of life is falling apart in many places under relentless pressure from outside forces, oral (and many other) traditions are being forgotten. In such places, the indigenous people sometimes need to thank, however anathemic this might be, some of the earliest Europeans to visit them, who recorded at great length oral traditions of all kinds, particularly belief systems, traditional knowledge, kinship systems and genealogies, and myths. Had these records not been made, many Pacific communities would be much the poorer today for the concomitant lack of knowledge.[1]

An important consideration in assessing the authenticity or otherwise of particular details in oral traditions, both those recorded historically and those still known within particular communities, is the likely degree to which outside influences played a role in altering the content of these traditions. Clearly, the more sustained contacts a society had with people from elsewhere, particularly people from different cultures, the more likely it is that their oral traditions became diluted and are therefore less authentic than those from societies that remained comparatively isolated. For example, it has been argued that the oral traditions of isolated Niue Island in the central South Pacific are more authentic than those on other islands in the region because Niue's 2,000 years or so of human occupation was characterized by "isolation rather than interaction with other islands or archipelagoes."[2]

In the modern Pacific, there are many communities where one might reasonably assume, because of the degree of historical interaction with outsiders and the associated acculturation, that few authentic myths remain alive. Conversely there are islands and island groups where, because of continued relative isolation, many oral traditions remain little changed in essence and in the social context of their delivery from the way they were hundreds of years ago.

Pacific Islander oral traditions contain many tales about vanished islands. Typically such an island is recalled as a faraway place of origin for a particular group of people. Sometimes the geological and the environmental consequences of the island's disappearance are recalled, highlighting their impacts on particular communities and explaining the need for them to move. The cause of the island's disappearance is frequently attributed to a particular event, sometimes involving a god, although gods are more commonly claimed as being involved in island appearances, not disappearances. The most common of these myth-motifs is that involving the demigod Māui, who is reputed to have fished up islands across the Pacific using his magic fishhook.[3]

Oral traditions about vanished islands could be argued as having a higher probability of being authentic than traditions that concern other, less memorable, events. Instinctively it may be felt that the fact of an island disappearing, particularly if that disappearance was catastrophic and/or abrupt, is a subject worthy of embodiment into oral tradition, which may attract less subsequent embellishment than a less-remarkable story. Yet in Pacific Islander myths, many mentions of vanished islands appear as almost incidental details, leaving one to wonder whether they ever existed or whether their disappearance was so long ago that it has literally become a footnote in history.

This chapter deals exclusively with those islands in Pacific Islander traditions that are adjudged, on the basis of information available to me, never to have existed. Yet in many of the island groups from which such mythical islands have been reported, our knowledge of the natural environment is such that there may well at one time have been island disappearances, partial or total, from which details of these myths have been drawn.

Table 6.1 lists those vanished Pacific Islands for which (most) evidence suggesting that they could once have existed comes from oral traditions. Of the sixty-two listed, forty-one are regarded as unsatisfactorily authenticated and therefore classified in this book as largely fictional or mythical; they are discussed in this chapter. The remaining twenty-one islands are regarded as satisfactorily authenticated and are discussed in chapter 9.

A few Pacific Rim societies, commonly in East Asia, also have traditions about mythical islands. One such tradition from the AD 712 *Kojiki* (Record of Ancient Matters) concerns the origin of Japan and is considered to be an authentic tradition of the people living in those islands before the eighth century. In this tradition, the origin of habitable land is said to have been brought about by the consolidation of Izanami (Male-Who-Invites) and Izanagi (Female-Who-Invites). Specifically, it was Izanami who stood on the floating bridge of heaven and "plunged his jewel-spear into the unstable waters beneath, stirring them until they gurgled and congealed. When he drew forth the spear, the drops trickling from its point formed an island, ever afterward called Onokoro-jima, or the Island of the Congealed Drop."[4]

The island Awaji at the entrance to the Inland Sea is identified as Onokoro-jima.

TABLE 6.1. Vanished Islands in the Pacific.

(Note that none of the shallow underwater volcanoes that periodically erupt and form short-lived islands is included in this list. No islands known to be superficial islands [such as sand cays] are intentionally included in this list.)

Island	Island Group	Principal References
Unsatisfactorily authenticated islands (probably largely mythical)		
Arapa	Central Pacific (10° S to 30° N)	Henry (1928)
Ato Usune	Near Ulawa, Solomon Islands	This book
Davis' Land	Near Easter Island	Wafer (1934)
'Ereni Kaule	Near Ulawa, Solomon Islands	This book
Fasu	Yap, Federated States of Micronesia	Ashby (1983)
Fatu-pu	Central Pacific (10° S to 30° N)	Henry (1928)
Hiti-marama	Tuamotu Group, French Polynesia	Henry (1928)
Hoahoamaitu	Tuamotu Group, French Polynesia	Beckwith (1940)
Hotu-papa	Central Pacific (10° S to 30° N)	Henry (1928)
Huto Mwasa	Near Ulawa, Solomon Islands	This book
Iumurafa	Off Makira, Solomon Islands	This book
Kahiki	Hawaiian Islands	Fornander (1878)
Kāne-huna-moku	Hawaiian Islands	Lyons (1893), Beckwith (1940)
Ma-ahu-rai	Central Pacific (10° S to 30° N)	Henry (1928)
Maro'rouhu	Near Maramasike, Solomon Islands	This book
Matai-rea	Central Pacific (10° S to 30° N)	Henry (1928)
Matang	Kiribati	Grimble (1989)
Matennang	Banaba, Kiribati	Grimble (1972), Maude and Maude (1984)
Mone	Butaritari-Makin, Kiribati	Grimble (1989)
Mwarohura'a	Near Ugi, Solomon Islands	This book
Nono-kia	Tuamotu Group, French Polynesia	Stimson (1937)
Nuku-tere	Tuamotu Group, French Polynesia	Buck (1938a)
Oeni	Off Malaita, Solomon Islands	This book
O'o-va'o	Marquesas, French Polynesia	Handy (1930)
Outu-taata-mahu-rei	Central Pacific (10° S to 30° N)	Henry (1928)
Pali-uli	Hawaiian Islands	Beckwith (1940)
Pirokeni	Near Maramasike, Solomon Islands	This book
Pisiiras	Chuuk, Federated States of Micronesia	Ashby (1983)
Pororourouhu	Near Maramasike, Solomon Islands	This book
Raerii	Near Maramasike, Solomon Islands	This book
Raparapa	Central Pacific (10° S to 30° N)	Henry (1928)
Sipin	Yap, Federated States of Micronesia	Ashby (1983)
Ta'aluapuala	Off Malaita, Solomon Islands	This book
Taiero	Tuamotu Group, French Polynesia	Montiton (1874)
Tai-nuna	Central Pacific (10° S to 30° N)	Henry (1928)
Te-vero-ia	Central Pacific (10° S to 30° N)	Henry (1928)
Toko-eva	Marquesas, French Polynesia	Christian (1895, 1910)
Tonaeva	Marquesas, French Polynesia	Christian (1910), Luomala (1949)
Tongareva	Tuamotu Group, French Polynesia	Stimson (1937), Langridge and Terrell (1988)
Uririo	West of Samoa Group	Newell (1895)
Wawao	Near Ugi, Solomon Islands	This book

TABLE 6.1. *(Continued)*

Island	Island Group	Principal References
Satisfactorily authenticated or partly authenticated islands (probably real islands)		
Bikeman	Tarawa lagoon, Kiribati	Howorth (2000)
Bikenikarakara	Western Kiribati (Gilbert Group)	Grimble (1972), Ward (1985)
Burotu	Southeast Fiji	Geraghty (1993), Nunn et al. (2005)
Calafia	Channel Islands, California	Keller et al. (2001)
Fatu Huku	Marquesas, French Polynesia	Christian (1895)
Karawanimakin	Near Makin, Kiribati	This book
Kuwae	Central Vanuatu	Eissen et al. (1994), Monzier et al. (1994), Clark (1996)
Los Jardines	Northwest Pacific	Beaglehole (1967), Stommel (1984)
Malveveng	Off Malakula Island, Vanuatu	Nunn et al. (2006a)
Santarosae	Channel Islands, California	Porcasi et al. (1999)
Te Bike	Near Makin, Kiribati	This book
Tebua	Tarawa lagoon, Kiribati	Moore (2002)
Teonimanu	Central Solomon Islands	Fox (1925), Ivens (1927), Mead (1973), Nunn et al. (2006b); see also Appendix 1
Tolamp	Off Malakula Island, Vanuatu	Nunn et al. (2006a)
Tuanaki	Southern Cook Islands	Wi. Gill (1856), Smith (1899), Wy. Gill (1911), Te-ariki-tara-are (1920), Crocombe (1974, 1983); see also Appendix 2
(Unnamed)	Off Ambae Island, Vanuatu	Bonnemaison (1996), Nunn et al. (2006a)
Vanua Mamata	Central Vanuatu	Nunn et al. (2006a)
Victoria	Northern Cook Islands	Percival (1964)
Vuniivilevu	Central Fiji	Nunn et al. (2005)
Warapu (?)	Off New Guinea, Papua New Guinea	Churchill (1916), Beckwith (1940)
Yomba	Papua New Guinea	Mennis (1978, 1981), Blong (1982)

Mythical Islands of the Marquesas

Three characteristics of the Marquesas Island Group (Figure 6.1) in northeastern French Polynesia make it especially susceptible to large-scale collapse and large-wave impact, giving rise to an uncommonly large body of relevant geomythology.[5]

First is the steep-sided nature of the Marquesas Islands, which, like many similar linear intraplate island groups, have a hot-spot origin. The huge sediment apron that exists around these islands formed by perhaps hundreds of catastrophic collapses. Second, there are no coral reefs around the Marquesas, even though water temperatures are suitable for coral growth. This observation, although puzzling, is best explained by the steepness of the underwater island slopes. Finally, owing to the bathymetry of the ocean floor to the east, the Marquesas are especially vulnerable to long-range tsunamis generated along the ocean trench off the west coasts of Central and South America.[6]

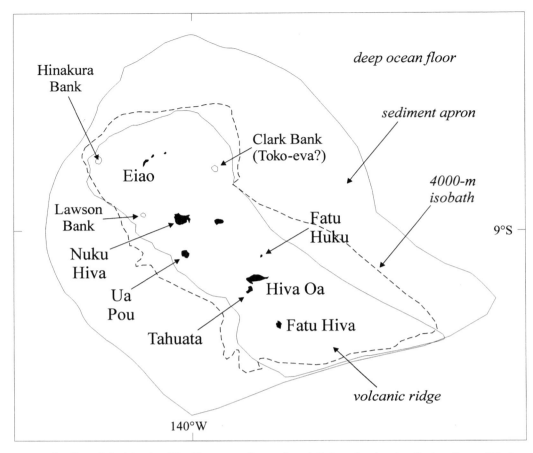

FIGURE 6.1. Map of the islands of the Marquesas Group, French Polynesia, showing the locations of Clark Bank (Toko-eva?) and Fatu Huku. Details of the area's geology include the volcanic ridge, formed by hot-spot volcanism, and the sediment apron, formed by the erosion of the island edifices over the past few million years. Note the location of both possible vanished islands (and Hinakura and Lawson banks) on the edges of the volcanic ridge where slopes are steep and where it is plausible to suppose that large-scale collapses have occurred. (After various sources, principally Filmer et al. [1994].)

There are numerous oral traditions in the Marquesas that tell of land sinking, often cata-strophically. It is likely that landslips of catastrophic proportions have occurred throughout the human occupation of the Marquesas and that these incidents have regularly rejuvenated older dying traditions about vanished islands in the area. Many such stories have been be-lieved in literally, so that at various times, often times of famine, people have searched for them. A 1930 account told of more than 800 canoes that had at some time set off in search of these fabled islands of plenty but only one canoe was ever heard from again.[7] Some of these expeditions for *he fenua imi* (land seeking[8]) may have been looking for the islands reputed to have existed in the voyaging corridor between the Marquesas and the Hawaiian Islands (discussed later in this chapter).

A likely mythical island in the Marquesas is Toko-eva. In his account of the tribal groups of the Marquesas and the geographical areas they occupied in the early twentieth century, Frederick Christian noted that northeast of the uninhabited island Eiao lies the "sunken land" of Toko-eva, which "once was a populous land."[9] The story was told to Christian by Titi-Ouoho, the aged chieftainess of Taipi Valley on Nuku Hiva Island, who recalled that Māui had once fished up Toko-eva from the underwater realm of Mahuike, the god of volcanoes and earthquakes. Angered by Māui's impunity, Mahuike caused Toko-eva to sink once more. Christian identified Toko-eva as Clarke's Reef, shown on more recent maps as Clark Bank.[10]

It seems quite plausible to suppose that Clark Bank, which is east of Eiao, not northeast, marks the site of an island that disappeared, perhaps sank in a catastrophic collapse event, but the possibility has not been investigated. Geophysical surveys in the area show that Clark Bank is on the eastern edge of the Marquesas volcanic ridge, where it slopes down into the sediment apron. It is more likely that a catastrophic flank collapse has occurred in this location, where slopes are steeper, than on the ridge crest.

Among the other islands that myth claims to have disappeared in the Marquesas Group is Oʻo-vaʻo, reported as a "land under the sea."[11] The sunken island named Tahu-uku by Christian[12] is probably Fatu Huku, described in chapter 9. There is a tradition of an island named Tonaeva near Tahuata Island said to have been fished up, as with many others, by the demigod Māui but then, less usually, allowed to sink again.[13] This may refer to Toko-eva[14] (see earlier in this section) or may recall a flank collapse of Tahuata itself or that of a now-submerged edifice nearby, not shown in Figure 6.1. As discussed later in this chapter, in Pacific Island myths from eastern Polynesia the place under the ocean to which the souls of the dead were said to go was often called Hawaiki. Most stories from tropical Pacific Islands state that Hawaiki was in the west, but in most of the Marquesas this is not so. Marquesan traditions explain that dead souls travel southeast along the island chain until they reach Hiva Oa Island, whereupon they leap off the high cliff at Kiukiu in the west of the island toward sunken Hawaiki. Many ethnologists have wrestled with this apparently contradictory account of the location of Hawaiki in the southeastern Marquesas,[15] but the simplest explanation may be that the pan-Pacific Hawaiki tradition here became enmeshed with a story of a sunken land in the southeast of the group, perhaps an island that indeed disappeared off Hiva Oa.

Alternatively it may even be that it was part of Hiva Oa itself that disappeared, for at Kiukiu there was a "high steep cliff" at the foot of which, Williamson reported (in a translation of a nineteenth-century oral tradition), "the breakers roared, and . . . closed the entrance to the nether-world. It opened or moved aside when the souls leaped upon it from above."[16] Perhaps it was here, early in the human occupation of the Marquesas, that a huge chunk of the western flank of Hiva Oa Island slipped into the sea, carrying people with it and becoming thereafter the abode of all dead souls.

Mythical Islands of the Tuamotus

The islands of the Tuamotus in French Polynesia are mostly low atoll islands, although some have experienced slight emergence as a result of passing across a low-amplitude intraplate swell known as the South Pacific Superswell.[17] Like the atoll islanders of Kiribati, the people of the Tuamotus have many stories about vanished islands. Some of these, as we shall see in chapter 9 when vanished islands in Kiribati are discussed, may refer to real islands that periodically form on shallow-water reef platforms when sediment is driven onto them during storms. The life of such islands (cays) is invariably short because the sediment of which they are composed is unconsolidated and may be readily removed from such platforms by waves subsequent to those that deposited them, giving rise to the observation of a vanished island.

Many Tuamotuan traditions refer to transitory islands of some kind, often described as floating or wandering, which reinforces the belief that they may be founded in part on observations of short-lived sand cays. The stories of wandering islands from the Tuamotus include those of Uporu and Havaiki (possibly the names of Upolu and Savai'i islands in Samoa) that are said to be visible to those on a boat halfway between them.[18] In another story, recalled in a traditional chant, the upstart god Tane is sent to pursue the swiftly moving island Nuku-tere, the name of which means "floating island."[19] The relevant part of the chant, translated by ethnologist J. F. Stimson, is as follows:

> Here is the Sailing Island, the swiftly fleeing land,
> Poised to depart on the long voyage to the far shore of Hiva-nui,
> Great land of darkness,
> Flocking birds, wheeling above the clouds, trail their fleeting shadows on the land,
> The Vanishing-Isle is as a migratory bird flashing in undeviating flight, now
> launched upon the wind.[20]

It is important not to interpret the idea of a wandering island too literally. It may simply be a metaphor for a group of itinerant people or the memory of a group displaced from a real island.

There are stories from several islands in the Tuamotus about the island named Tongareva; the suffix *reva* means floating, thus Tongareva is "floating Tonga."[21] It cannot be proved beyond doubt, but it seems likely that the myth of Tongareva predated the naming of Tongareva Atoll (also known as Penrhyn) in the northern Cook Islands.[22] The mythical Tongareva is typically mentioned merely in passing, as in a traditional tale from Vahitahi Atoll, where a man-fish character named Tongamaututu meets a school of fish while he is swimming who tell him that they are traveling to Tongareva, "underneath the deep bottom of the sea just beyond the reef."[23] From Anaa Atoll, the familiar story of the demigod Māui fishing for islands includes the detail that an island named Tongareva existed on the bottom of the sea.[24]

Other allegedly vanished islands in the Tuamotus include an island named Hoahoamaitu that is described as an ancient homeland of many Tuamotu Islanders that sank "beneath the waves."[25] The island Nono-kia is described as "a land flung down in jumbled ruins, - long since effaced from the memory of man."[26] The island of Taiero and its neighbors were "submerged" by the spirit of Temahage in revenge for his murder.[27] The island Hiti-marama has been "long since swallowed in the sea."[28] It is interesting that Hita-marama was located north of Pitcairn Island, where a number of shallow submerged islands have recently been mapped (see Figure 3.8).

Mangareva: The Myth of an Island That Sank?

Mangareva is the largest island in the Gambier Islands, located to the southeast of the Tuamotus and northwest of the Pitcairn Group. Throughout the Pacific Islands there are myths concerning the demigod Māui who fished up islands that exist still, but the story from Mangareva is somewhat different. In this story, while out fishing with his brothers, Māui tore off his ear and used it to bait his fishing line, upon which he eventually hooked an island. The following translation is by Sir Peter Buck (Te Rangi Hiroa), one of the most respected chroniclers of Pacific Islander myths.

> Maui called, *Haul up the anchor and paddle for the shore.* They made for shore, hauling the fish which loomed up like a mountain. The brothers made a great noise because the fish caught by Maui was land.
>
> Observing the commotion in the water, their mother came to see what was happening. She saw a mountain with trees on it and began to dance with excitement. The people arrived on the beach; the land was well elevated above the water.... As the land got close, Maui prepared to leap onto it to tether it, but the hook loosened and the land sank below the water.[29]

Mangareva is in a seismically and volcanically quiet part of the Pacific so it is unlikely that there have been any observations locally from which the fishing-up component of this myth originated; it is likely that it diffused here from elsewhere, perhaps the Tonga-Samoa region, where it is believed that fishing-up myths of the Pacific were conceptualized.[30] Yet the component of this myth that involves the land sinking after it had been pulled up may well recall a flank collapse of one of the islands in this group, quite possibly Mangareva itself. The northwest-facing coasts of Mangareva are deeply embayed, possibly evidence for island-flank collapses, one of which may once have carried away "a mountain with trees on it" (Figure 6.2).

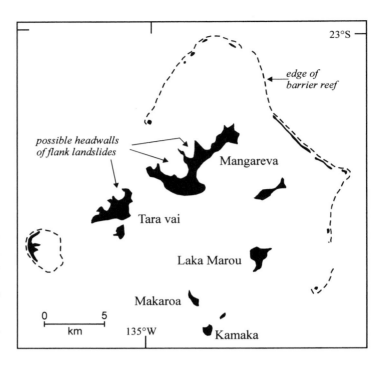

FIGURE 6.2. Map of the Gambier Islands (Mangareva Group) showing the deep embayments on the northwestern coasts of some of the islands that may be evidence of former large flank collapses.

Another Gambier Island tradition lends support to this suggestion. From the people of Tara vai Island, a story was collected in the late nineteenth century that recalled that their island was once "large and high," but the god Pere (cognate with Pele, the Hawaiian goddess of volcanoes) carried it away, leaving them only "this low island." The involvement of Pere may have been stimulated by contact with Hawai'i, where Pele is well known, but the component of this myth whereby part of the island was carried away may be autochthonous.[31]

Traditions of Vanished Islands from Easter Island

The issue of how and why Easter Island was discovered and settled has been much discussed and debated, often in the absence of much hard fact. Given the remote location and comparatively small size of Easter Island and, at least in recent times, the poverty of its natural environment, it is no surprise that many pseudoscientists have argued that the island was settled by refugees from a nearby island (or archipelago) that sank. This notion appears to have been conceived by John Macmillan Brown in his 1924 *The Riddle of the Pacific,* one of the most frequently cited sources of supposed Pacific Islander origins before these became the subject of systematic scientific study in the 1960s and 1970s. Brown supported his hypothesis by quoting the 1687 account of John Davis' apparent sighting of a small sand island in the area.[32] Brown believed that this island must have been the remains of a larger, higher one, which in turn signaled the presence of a sunken archipelago in the area.[33] If such a sand island ever truly existed in this area, a more plausible explanation is that it was a pumice mat or a sand cay, thrown up on a reef by a storm surge, that disappeared shortly afterward. Similar explanations have been given for sightings of wandering or vanished islands in the Tuamotu Group (see earlier in this chapter) and Kiribati (see chapter 9).

Brown's inference that the probably nonexistent Davis' Land disappeared sometime after Davis saw it in 1687 has spawned a rash of pseudoscientific accounts of the sunken homeland of Easter Islanders. In one, they are said to have come from the continent named Hiva that sank;[34] in another, Hiva was in the process of sinking when the people abandoned it.[35] The details of these stories appear to lie in the alleged traditions of the people of Easter Island themselves, as supposedly uncovered by Thomas Barthel in 1957–1958 in a series of written accounts "kept secret from outsiders until recently."[36] These stories are remarkable because they are significantly different in key details from those collected decades earlier on the island. For example, none of the origin stories recorded around the time of the first sustained contact between Europeans and Easter Islanders speak of their homeland [named Marae-renga] as having sunk,[37] whereas two of the traditions collected by Barthel do indeed state that they left their homeland [named Maori] because it was "disappearing into the sea."[38] For such reasons it seems prudent not to place too much credence on the Barthel collection of stories, which are "almost too neat to be anything but very valuable literature."[39]

It is unlikely that there are any authentic myths about vanished islands close to Easter Island preserved in the oral traditions of the island's inhabitants. As ethnographer Alfred Métraux bluntly pointed out, "there is no other Polynesian island whose past is so little known as that of Easter Island. This is largely due to sudden reductions of population . . . and the complete acceptance of European culture and belief."[40]

Mythical Islands of Hawai'i

It seems plausible to suppose that many accounts of supposedly vanished islands in Hawaiian traditions in fact recall the gradual disappearance of island homes over the horizon as they appeared to people voyaging to Hawai'i who thought it unlikely that they would see those homes again. Certainly the ocean corridor between Hawai'i and the tropical Pacific (Cook Islands–French Polynesia), which is today largely void of islands, has been claimed as the location for many lost islands, possibly because they marked the route that early voyagers followed.[41] In his 1878 compilation of Pacific Island myths, Fornander claimed that there are references to vanished islands "great and small" in this area in Hawaiian folklore,[42] just as they are mentioned in oral traditions from the other end of this ocean corridor. To make the point explicit, it is worth quoting part of a chant recited in 1817 by the Raiatean scholars Ara-mou'a and Vara titled *The Birth of New Lands*.[43] The part of the chant quoted describes how a voyager should proceed and what is encountered between Pukapuka Island (named Pupua) in the northern Cook Islands and Hawai'i (Aihi). The names of islands that could not be identified and may have vanished are underlined.[44]

Areare te tai, o Pupua	The sea casts up Pupua [Pukapuka]
Rutu a'e i toerau roa!	Strike far north!
Areare te tai	The sea casts up
O Nuuhiva roa,	The distant Nuuhiva [Nuku Hiva, Marquesas],
I te are e huti	Of the waves that rise up
I te tai o vavea.	Into towering billows.
Tari a oe i toerau i tooa!	Bear thou on to the northwest!
Rutu i hea?	Strike where?
E rutu i vavea!	Strike the towering wave!
Areare te tai, o Hotu-papa	The sea casts up Hotu-papa [surging rock]
O te vavea.	Of the towering wave.
Tari a rutu a oe i tea vavea!	Bear thou on and still strike the towering wave!
Areare mai, o Tai-nuna, fenua	There comes Tai-nuna [mixed-up shoal] land
I o atu i Hotu-papa.	Beyond Hotu-papa.
Areare te Tai o Putu-ninamu,	Sea of Putu-ninamu [sooty tern] casts up,
O Ma-ahu-rai te fenua;	Ma-ahu-rai [cleared by the heat of heaven] is the land;
Areare a,	There is cast up again,
O Outu-taata-mahu-rei.	Outu-taata-mahu-rei [the people's verdant headland].
Areare te Tai o Nu'u-marea,	The Sea of the Nu'u-marea [host of parrot fish],
O Fatu-pu maira.	Casts up Fata-pu [clustering pile].
Areare te Tai-o-manunu	Tai-o-Manunu [sea of cramps] casts up
O Te-vero-ia te fenua.	Te-vero-ia [fish-producing storm] Island.
Tari a oe!	Bear thou on!
Tari a rutu mai i hea?	Bear on and strike where?
I toerau.	Strike north.
Areare te tai, o Matai-rea,	The sea casts up Matai-rea [breeze of plenty],
Te fenua o te pahu rutu roa:	Land of the long beating drum:
O Taputaputea te marae hoho roa.	Taputaputea is the temple with long courtyard.
E rutu i hea? E rutu i toerau.	Strike where? Strike north.
Areare te tai, o Arapa iho;	The sea casts up Arapa [basket, island] alone;
O Raparapa iro.	Raparapa [angular, island] alone.
Tei tai atu o Tai-Rio-aitu.	Just over the sea is Tai-Rio-aitu [the star Aldebaran].
Tari a oe! E au tia e hea?	Bear thou on! And swim where?
E au tia i te taha o te ra,	Swim towards the declining sun,

E au tia i te Uru-meremere.	Swim towards Orion.
O atea te maunu a tae 'oe,	Distance will end at thine approach,
Tupu o ura.	Redness will grow.
E tupu i rei o te moua	It will grow on the mountain figurehead [a landmark]
A tae oe.	At thine approach.
Moti mai ai te moua i o atu e!	Where the mountain is the boundary over there, O!
A huti te vera hiehie;	Angry flames shoot forth;
Tupu o 'ura, tupu i rei	Redness grows, grows on the figurehead
Moti mai ai	Bounding in
Te moana i o atu e!	The ocean over there!
Oia o Aihi	That is Aihi [Hawai'i]

It is no surprise to also find traditions of these islands in the Marquesas Islands; one 1823 account mentioned "several islands which are supposed . . . to exist" and there have evidently been many fruitless expeditions to find these islands within the last few hundred years.[45]

In recent Hawaiian tradition, some of these "islands hidden by the gods" between French Polynesia and Hawai'i can sometimes apparently still be seen on the horizon at sunset or sunrise. At the command of the gods, they can emerge or submerge, approach close to land, and even float through the air. Twelve in number, the islands are sacred and must not be pointed at.[46] Although such traditions may mix a belief in vanished islands with various manifestations of volcanic eruptions, including mats of floating pumice, and other more enduring ocean marks, the stories about two of these islands, Kāne-huna-moku and Kahiki, are especially widespread and are discussed next. Other floating islands are known in Hawaiian traditions, the most common being Pali-uli, a paradisical island that is sometimes said to be deep under the sea, at other times floating above the clouds. More recently it has been claimed that Pali-uli is in fact hidden on Hawai'i Island itself.[47]

Kāne-huna-moku or the *'āina huna a Kāne* (hidden land of Kāne) is perhaps the most commonly mentioned of the vanished islands in Hawaiian mythology. It has been characterized as a "fairy island, to which the souls of good chiefs went after death."[48] Kāne was one of the creator gods of Hawai'i who for a time in late prehistory was evidently worshipped as "the great god" of the Hawaiian people.[49] Kāne rules Kāne-huna-moku, to which no land compares "in excellence,"[50] where the spirits of people go and live well, either to return to earth alive or to become gods and dwell in the clouds.

A literal belief in the periodic reappearance of Kāne-huna-moku has many parallels with those of the Fiji island named Burotu, discussed in chapter 9. In the early decades of the twentieth century, a family living at Hāna on Maui Island predicted that on a particular day Kāne-huna-moku would pass by and carry them away; strange shapes in the clouds seemed to presage the island's appearance but nothing happened.[51] Another account from Ka'ū on Hawai'i Island tells that when Kāne-huna-moku passed by, it was possible to hear cocks crowing and pigs grunting on the island, to see lights flickering and people moving about.[52] And on another occasion, two fishermen from Pu'uloa were blown to Kāne-huna-moku and brought breadfruit back to the Hawaiian Islands.

It appears that Kāne-huna-moku has been seen in many places within the Hawaiian Islands and beyond, which suggests that many instances of coastal submergence in these islands may have been enshrined in oral traditions, sometimes by modifying existing stories. For example, the occasional reappearances of Kāne-huna-moku in Hawai'i may be the outcome of a mirage or the appearance of a pumice mat, combined with a perceived need to reinforce some element of tradition.

Much has been written about the interpretation of a seventeenth-century Hawaiian epic poem called the *Chant of Kuali'i* in which a land named Kahiki is prominent.[53] In Andersen's translation, the relevant section begins as follows:

> O Kahiki, land of the far-reaching ocean,
> Land where Olopana dwelt!
> Within is the land, outside is the sun;
> Indistinct are the sun and the land when approaching.

The chant continues and describes Kahiki as a land where people are not like those of the Hawaiian race and speak a strange language. These details may preclude Kahiki from being Tahiti, as sometimes suggested, and it may be that Kahiki was one of the islands that once signposted the ocean route between Hawai'i and French Polynesia but has now vanished (see earlier in this section). Fornander interpreted the "within is the land, outside is the sun" detail in the chant as meaning that Kahiki lay east of Hawai'i, where islands were certainly once claimed to lie but probably never did, and it is possible that it refers to a place on the American mainland.[54] Andersen sketched a speculative scenario whereby a Spanish galleon voyaging between the Philippines and the Americas in the mid-sixteenth century came across Kuali'i fishing at sea, picked him up and took him to Acapulco (Mexico), then brought him back on a return trip.[55] In this interpretation, the observations made by Kuali'i became enshrined for posterity in his chant.

This may be altogether too elaborate an explanation. Kahiki may have simply been the Society Islands,[56] the name of a place that was fading in Hawaiian traditions and therefore had its character embellished to make the tales more memorable.

Mythical Islands of Western Kiribati

The island nation of Kiribati is one of the largest in the Pacific yet is mostly ocean. With the exception of the uplifted coral island of Banaba (formerly Ocean Island), all the islands in Kiribati are *motu,* low islands built on atoll coral reefs from reef fragments cemented together. Unlike many *motu* and atoll reefs in the Pacific, some of those in Kiribati are quite large and may have been inhabited for 2,000 years or more.

Some oral traditions from Kiribati talk about islands beneath the ocean as almost incidental details, perhaps a reflection of the comparatively high frequency of the disappearance (and creation) of sand cays on the reefs of these low islands by the action of large waves. One such account notes that an island named Uririo once existed between Savai'i Island in Samoa and Tarawa Island in Kiribati. Uririo was trodden on by the god Nareau as he traveled from Savai'i to Tarawa, and his weight caused it to sink.[57]

From Banaba comes a fragmentary tradition of a sunken island in the west named Matennang where the souls of dead fishermen go. Although Banabans insist that Matennang is "under the sea," it has been suggested that this statement should be interpreted figuratively to mean "over the horizon."[58]

Matennang is not the same place referred to as Matang. Inhabited by the old gods (*antin ikawai*), stories about Matang are common in Kiribati—it was the place from which the ancestors came and to which dead souls travel.[59] Matang is said to lie west of the Gilbert Islands of western Kiribati and is thought to be a tradition based partly on recollections of a homeland for at least one group of migrants. Like some other mythical islands, Matang is said to fly to heaven or sink beneath the sea when approached by the living.

There is another vanished-island tradition from northwestern Kiribati that refers to an island named Mone,[60] said by the people of Butaritari and Makin to be *i nano.* This term can either mean "beneath the ocean" or "in the west," and it seems probable that Mone was originally a homeland in the west but that it passed into myth as an island beneath the sea.[61] Better-authenticated myths about vanished islands from Kiribati are discussed in chapter 9.

Mythical Islands of the Federated States of Micronesia

The long ocean distances between the islands of the Federated States of Micronesia (FSM) and the close contacts that islanders regularly renewed across these distances after European contact provide the ingredients for a mass of stories about vanished islands in this part of the Pacific.[62] Unable to otherwise explain these regular long-distance contacts, John Macmillan Brown, in his 1924 *The Riddle of the Pacific,* was sure that islands and island groups (that acted as stepping-stones) must have disappeared in the region. More recent study of oral traditions recalling the history of these contacts suggests that they were sustained without now-vanished island intermediaries.[63]

Many of the islands in the FSM are high and rise steeply from the surrounding ocean depths. Many are also subsiding. The islands composing Chuuk (Truk), for example, are a classic almost-atoll. High volcanic Pohnpei (Ponape) Island has remarkably steep slopes but not so steep as in the Marquesas, where this factor explains the absence of fringing coral reefs. Many island groups in the FSM were settled about the same time as the Lapita people farther south were moving out into the tropical western Pacific, but the colonization of FSM appears to have been directly from the western rim of the Pacific.[64]

Like those of the Marshall Islands and other island groups in the northwest quadrant of the Pacific, Europeans made contact with the island peoples of the FSM far earlier in their history, typically in the early to middle sixteenth century, than those elsewhere in the Pacific, with the result that many authentic oral traditions have been lost or modified through acculturation.[65] Consequently there are doubts about the authenticity of the islanders' myths, which is why they are exclusively in this chapter and classed as unsatisfactorily authenticated.

Most parts of the FSM are in parts of the Pacific Basin where there is no significant seismic or volcanic activity. The exception is the islands of Yap State in the west of the FSM that straddle the convergent plate boundary along which the Pacific Plate was once being thrust beneath the Philippine Plate; the associated ocean trench, the Yap-Palau Trench, is thought to have ceased accommodating plate convergence around two million years ago (Figure 6.3). Despite this inactivity, there is likely to have been some residual movements, including those associated with instability along the steep trench walls. Islands like Ngulu and Yap itself are part of the associated island arc west of the trench, but emerged limestone islands like Fais and Ulithi lie on the eastern side, part of the upbent Pacific Plate. It is probably no coincidence that the stories about vanished islands from FSM come mostly from Yap State.

This corner of the Pacific Basin is considerably more complex geologically than once envisaged, with the identification of the Caroline Plate south of the main islands of Yap State.[66] The northern boundary of this plate, which may comprise elements of both seafloor spreading (plate divergence) and thrusting (plate convergence), passes through southern Yap State with islands ranged about it. This again provides good reason for understanding why most vanished-island traditions in FSM come from Yap. Four traditions are recounted here: about Fais, Fasu, Sipin, and Pisiiras.

The island named Fais in Yap State is an emerged limestone island and, like many of this kind throughout the Pacific, is regarded in myth as having been fished up. One myth explains that Fais was once "under the sea," and visited regularly by a woman named Lorob to gather "fruits and large baskets of food." This hints at submergence, which might be expected in this location. This particular myth about Fais represents the fusion of the popular idea of a limestone island being fished up with details observed locally of a fertile (part of an) island having been submerged, something that could have happened here within the few thousand years the islands have probably been inhabited.[67]

Fais lies on the Pacific Plate, in the area where it has been flexed upward before being thrust downward beneath the Philippine Plate to the west. As with islands like Niue that are in an equivalent geotectonic situation, there is ample reason to suppose that islands (or parts of islands) have slid downward to become submerged in the vicinity of Fais.

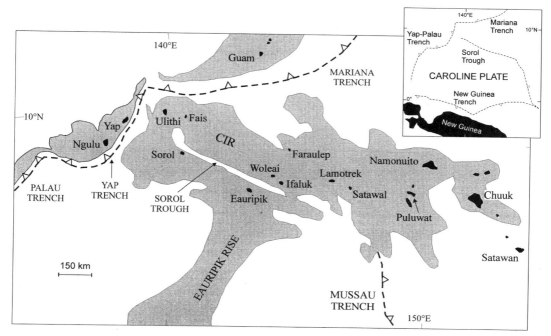

FIGURE 6.3. Map of part of the Federated States of Micronesia showing the locations of islands mentioned in the text. Note that, to improve clarity, atoll reefs have been filled in with black like other islands. Shaded areas in main map are ocean shallower than 4 kilometers. The ocean trenches in the area are shown with triangles pointing in the direction of crustal downthrusting. Inset map shows the boundaries of the Caroline Plate. CIR, Caroline Islands Ridge. (Map compiled from various sources, most geologic detail from Altis [1999] and Ali et al. [2001].)

The story of Fasu comes from Ifaluk (Ifalik) Island in Yap State.[68] Fasu is said to have been a high island east (or northeast) of Ifaluk that was caused to disappear by a visiting navigator who feared it was an approaching hurricane. Today the island is said to be barely submerged. That such a story is based on observations is believable if not currently verifiable, because both Ifaluk and neighboring Woleai Atoll rise from a former seafloor spreading center marked by the Sorol Trough (see Figure 6.3).

The Sorol Trough is interpreted as a former seafloor spreading center that has been inactive for the past million years or so, during which time rifting associated with plate convergence in the area has changed it to a strike-slip fault, part of the developing plate boundary marking the northern limit of the Caroline Plate.[69] Ifaluk and Woleai lie on the northern side of the Sorol Trough, rising from the Caroline Islands Ridge. It is possible that Fasu slipped off the crest of this ridge, perhaps in an event triggered by an earthquake along the Sorol Trough, but there are insufficient data to test this idea.

In a myth that is well known in Yap State, the island named Sipin one day disappeared suddenly.[70] The lifestyles of the people of Sipin had been among the most traditional in all the islands of Yap, but, as more outsiders came to live there, the chief of Sipin felt that his people's way of life was threatened. So he caused the island to disappear under the sea. At first, the myth tells, the island did not sink far so that people sailing across the area during the day could still see it. And at night it was possible to see the torches and hear the voices of the people below. But now the island has sunk too deep for it to be seen or its people heard.

Sipin was reportedly located east of Rumung Municipality in the north of Yap Island, close to the steeper side of the Yap Trench. Although such a location is a likely site for an island to have disappeared, slipped down the trench wall, the absence of any other corroborative information suggests that Sipin probably never truly existed.

In the lagoon of Chuuk (Truk) Atoll, well away from any plate boundary, there is an island called Moen near which lies the reputed "lost island" of Pisiiras. There is a somewhat anticlimactic myth that tells of five brothers who searched unsuccessfully for Pisiiras; all but the youngest gave up, and he was guided there by a shark.[71]

This story cannot be readily authenticated, although Chuuk is a classic example of an almost-atoll in which a volcanic island is being gradually transformed through subsidence into an atoll. In such a scenario, part of the Subsidence Theory of Atoll Formation illustrated in Figure 2.6, one expects the islands to disappear slowly, not so fast that any sizeable island could have disappeared within the past few thousand years. It is possible that generations of islanders came to recognize the signs of submergence of the existing islands and linked them, correctly, to the origin of shallow underwater banks, perhaps favored fishing grounds, as Pisiiras may have been.

Hawaiki: The Generic Myth of the Pacific

Many Pacific Islanders, both today and in the past, believe that they reached the places where they (or their immediate ancestors) lived from a land known as Hawaiki. Sir Peter Buck explained that Hawaiki had once been a "land below the sea." The "Great Source" Tumu-nui raised this land to become a coral reef and then an island.[72]

> Then the Heavens became light, the early dawn, the early day,
> The mid-day. The blaze of day from the sky.
> The sky which floats above the earth
> Dwelt with Hawaiki.[73]

This section does not attempt a comprehensive review of ideas about Hawaiki and its functions, which has been done well elsewhere,[74] but focuses on arguments as to whether it is today a real "sunken land."

In most islands in Polynesia east of Tonga and Samoa, Hawaiki is the homeland of the people, and it lies to the west. In many Maori traditions, Hawaiki is the place from which their ancestors who settled New Zealand originated. In these traditions, as in the scientific record, particularly of obsidian circulation, there is evidence for return voyaging from New Zealand to Hawaiki. For example, some of the earliest Maori settlers of the Bay of Plenty are said to have returned to Hawaiki to bring sweet-potato (*kumara*) seeds for planting. The stories they told about New Zealand upon reaching Hawaiki are said to have inspired a great migration from there about AD 1350.[75] One explanation that has been given for the colonization of New Zealand about AD 1250–1300 is that the homeland named Hawaiki had become overpopulated, stimulating the migration in ten canoes — the "great fleet" of Maori myth.[76]

In many versions of the myth of Hawaiki, it is the abode of the gods. This accords with those traditions of Hawaiki as the nether world inhabited by the spirits of the dead: a "veritable Hades, the shadowy Under-world of death, and even of extinction."[77] The combination of these two explanations with that of Hawaiki as homeland is natural, recalling not only the direction and place the ancestors occupied but also that to which the dead return to join the souls of their ancestors.[78] Many Pacific Islands (or island groups) have well-known pathways that dead souls follow that typically lead to a coastal promontory, commonly facing west (the direction of Hawaiki), off which they leap.

For most Pacific Islanders in the past, perhaps the most important function of Hawaiki was as a symbol: a symbol of shared ancestry, of seafaring achievements, and of cultural antiquity.[79] And it is for that reason that Hawaiki has been invested with a range of functions, including mystical ones that have endeared it as a real place to many new-age thinkers. Conversely, it has also led to the suggestion that the entire story, along with other such Maori traditions, is symbolic rather than literal.[80]

Hawaiki is linguistically cognate with 'Avaiki, which in many Cook Island myths is said to be below the ocean surface. A myth from Mangaia Island in the southern Cook Islands, for example, talks of 'Avaiki as the underworld inhabited by the gods:[81] an account that laid the foundations for more recent ideas about 'Avaiki as a land beneath the sea. For many writers, the fact that dead souls heading for Hawaiki are last seen leaping into the ocean was enough to confirm that Hawaiki was below the ocean surface.[82] The inference became a truth. Writing in 1920, John Macmillan Brown described Hawaiki as "now a mythical and shadowy land beneath the waters."[83]

It is not possible to identify with certainty the origins of the idea that Hawaiki had sunk, but it might lie with a Maori myth, told to Elsdon Best by the Takitumu people, that

> about eight generations before the coming [to New Zealand] of the Takitumu canoe from eastern Polynesia, a volcano named Maunga-nui, at or near Rangiatea [Ra'iatea Island in French Polynesia], was destroyed by a terrific explosion. At the same time an extensive tract of land called Whainga-roa was submerged by the sea, in which disaster whole tribes perished.[84]

Although such a disaster may indeed have occurred in the Ra'iatea area during the early period of its human occupation,[85] the critical things are that the tradition does not refer explicitly to Hawaiki nor does it suggest that the events described led to the migration of people from the Ra'iatea area to New Zealand. Yet many stories about Hawaiki refer to it as sunken or submerged today and describe the people who left it as refugees. One story, echoing those about Atlantis, refers to Hawaiki as "a veritable Paradise before its destruction by a huge volcanic cataclysm."[86]

Some pseudoscience writers have suggested that people left Hawaiki because it was sinking.[87] This does not appear to be a widespread detail in authentic Pacific Island myths and may have been invented by unscrupulous theorizers to justify linking Hawaiki with supposedly sunken continents like Lemuria and Mu in the Pacific. But these continents never existed (see chapter 7), and to connect them to Hawaiki, which probably was a real place (or places), is to trivialize a revered tradition.

Whether as a homeland or the place where dead souls went, almost all traditions state that Hawaiki was in the west. The belief that this is consistent with the migration pattern of Pacific Island people (shown in Figure 5.1) may appear uncontroversial, although it could simply manifest, as in many other cultures, an association between death and the setting sun.[88]

It is most plausible to assume that Hawaiki was (or is) an island — probably not the same island for every group of people possessing relevant oral traditions — from which one or more memorable migrations took place, probably within the last 2,000 years. One of the first people to propose that Hawaiki referred to a real place was Horatio Hale, attached as ethnologist and linguist to the 1838–1842 United States Exploring Expedition. Hale argued that the name (if not the place) Hawaiki referred to Savai'i, the largest island in the Samoa Group, and few have found reason to dispute this association.[89] It is possible that Savai'i was just one of the places named as Hawaiki, perhaps because the people of eastern Polynesia had arrived there from multiple homelands, each of which came retrospectively to be recalled as Hawaiki. It is also possible that the name Hawai'i (for both the island group and the island Hawai'i) came from people from eastern Polynesia who had sought to return to Savai'i but were blown off course to Hawai'i, naming it such because they believed it at first to be Savai'i.[90]

Another prime candidate for Hawaiki appears to be the island of Ra'iatea in French Polynesia, which was identified as such in a number of accounts by New Zealand Maori collected during the middle and late nineteenth century.[91] Also compelling is the idea that Hawaiki referred to a group (or groups) of islands rather than a single island.[92]

For linguists like Percy Smith, writing around the start of the twentieth century, who believed that "Polynesians" had come from southern Asia, Hawaiki was in India:[93] an idea that is no longer regarded seriously. Others regard Hawaiki as having been the "sunken continent" of Southeast Asia, one even trying to justify this by claiming that Hawaiki is cognate with Jawaiki (it is not) and therefore regarding it as synonymous with Java Island in Indonesia.[94]

The Maori belief that *kumara* (sweet potato) was introduced to New Zealand by people from Hawaiki, alluded to earlier in this section, might also sound relatively innocuous but, when combined with the probability that *kumara* was introduced to the Pacific from South America (where it originated), has provided a fertile mix for pseudoscience theorizing. Some state that "Polynesians" must have come from South America.[95] As explained in chapter 5, they did not. Others find in this support for the idea of a sunken continent in the Pacific that had connections with South America, but such ideas are outlandish, however appealing they might sound.

Pulotu

The story of Pulotu is one that is deeply immersed in many Pacific Island cultures, and almost certainly much more information about the island and the circumstances of its disappearance lies out there in the unwritten oral traditions of Pacific Island peoples. Yet, like Atlantis and possibly Hawaiki, the story of Pulotu appears largely allegorical, so it is considered here as a mythical island. Yet with that in mind there is strong evidence that many stories about Pulotu were built around the story of an island named Burotu that did indeed disappear, probably somewhere in southern Fiji (see chapter 9).

This account of Pulotu is not intended to be all-encompassing; it is focused only on those details that represent it as an island that disappeared. At the outset, it should be noted that many writers have emphasized how varied are oral traditions about Pulotu, compared with those of Hawaiki, for instance, and how they often contradict one another.[96] Yet, in general it appears that it is the indigenous people of eastern Polynesia who referred to Hawaiki as their homeland; those in western Polynesia (principally Samoa and Tonga) spoke instead of Pulotu as their homeland.

The earliest written account of Pulotu, collected from Tonga by William Anderson in 1777 during the third voyage of Captain James Cook, stated that Tongans

> say that immediately upon death [the soul of a dead chief] . . . separates from the body and goes to a place calls Boolootoo [Pulotu], the chief or god of which is Gooleho [Hikuleʻo]. They feign that this country was never seen by any person, is to the westward beyond Feejee [Fiji] . . . and that there they live fore ever, or to use their own expression, are not subject to death again and feed upon all the products of their own country with which it is plentifully storʼd.[97]

In 1789, James Morrison, second mate of HMS *Bounty,* wrote the first description of the island of Tubuai in the Austral Islands (French Polynesia), remarking that "some of the people had been driven [to Tubuai] from an island far to the westward which they had called ʻParoodtoo.ʼ"[98]

An important written source of information about Pulotu is the journal of William Mariner, who was in the Kingdom of Tonga from 1807 to 1811, long before there had been any significant European influence on Tongan traditions. Mariner recorded the Tongan belief in Pulotu (Bolotoo), a large island to the northwest

- where gods resided, including Tangaloa, the father of Māui, who had caused the islands of Tonga to emerge,
- where the spirits of the dead went, and
- where their ancestors had lived before they reached Tonga.

This account neatly identifies the three common functions of Pulotu and Hawaiki (see previous section) in myths from throughout the Pacific Islands region. Mariner also echoed the commonly held beliefs that Pulotu lay to the west of Tonga and was both invisible and inaccessible to mortals.[99] Numerous other legends about Pulotu that bear out these beliefs are known.[100]

A radical new interpretation of the Pulotu myths was proposed by linguist Paul Geraghty in 1993. He argued that although the Pulotu-as-homeland myth is widespread throughout Polynesia, it did not arrive on most islands with their first inhabitants but only with later groups, refugees from a conflict in Fiji sparked by the dwindling supply of red parrot (*kula*) feathers, a prized cultural symbol. Geraghty proposed that it was the island Matuku in southeastern Fiji where this conflict erupted and from which many of the place-names and familial names associated with the Pulotu legends derived. Pulotu-as-Matuku may perhaps have been described as disappeared in some legends because of the island's sharply reduced importance to the regional *kula* trade. Matuku may have had its reputation as an island of mystery enhanced "by a natural catastrophe of some description,"[101] a subject that is discussed further in chapter 9.

Mythical Pacific Islands of Modern Times

Islands were invented by some of the earliest European explorers of the world's oceans for reasons that are much easier to comprehend, commonly associated with an individual's self aggrandizement.[102] In the Pacific, persistent rumors of a cluster of islands (named the American Group) along the shipping route between Panama and Hawai'i were quashed during the nineteenth century by the systematic surveys of countless ships' captains. But then on March 17, 1902, the question of the existence of the American Group suddenly made headlines when the SS *Australia,* captained by Robert T. Lawless, reported shallows in the area where these islands had reputedly existed. A strong supporting statement was made by Captain John DeGreaves, self-styled science adviser to King Kamehameha of Hawai'i, who claimed to have picnicked on the island some years earlier in the company of actress Lola Montez. Had DeGreaves' statement been less than authoritative, it is unlikely to have been quoted in an article on May 4, 1902, in the influential *New York Herald;* his island was last seen on the map in the December 1921 issue of *The National Geographic Magazine.*[103]

There have been many genuine mistakes made in the identification of islands where none ever existed. These vigias, or reported shoals, have been attributed to schools of breaching fish, rushing up to the surface and disturbing it in the same way as when a wave meets a shallow coast or the edge of a coral reef. Likewise the white scum that floats away from the seasonal rising of the *palolo* (*Eunice viridis*) worms has been mistaken for islands.[104] And of course there are many more solid yet transient islands, such as those made from masses of floating pumice or mats of vegetation, that periodically appear on the ocean surface and are acknowledged as important agents of cross-ocean dispersal of biota.[105]

7 Mythical Continents of the Pacific

The best-known island to have vanished in the history of the world is named Atlantis, which according to the Greek philosopher Plato sank around 9600 BC in the Atlantic Ocean.[1] But for many people in the Pacific, as well as many in Asia and much of the non-English-speaking world, the name Atlantis conjures up no such excitement. It has no significance. Notwithstanding this, in terms of inspiring and informing the modern vanished island and hidden continent myth-motif, the story of Atlantis has been remarkably influential. Not only does it directly continue to generate a huge amount of interest, but it has also inspired an entire genre of mythmaking (ranging from vehicles for social comment and satire to literature and entertainment) and has permeated global consciousness in ways that we shall probably never fully appreciate. But, for all the hype, Atlantis never existed. The same is true for most of the continents suggested as having once been emergent in the Pacific.

Size: Does It Matter?

Many pseudoscience and new-age writers appear uncertain as to the difference between an island and a continent other than the supposedly greater size of the latter. Their accounts often describe something that is physiographically more insular than continental in character — for instance, having only a few major rivers draining the center radially[2] — yet the dimensions of which are stretched beyond what is geographically credible.

It is likely that many such writers have invoked the existence of former continents (rather than islands) because their greater size might be regarded as adding weight to the descriptions given and arguments outlined. Size does matter. The larger the stage for the human actors, the more believable it is that they actually accomplished all the things that are claimed. Another reason for such writers favoring the existence of former continents rather than islands is that they are targeting continental audiences, who might, they think, regard an island as being somehow less compelling as a place of human origin, for instance, than a continent.

Obviously, there is less potential for complex social interaction and change within a smaller island than on a larger one or on a continent, but island communities have their own idiosyncrasies that are worthy of study.[3] Yet, new-age and pseudoscience writers require larger areas for their invented scenarios. For example, how can one suppose that ancient Egyptian sailors made such a fuss about Nan Madol on Pohnpei Island in Micronesia had it been only the handful of tiny islands that it is today?[4] How can the temples (*marae*) on remote Malden Atoll have been built (as they were) by this island's former inhabitants rather than being interconnected through a grand system of roads with similar features on other islands nearby[5]? It is precisely this kind of embellishment that made it necessary for Plato's island of Atlantis to evolve into a continent, for how else could Atlantean society have developed all its fabulous accoutrements, unless it was exceptionally large?

Exploring the Pacific: The Lure of Hidden Continents

The first people known to have explored the Pacific Islands entered the region from its western rim around 1330 BC, having honed their long-distance sailing and navigational skills in the voyaging nursery of the Bismarck Archipelago of Papua New Guinea. As discussed in chapter 5, these early people, the Lapita people, sailed against the prevailing wind and thus came to discover the archipelagos to the east: the eastern outer Solomon Islands, Vanuatu, New Caledonia, Fiji, Tonga, and Samoa. These remarkable voyages of colonization are thought to have been intentional because the Lapita people appear to have carried with them all the things they believed they would need to re-create their homelands in the places they expected to find.[6]

But what did they expect to find? They and their immediate ancestors had lived in archipelagos where islands were, if not visible from one another, only just beyond the horizon. In the absence of a world map, it was perhaps not unreasonable for Lapita people in the Bismarck Archipelago about 3,400 years ago to believe that they would encounter another island over the eastern horizon, even before they had explored in that direction.[7]

It is possible that they expected to find a landmass even larger than an island. As explained in chapter 5, we are fairly sure that the Austronesian-language-speaking ancestors of the Lapita people lived along the continental fringes of East Asia. Certain groups are inferred to have passed through the Philippines, some of the islands in eastern Indonesia (including Halmahera), and thence east into the outer islands of Papua New Guinea, including the Bismarck Archipelago, where around 1350 BC they became the Lapita people.

Could it be that the Lapita people sailed east into what we now know as the island Pacific — Remote Oceania — believing that they would encounter a continent? It may seem a farfetched notion, but we know nothing of their worldview, and ideas of balance between large and small landmasses in space were central to Greek global cosmogony 1,000 years later. And after all, their ancestors had left a continent for seas full of islands, so it may not have seemed unreasonable for them to posit the existence of another continent over the next unexplored horizon.

If the story about the Lapita people in search of an unknown continent nearly 3,400 years ago seems far-fetched, then it may come as a surprise to learn that this vision was ultimately fulfilled. For although the Lapita diaspora lost its momentum about 950 BC in Tonga and Samoa (an astonishing 4,500 kilometers from its origin in the Bismarck Archipelago), the descendants of the Lapita voyagers, who subsequently continued the eastward migration across the Pacific, appear to have reached three places along the west coast of the Americas, including Panama (almost 14,000 kilometers in a straight line from Papua New Guinea), long before the first European did.

The European in question, Vasco Nuñez de Balboa, saw the Pacific on September 26, 1513.[8] His contemporaries found coconuts in the area and in neighboring Columbia; there were also reports of Melanesian-looking people living there. Coconuts originated in the Southwest Pacific, possibly in Queensland (Australia) or New Caledonia, yet succeeded in colonizing many Pacific Islands before humans arrived. Experiments show that a coconut can remain viable bobbing around in seawater for almost eighteen months and will germinate once it lands on a suitable substrate (such as an island beach). Ocean currents could have carried coconuts throughout the tropical Pacific, but not to Panama. For the ocean current (named the Peru or Humboldt Current) that runs northward off the west coast of South America and then westward across the tropical Pacific has not shut down since it started, around the time of closure of the Panama Isthmus nearly three million years ago. It is simply not possible for a coconut to have crossed the easternmost Pacific unaided between an island in French Polynesia, for example, and Panama. The only way a coconut could have crossed that gap would be if it had been carried across it by humans. Therein lies the prima facie case for humans having crossed the entire Pacific, from west to east, long before this vast ocean became known to Europe.

Other lines of evidence for pre-European contact by Pacific Islanders with the west coast of the Americas have also been presented. The most long standing of these is the evidence of the design and name of the polished stone axes of the Mapuche people of Chile, which have undeniable affinities with those of New Zealand Maori.[9] This evidence has recently been supplemented by the discovery of caches of Polynesian chicken bones in Mapuche sites, something that puts the issue of this contact event beyond doubt.[10] Farther north, there is linguistic evidence for at least one prehistoric contact event between the people of Polynesia and the Chumash of southern California.[11]

The Pacific Islanders who repeatedly traversed large parts of the Pacific Ocean acquired a massive body of lore about its geography, some of which will be explored later. But there was no belief amongst them that has come down to us today suggesting that they ever thought a huge continent might exist (or have once existed) in this vast region.[12] Yet, astonishing as it may seem today, the belief in just such a continent was one of the main reasons for Europeans to come to the Pacific centuries later.

The Search for *Terra Australis*

The ancient Greeks believed that the world was round[13] and, from their knowledge of the size of landmasses in the Northern Hemisphere, considered that there must be an equal amount of land in the Southern Hemisphere to balance them.[14] This was one of a number of philosophical expressions of the Greek concept of equilibrium. The Greeks' idea of a great southern continent was passed on to Roman and Egyptian thinkers by several key people, including Claudius Ptolemy, regarded by many as the father of modern astronomy and geography. Neither the Greeks nor anyone else at that time is known to have traveled far enough south to confirm the existence of this supposed great southern continent, or *Terra Australis* as it came to be known.

The earliest records of the intentional search for *Terra Australis* come from the European explorers of the Pacific. When Ferdinand Magellan passed into the Pacific through the islands at the southern tip of South America in 1520, he thought that the land to his south, Tierra del Fuego, might in fact be the tip of *Terra Australis*. The awful weather that is common in that part of the world, particularly around Cape Horn, caused many ships to be blown off course and to end up sighting land that was thought might be *Terra Australis*, thereby reinforcing a belief in its presence there.

But there were more pragmatic reasons for wanting to find *Terra Australis*, especially before it was found and laid claim to by another country. Reports of fabulously wealthy islands in the South Pacific collected from the Incas by the Spanish in Peru in the decades following their conquest in 1533 spurred the European search for *Terra Australis*. For example, an Inca named Tupac Yupanqui was reported to have visited the islands Hagua Chumbi and Niña Chumbi and returned with "gold and silver, a copper throne, black slaves and the skin of an animal like a horse."[15] Initial thoughts were that these and similar islands[16] were the Islands of Solomon, where the biblical king had acquired his riches. It was a Spaniard, Pedro Sarmiento de Gamboa, who first became convinced that these islands were part of a continent stretching from Tierra del Fuego north to within 15 degrees of the equator in the Pacific. The first Spanish expedition charged with discovering this continent left Callao in Peru under the command of Alvaro de Mendaña in November 1567 and found the several large islands that became known as Solomon Islands. Mendaña believed that the islands he had visited were outliers of the great continent he sought and returned there in 1595. Later Spanish expeditions to search for *Terra Australis* included that of Quiros in 1605.

The reputed wealth of *Terra Australis* led to other countries and companies taking an interest in finding it. The Dutchmen Schouten and Le Maire left Holland in June 1615 with the express intention of landing on *Terra Australis*. They headed for the Juan Fernández Islands, 670 kilometers off the Chile coast, and then northwest and west where the continent should have lain. Schouten wrote that at first they "were looking out eagerly for the southern land" but that later they were "almost despairing and fearing there was no such land,"[17] a tale similar to that of many others.

By the mid-seventeenth century, the numbers of Spanish galleons regularly plying their way across the Pacific Ocean carrying valuable cargoes between the Philippines and Mexico caught the attention of British privateers. They attacked the galleons off the South American coast and sometimes ventured farther out into the Pacific. One of the privateers was Edward Davis, who in 1687 apparently discovered land some 2,800 kilometers west of Copiapo on the coast of Chile. This became known as Davis' Land but it has apparently never been found again. Although the latter fact is relatively unimportant — many islands were not properly observed, invented by unscrupulous ship's captains, or wildly mislocated in the era before longitude could be calculated accurately — it is important to understand that the report of the discovery of Davis' Land (quoted at the start of chapter 8) revitalized the flagging search for *Terra Australis*.

One of the first to continue the search for the southern continent was Arnold Roggeveen, a Dutch mathematician who sailed from Holland in August 1721. He found Easter Island (Rapa Nui) and spent some time in the area searching for Davis' Land. Having lost one of his three ships and failing to fulfill any of the aims of his voyage, Roggeveen's voyage effectively disabused the Dutch of any further exploration of the Pacific.[18]

On August 7, 1768, Lieutenant James Cook set sail from London for the Pacific in HMS *Endeavour* destined to make the first of his round-the-world voyages.[19] One of the aims of his mission was to search for *Terra Australis,* the great southern continent that the British Admiralty was convinced must lie in the South Pacific. This belief was based largely on the analytical works of Augustus Dalrymple, who regarded the continent-ocean ratio in the Northern Hemisphere as normal. Influenced by Greek thinking, he argued that there could not be an imbalance in the Southern Hemisphere otherwise the earth would not be able to rotate uniformly and so concluded that there had to be a large southern continent that had not yet been discovered.[20]

After his first round-the-world voyage, Cook was sceptical that *Terra Australis* existed. But it was his second voyage (1772–1775) that convinced him of this beyond any doubt. For in HMS *Resolution* he penetrated more than 71 degrees south of the equator, becoming the first ship known to have passed within the Antarctic Circle and crisscrossed the Pacific, successfully relocating many places where sightings of *Terra Australis* had been made by earlier visitors and finding nothing to fit the description.

As far as we know, *Terra Australis* has the distinction of being the first mythical continent in the Pacific whose presence was inferred by science, rather than being simply a product of wishful thinking. The ideas of the ancient Greeks and, later, those of Dalrymple may seem woefully outdated today, but at the time they were viewed very seriously. The failure of the search for *Terra Australis* has not discouraged some people from believing in a large continent in the Pacific. Yet since this continent clearly does not exist above water, then it must, they argue, have become submerged. Often it is termed a lost continent, a number of which have been suggested as lying beneath the surface of the Pacific (see later in this chapter). Yet no such continents do exist. They never did. Nor indeed is it theoretically possible that a continent could sink.

Ironically, although some of Cook's observations led to the rejection of belief in *Terra Australis,* others of his observations were used to build the case for a sunken continent in the Pacific. Along with many others, Cook reported the similarities, particularly in language and culture, of the peoples inhabiting Pacific Islands separated by vast ocean distances. "How shall we account for this nation spreading itself so far over this vast ocean?," asked the bemused Cook.[21] Notions of racial superiority that were current in the West at the time, but that are no longer tenable, influenced the answer. The idea that darker-skinned people who apparently lacked the ships and the navigational skills of lighter-skinned European explorers like Cook could possibly have colonized the far-flung islands of the Pacific Ocean seemed impossible.[22] The only plausible explanation, it appeared, was that those islands represented the peaks of mountains that had once risen from a huge continent. Then, at some time in the past, that continent had sunk, perhaps abruptly, and only the more fortunate of its inhabitants had escaped drowning by fleeing to the tops of the mountains. This scenario was regarded as plausible by many scientists up until about 150 years ago.

We now know beyond any doubt that this scenario is incorrect.[23] We know that pre-European-contact Pacific Islanders were seafarers and navigators of unprecedented accomplishment who successfully crossed ocean distances many times longer than those of their Neolithic contemporaries elsewhere in the world. There is no reason to infer that within the past 3,000 years or so a continent (or any large landmass) ever existed in the Pacific that made sailing between these far-distant islands significantly easier.

Modern Myths

As far as we know, the first occupants of the Pacific Islands had no traditions about hidden continents within this great ocean basin (see earlier in this chapter), although some pseudoscience writers claim otherwise.[24] It seems that modern discussions of hidden continents in the Pacific are extremely recent in the human history of the region, postdating not just European arrival in the Pacific but also the publication of sufficient information about the Pacific to make such views appear credible. Most contemporary discussions of hidden continents in the Pacific have been invented or are in other words, perhaps more generous, the subject of modern myths. Such myths have been designed for much the same reasons as they were by earlier people, including those in the Pacific Islands: sometimes to entertain, sometimes to instruct, sometimes to reassure. One difference is that many of the most successful purveyors of modern myths about hidden continents in the Pacific (and elsewhere) are making money, often huge amounts of money, by callously playing on the insecurities and ignorance of their readers.

Readers of a scientific bent may well wonder why I, while professing to write a scientifically rigorous book, should choose to go into such detail about hidden continents for which no scientific evidence has been found and that, when you scratch off the clumsily applied surface gloss, are exposed as unsophisticated creations of human imagination. In response, I argue that this is necessary for three reasons:

- First, many people believe in these new-age and pseudoscience creations and all the associated gobbledygook, particularly people who do not have sufficient scientific training to see past the dupe and/or who are readily beguiled by the often-accomplished spinners of these yarns. Many of us in the modern world have insecurities that we yearn to overcome but cannot readily do so within the routine context of daily life. So to believe in a place that no longer exists, where life was much better, contributes to the personal solace that many people seek.
- Second, for all their recent invention, some of the ideas associated with stories about mythical continents in the Pacific (and elsewhere) have pre-nineteenth-century antecedents that their inventors commonly amplify or distort to add credibility to the entire proposition. Some of these early ideas are worth focusing on, not just in their own right, but also to understand the influences on modern myths.
- Finally, it is regrettable yet understandable that few reputable scientists have chosen to challenge the idiocies of modern mythmakers concerned with hidden continents in the Pacific. Such is not intended to increase one's popularity, nor help one's career. Perhaps the only benefit is within the context of general awareness-raising. Yet there is a danger that one day belief in such a myth will greatly outweigh belief in a contrary scientific truth, not just for the subject of hidden continents, and therefore scientists have an ethical responsibility to contest mythical explanations when they are in danger of overwhelming scientific truth.

Pacific Islands as Evidence of a Sunken Continent?

To some disingenuous modern writers, "Pacific Islands do look like mountain peaks of a much larger area."[25] This is not an objective view. After all, none of the same writers regards the North American continent as a vast emerged area of islands from which the water has recently receded. Rather this view seems to be another ploy to attempt to validate the whole idea of there being a sunken continent in the Pacific. But it was not always so.

To the first Europeans to think about the Pacific, having spent most of their lives on continents, the idea of a vast ocean dotted with islands was quite alien. So it is not surprising that some such writers came to regard the Pacific as the site of a formerly emergent continent. But there is no excuse for any modern writer to regard the Pacific in that way — the distinction between continents and ocean basins (and the islands they contain) is taught at elementary levels in most countries.

The idea of regarding the Pacific Islands as part of a continent undoubtedly had antecedents in the belief in the existence of *Terra Australis* (see earlier in this chapter). In this regard, the inclination of eighteenth-century Europeans to regard Pacific Islands as outliers or even promontories of *Terra Australis* is understandable. Hence, when at 3:00 p.m. on June 19, 1767, the crew of HMS *Dolphin* saw Tahiti in the distance, the sight "made us all rejoice and fild us with the greatest hopes Imaginable" because they believed they had found *Terra Australis,* the long-sought-after southern continent.[26] Only when James Cook repeatedly traversed the Pacific during his second voyage (1772–1775) did it start to become finally accepted amongst European scholars that there was in fact no emergent continent hidden in the Pacific. As a result, the focus shifted away from ideas of crustal balance to speculations about the origin of this huge, largely landless, ocean basin.

From this point onward, any speculations about Pacific continents had to focus on what was below, not above, the ocean surface.[27] And indeed, it was at this point in time, around the start of the nineteenth century, that the first accounts of sunken continents in the Pacific began to be published: not as fiction, but as fact, a trend that has continued subsequently.

One of the earliest, perhaps the earliest, person to speculate that the Pacific Islands had once been the peaks of a now-sunken continent was Dumont d'Urville, an assiduous explorer of the Pacific on behalf of the government of France.[28] But it was Jacques Moerenhout who in 1837 was the first to lay out a detailed account of the probable nature of this continent and the cultural associations of its inhabitants. Reacting against earlier ideas that Pacific Island peoples were the descendants of voyaging Malays or American Indians, Moerenhout considered that the apparent uniformity in culture and language between Pacific Island peoples could be explained only by them being the remnant inhabitants of a great continent that had once existed in the Pacific.[29] Given our modern understanding of Pacific Island cultures, Moerenhout's view, uncritically echoed by many more-recent pseudoscience and new-age writers, is as untenable as that which exclaims that "Pacific Islands do look like mountain peaks of a much larger area" quoted earlier in this section. If Europe or South America were drowned today, the survivors on the mountaintops would have little shared culture or language in the way that Pacific Islanders do.[30]

The idea that there was a sunken continent in the Pacific was certainly pondered by many nineteenth-century geologists, even espoused by some like Jules Garnier,[31] but majority scientific opinion was against it. James Dwight Dana, the geologist with the United States Exploring Expedition (1838–1842) that spent a long time in the Pacific Islands, wrote succinctly on the matter: "We should beware of hastening to the conclusion that a continent once occupied the place of the ocean, or a large part of it, which is without proof. To establish the former existence of a Pacific continent is an easy matter for the fancy; but geology knows nothing of it, nor even of its probability."[32]

Another influential exponent of the view that the islands in the Pacific were the extremities of a submerged continent was Louis Jacolliot. In his 1879 book, Jacolliot gave the name of this continent as Rutas (see later in this chapter) and contended that "its existence rests on such proofs that to be logical we can doubt no longer."[33] He gave four lines of alleged proof.

First, he pointed to the great distances between these island groups, echoing Moerenhout by stating that "all navigators agree in saying that the extreme and the central groups could never have communicated in view of their actual geographical position, and with the insufficient means they had at hand. It is physically impossible to cross such distances in a pirogue [dugout canoe] . . . without a compass, and travel months without provisions."[34]

If you assume that Pacific Island people could not have crossed such distances, then clearly your explanations for the settlement of the farthest-flung Pacific Islands are hugely constrained. To writers like Moerenhout and Jacolliot, inculcated with nineteenth-century European notions of racial (Caucasian) superiority buoyed by the first written accounts of Pacific Islanders, this assumption was self-evident.[35]

Second, Jacolliot supported his interpretation by asserting that, before the arrival of Europeans, the peoples of the various Pacific Islands had no knowledge of each other's existence, something that is abundantly contradicted by their oral traditions.

Third, and in apparent contradiction of his second line of proof, Jacolliot averred that the occupants of each Pacific Island archipelago unanimously declared that at one time, their islands "had at one time formed part of an immense stretch of land which extended towards the West," something that is contradicted by the conspicuous lack of hidden-continent myths among Pacific Islanders. He strengthened this assertion by noting that, for all Pacific Islander groups, when asked the question "where is the cradle of your race," all responded by extending their hand toward the setting sun.[36] This could be an invented detail, used by Jacolliot to support his theory. But it may be otherwise, for if everyone in the Pacific Islands pointed west when asked where the "cradle" of their race is, that could be interpreted either as meaning that their *island* homeland is in the west (as with Hawaiki or Pulotu, see chapter 6) or it could mean that all their islands were once connected by a now-sunken continent and that they originally crossed this from west to east.

From what we know today about Pacific Islander traditions, it is clear that the first of these two explanations is correct — the cradle was an island homeland in the west. Yet Jacolliot chose to subscribe to the latter explanation and so gave birth to the mistaken belief that Pacific Islanders had themselves explained that their ancestors had occupied a now-sunken continent.

Finally, Jacolliot noted correctly the similarities in language and cultural traits such as customs and religious beliefs among scattered Pacific Islander groups and interpreted these similarities as evidence that they had all come from the same ancestral group.

The views of Jacolliot are quoted at length here because they appear to have been his genuine beliefs deduced from his collection of oral traditions in India and elsewhere, rather than ideas he developed to support a particular worldview. As such, these ideas may be seen as representative of some respectable scientific views during the second half of the nineteenth century in Europe. Yet, because these views were widely regarded as irreproachable they became vulnerable to the interpretations of charlatans and others competing to foist their own peculiar worldviews on an unsuspecting public.

A good example of how Jacolliot's deductions were subsequently misused comes from the writings of Helena Blavatsky, the founder of the mystical belief system known as Theosophy.[37] Although she relocated the sunken continent of Rutas in the Indian Ocean, she quoted much of Jacolliot's evidence for its former existence approvingly. She noted that, although his description was not quite the same as the "truth" she had been given by her spirit guides, it nevertheless "shows the existence of such traditions, and this is all one cares for. For, as there is no smoke without fire, so a tradition must be based on some approximate truth."[38]

As a respected academic, well traveled in the Pacific Islands, John Macmillan Brown became a voice of authority in the 1920s about the origins of both the islands and their inhabitants. Brown's fundamental belief in the former presence of a huge continent in the Pacific was spawned by his inability to believe that Pacific Islanders could have sailed the vast distances between archipelagos necessary to colonize them, but it was also propped up by two scientific notions that Brown accepted.

First, Brown believed that most geologists regarded the Pacific as the site of a sunken continent, but he was misled in this, probably giving undue weight to the influence of a few writers such as Jules Garnier.[39]

Second, Brown drew support for his beliefs from Charles Darwin's Subsidence Theory of Atoll Formation, in which the presence of atolls can be taken to indicate subsidence (see Figure 2.6). But this subsidence refers only to that of an individual edifice, not an entire region, although Darwin's ambiguity on this point might also have misled Brown.[40]

Once locked into the idea that the Pacific was the site of a sunken continent, Brown could not escape it even though the additional evidence that he accumulated in support of the idea must have appeared so contradictory that one wonders whether he struggled in the face of increasing self-doubt.

For example, in his influential 1924 book *The Riddle of the Pacific*, Brown argued that support for the subsidence of the Pacific continent comes from "occasional compensatory elevations and volcanic spurts."[41] In other words, it seemed to Brown that the principal process (subsidence) must be occurring because of infrequent manifestations of the opposing process! Brown also argued that the development of infanticide and cannibalism in some central Pacific Islands came about as a result of the reduction in cultivatable area and associated societal stress after the continent sank. And, perhaps most regrettably considering its enthusiastic uptake by the current generation of new-age and pseudoscience writers, Brown also averred that the megalithic site of Nan Madol on Pohnpei Island in Micronesia was the emergent remains of an area twenty times larger and that "the houses and huts of the nobles and commoners and slaves must have vanished centuries ago."[42] The reality is quite different because, although Nan Madol has been affected by slight submergence, this has probably not significantly reduced the area of the original settlement, a settlement that was built on ninety-two artificial islands (see Figure 7.2c).

The legacy of Jacolliot and Brown has endured. In a 1931 description of the end of the continent Mu, which he imagined to have been in the central Pacific (see later in this chapter), James Churchward used the evidence of Pacific Islands, supposing that these had become "pathetic fringes of that great land, standing today as sentinels to a silent grave."[43] And today, such evidence for a sunken continent in the Pacific is widely quoted in pseudoscientific writings despite the overwhelming amount of real scientific evidence showing it to be false.

Sensing Mythical Pacific Continents

Many people, both today and in the past, believe that communication with the dead is possible. The issue here is not specifically whether that is indeed true, but, because sometimes the dead apparently give us details about vanished islands and hidden continents, the question arises as to how much weight should be attached to such forms of received wisdom.[44]

Most accounts of vanished islands and hidden continents allegedly obtained from the dead reach us through a number of pathways that may be broadly classified as channeling, whereby a human medium is temporarily possessed by the spirit of a dead informant. The spirit is alleged to recount details about these islands that have somehow largely escaped the more conventional processes of historical enquiry. For example, the famed Edgar Cayce (1877–1945), whose diagnoses and recommended treatments for a variety of illnesses while in a trance led many people to believe that he was a true psychic, had much to say about Atlantis.[45] More recently, the numbers of mediums divulging information about vanished islands and hidden continents has increased sharply, in line with the rise in popularity of the New Age in the English-speaking world. For example, Mark Williams, with his "two university degrees in history," became convinced that the sunken (Pacific) continent of Lemuria had indeed once existed after attending a channeling session by Jach Pursel, who spoke as the Lemurian spirit Lazaris.[46]

Channeling specifically targeting vanished islands and hidden continents rose to prominence in the late nineteenth century, largely through the voluminous writings of Helena

Blavatsky. Like many of the authors who wrote in great detail on such topics, Blavatsky and her contemporaries were coy about their information sources and how exactly they accessed them. One of the largest bodies of secret information about the lost continents Lemuria and Mu is alleged to be the Akashic Records, which apparently do not exist in written form but rather in some ethereal state, accessible only to select people like Helena Blavatsky and Rudolf Steiner.[47]

Automatic writing has been a popular form of supposed communication with the spirit world, leading to many accounts of lost continents. For example, J. B. Newbrough claimed that his lengthy account of the mythical Pacific continent named Pan was obtained in this way, a result of dictation by spirits while he was in a trance.[48]

Somewhat less common than information allegedly obtained by channeling is a series of techniques of what might loosely be classified as inspiration, which has been particularly important in producing accounts of lost continents. For instance, Lewis Spence (1874–1955), who wrote five books about Atlantis, claimed to have received most of his information about the lives of the ancient Atlanteans subliminally, which led him, amongst other things to claim that Cro-Magnon Atlantean refugees who had survived the island-continent's sinking produced the cave paintings of Europe.[49] Spence was scornful of traditional scientific methods, writing that

Inspirational methods, indeed, will be found to be those of the Archaeology of the Future. The Tape-Measure School, dull and full of the credulity of incredulity, is doomed.[50]

James Churchward, who wrote a number of influential books about the lost continent of Mu in the 1920s and 1930s, appears to have employed a variation on the inspiration theme although he was less forthcoming than Spence about his methods and therefore needed to be less combative in their defense. But it appears that he believed that persons like himself who had the appropriate gift could discern the meaning of ancient symbols simply by staring at them long enough.[51] If only science were really that simple!

Blavatsky and Churchward probably realized that little credibility would be given to their accounts if they claimed they received their information only by esoteric means like channeling or inspiration, so they were careful to point out that this information had been supplemented by that obtained by more conventional means. Blavatsky claimed to have viewed the *Book of Dzyan* in Tibet, and Churchward claimed to have been shown tablets with writing on them in India (whose meaning he clearly apprized!). There is not one piece of verifiable scientific evidence to suggest that either of these claims is true, nor in fact to suggest that Blavatsky and Churchward, like most of their ilk, were anything other than prolific science-fiction writers at a time when this literary genre was not well established.

The roots of modern channeling probably arose with Spence, Churchward, and theosophists such as W. Scott-Elliot who claimed to have received substantial amounts of information from his spiritual masters by the process of astral clairvoyance. Today many subscribe to the idea that "through the power of thought alone, we can and do create our own reality in a multi-leveled spiritual universe,"[52] something that may be suited to exploration of metaphysical worlds but that is no help at all in understanding the tangible world and its geological history.

Most of the more elaborate mythical accounts of hidden continents in the Pacific were constructed for what their inventors and adherents claimed were religious reasons. Detached observers, confronted by some of the outrageous claims of such religions, might reasonably question their authenticity as expressions of deeply pondered theological beliefs.[53] And yet, against such rationalist views, it can be legitimately argued that a person's religion is exactly what that person chooses it to be and, if it happens that your personal belief system involves a mythical continent inhabited by four-armed egg-laying hermaphrodites, each with an eye in the back of its head,[54] then so be it. *Chacun à sont goût.* Danger appears only when such personal beliefs are foisted on people who are less critically aware and therefore more vulnerable as part of a package ultimately intended to profit those who promulgate the belief system. It is in this spirit that the criticisms in this chapter of those snake-oil peddlers who masquerade as lost-continent gurus with a view to deceiving the unwary are offered.

The easiest way to structure this discussion is under the names of each of the allegedly sunken continents in the Pacific: Rutas, Lemuria, and Mu.

Rutas

A lost continent named Rutas is often referred to in theosophical writings as having been in the Indian Ocean. According to Helena Blavatsky, the founder of Theosophy, Rutas was a "huge land" that once lay east of India and that was described in "Brahminical traditions." But one day it was destroyed in a volcanic cataclysm and "sent to the ocean depths," leaving behind only Indonesia.[55] This story, probably concocted by Blavatsky just after the truly cataclysmic 1883 eruption of Krakatau Island volcano,[56] has its earlier roots in the writings of Jacolliot. Jacolliot's 1879 book included some myths referring to Rutas that he claimed to have collected in India.[57] His work was certainly a major influence on Blavatsky, particularly her use of sunken continents to support her theories of human origins. But Jacolliot's own interpretation of the legend of Rutas was quite different, for he believed that it had existed in the Pacific (not the Indian Ocean, as Blavatsky stated) and had included all the main oceanic island groups. When Rutas sank, Jacolliot claimed, an elite group survived and settled in India, introducing the Sanskrit language.

It is unclear why Jacolliot insisted that Rutas had been in the Pacific. The traditions that he collected apparently stated only that it had been a huge land east of India, so possibly Jacolliot could think of no place other than the Pacific where a "huge land" could have been. And because Jacolliot was writing before the 1883 eruption of Krakatau, he probably did not pause to consider that this region may have been the site of Rutas. Yet no other writers have located Rutas in the Pacific rather than the Indian Ocean, most contending that it was simply another name for Lemuria in the Indian Ocean.

There are no certain ancient traditions that refer to a land called Rutas, and it is likely that Jacolliot simply amplified an insinuation in the traditions he collected into a fully fledged legend, given that his book was written to be popular rather than a work of science per se. Yet in doing so he misled many later readers into believing that such a place once really did exist.

Lemuria

A hidden continent named Lemuria was once proposed by scientists to explain biotic similarities between the western and eastern borders of the Indian Ocean. Since the advent of ideas about the mobile earth's surface, particularly the plate-tectonics theory, this Lemuria has become unnecessary.

The idea of a former continent in the Indian Ocean was first suggested by a German biologist named Ernst Haeckel in his 1874 *Anthropogenie oder Entwickelungsgeschichte des Menschen (Pedigree of Man)*. In a somewhat cavalier interpretation of the evidence, Haeckel proposed that this continent had been the cradle of human civilization before being submerged. In 1876, this continent was christened Lemuria by another eminent biologist, Philip Lutley Sclater, who conceived it to have been an enormous land bridge, home to the lemurs that he believed once inhabited it and its extant fringes.

Since that time, Lemuria has been largely appropriated by popular philosophers like Blavatsky, who, in her monumental 1888 work *The Secret Doctrine*, named it as the sunken continent once inhabited by the Third Root Race, the Lemurians: egg-laying beings with three eyes. Most people place Lemuria in the Indian Ocean, but some relocated it to the Pacific, including Lewis Spence.[58] Having flexed his intellectual muscles trying to make sense of Atlantis legends, in later life Spence turned his attention to Lemuria in the Pacific. His ideas are not particularly noteworthy, although he did claim that Davis' Land was the last part of Lemuria to sink beneath the waves, which is why it had apparently not been seen since 1687.

The Rosicrucians are another organization that place great significance on a Lemuria located in the Pacific (Figure 7.1B). Their key text on the subject, written in 1931, describes Lemuria stretching from the South Pacific to the area north of Hawai'i and extending southeast, like most reconstructions of Mu (see the next section), to Easter Island.[59]

Some advocates of new-age explanations of human origins on a lost continent in the Pacific believe that a small part of this continent, usually named Lemuria, did not sink but became accreted onto the rim of the Pacific as Mount Shasta in California, where a community of Lemurians is said to continue living inside the mountain. This story, popularized

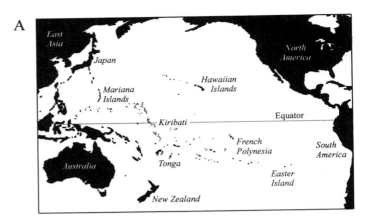

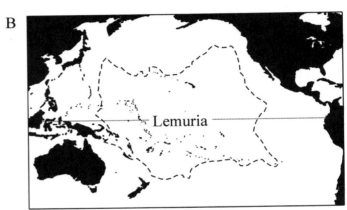

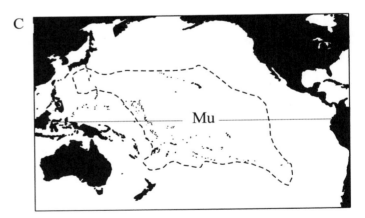

FIGURE 7.1. Some supposed "lost continents" of the Pacific. Neither Mu nor Lemuria ever existed. **A**. Index map showing main points of geographical reference. **B**. The lost continent of Lemuria in the Pacific according to Wishar Cervé. (Adapted and redrawn from Cervé [1931].) **C**. The lost continent of Mu according to James Churchward. (Adapted and redrawn from Churchward [1926].)

by Cervé[60] in the 1930s, has led to a massive increase in new-age interest in the Mount Shasta area. Mount Shasta is part of the Cascade Range, a line of young volcanoes that has formed above a zone of plate convergence. Yet it is interesting that, like many other parts of the western seaboard of the Americas, it lies close to the Klamath Mountains, parts of which are undoubtedly oceanic terranes that have crossed the Pacific Plate and become accreted onto the margin of the North American continental core (see chapter 2).[61] Fueled by the presence of Mount Shasta and the proximity of the Pacific Ocean, new-age interest in Lemuria and its peoples has flourished along the western seaboard of North America.[62]

Mu

The name Mu was used first to refer to a supposedly sunken continent in the Atlantic Ocean.[63] The idea of a lost continent named Mu in the Pacific was largely the creation of Colonel James Churchward, who, during service as an engineer in the British army in India, claimed to have been told the story of Mu by the sole survivor of the Naacals, a priestly elite that had presciently fled the island continent when it first showed signs of sinking. Churchward wrote six books about Mu and can be credited with giving rise in the middle twentieth century to a huge increase in a popular belief in lost continents.

In his first book (1926), Churchward claimed that Mu once stretched from Hawai'i in the north to Fiji in the south, and from Easter Island in the east to the Marianas in the northwest (Figure 7.1C). He believed that the continent was low and flat—the mountains had "yet to rise." For Churchward, Mu had all the attributes of the Garden of Eden, a place of harmony without savagery, where sixty million people were living 50,000 years ago, and where civilization was apparently more advanced than at present. Such nonsense continues to have wide appeal and has been given a boost by the revival of ideas such as that of a lost civilization in Egypt that was more advanced than today's, advocated by popular pseudoscience writers like Graham Hancock.

Churchward's Mu met its end about 12,000 years ago, when "cataclysmic earthquakes rent Mu asunder . . . she became a fiery vortex, and the waters of the Pacific rushed in making a watery grave for a vast civilisation and sixty million people."[64]

A widely published modern advocate of a lost continent named Mu in the Pacific is David Hatcher Childress. Echoing earlier ideas of cultural diffusion from the Mediterranean, particularly Egypt, to the Pacific (and elsewhere in the world), Childress listed the purported evidence of this cultural contact that he has found or interpreted in the Pacific. For example, he asserts that the *rongorongo* scripts of Easter Island are the same as those of the Indus Valley (they are not); that the megaliths on Pacific Islands were built through levitation using "harmonic sound" (they were not); that the people who lived on the continent of Mu had an advanced society, one government, and one language; and that the whole place disappeared in a great catastrophe and sank beneath the waves (none of which is true). The disappearance of Mu is very convenient because it means that theorists like Childress can say what they like and appear convincing to people who are comparatively uninformed, as many naturally are, of the huge body of scientific information on Pacific geology and cultures.

Mythical Pacific Continents Invented as a Reaction to Darwinism

The ideas of Charles Darwin about human evolution shook the foundations of nineteenth-century thinking in a variety of ways. As is common when radical ideas are first proposed, this led both to established scientists reaffirming their long-held beliefs with strong statements against the new thinking, and to the exposition of alternative, more palatable, explanations of comparable detail and complexity.[65] Darwin's ideas concerning evolution provoked strong statements defending the orthodoxy and also a number of supposed alternative models. The latter were fuelled largely by two popular (yet incorrect) interpretations of Darwin's ideas, repugnant to most people at the time, that humans were descended from apes and that divine intervention had not been required to produce humans.

Helena Blavatsky was fanatically anti-Darwin.[66] And it seems likely that the complex scenario she created concerning human origins and succession on lost continents in the Pacific and elsewhere was intended primarily as a counter to what she believed were Darwin's ideas about human evolution. Blavatsky built her ideas initially on Haeckel's work, which concluded that a continent (named Lemuria) had once existed in the Indian Ocean (see earlier section on Lemuria). Ironically, Haeckel was a strong supporter of Darwin's ideas who had embraced the idea of Lemuria to attempt to overcome criticism of them, particularly that involving doubts about how particular animal species had become distributed across large areas. But Haeckel overstepped the boundaries of deduction and suggested, almost as an afterthought, that Lemuria may also have been the cradle of humanity, the place from which all races of humans had come.

The essence of Blavatsky's views about human development is that, contrary to evolutionary theory, humans are the blueprint for sentient life, and most other species appear to have "evolved" from the union of humans and "beasts." For this to happen, and according to Blavatsky it happened repeatedly as one pure human "root race" eventually succumbed and mated with the part-human descendants of the earlier root race's much-deplored unions, it was helpful to Blavatsky to have places where the supporting evidence could not be located, such as a vast broken-up and submerged continent as she supposed Lemuria and others to be.[67]

Tangible Evidence Claimed for Mythical Pacific Continents

In his 1924 book *The Riddle of the Pacific,* John Macmillan Brown claimed that megalithic remains found throughout the Pacific Islands pointed to a "lost continent" that had been occupied by supposedly fair-skinned ancestors of modern "Polynesians." Brown's ideas were rooted in the unquestioning racism that still influenced some anthropologists in the first half of the nineteenth century and in the more recent diffusionist thinking of people like Franz Boas.[68] Yet Brown's ideas have been regarded as somewhat more authoritative, not only because he was a respected academic with training in both geology and literature, but because he did actually travel extensively in the Pacific Islands gathering what he supposed to be supporting evidence of his ideas.

Brown paid a brief visit in 1919 to the atoll of Malden in the Line Islands of what is now eastern Kiribati and saw for himself some of the megalithic structures found there. He was astonished by the presence of such elaborate megalithic remains on this low, uninhabited, resource-poor island.[69] Part of his account talked of "paved ways from the sea" to the central monuments, a detail that was subsequently inverted into the belief that "roads of basalt blocks extend from these temples in every direction only to disappear into the Pacific Ocean."[70] Similar trails on nearby Enderbury Atoll have proved no mystery because, facilitating foot travel, they "give access to interior ponds, and more generally to reefs and marine resources"[71] (Figure 7.2A).

Recent work on the so-called mystery islands of eastern Kiribati and the Line Islands (which include Malden) show that most of them have traces of human occupation, including megalithic structures, even though no one was living there when they were first visited by Europeans. The megalithic structures include many *marae* (temple platforms) that were built by the islands' early settlers, a sure sign that they were well-established, not overnight, visitors to these apparently resource-poor islands.[72] The megalithic remains on Malden must be viewed in the context of the region as a whole in which there were parent communities on larger islands and satellite communities within regular sailing distance that people visited to hunt birds in particular. Then there are isolated communities including Malden that probably had a more productive resource base in the past, which once made them satellites, but where conditions deteriorated as resources became scarcer and/or they were less frequently visited leading to the appearance of abandonment.[73]

The huge stone *moai* (statues) of Easter Island (Rapa Nui) in the Southeast Pacific have come in for special attention from lost-continent theorists (Figure 7.2B). Both Churchward and Spence saw the *moai* as examples of Mu/Lemuria artwork, unable to have been executed by ancestors of the known inhabitants of the island. Such interpretations were given credibility for many other new-age writers by the assumed proximity (actually 3,800 kilometers) of Easter Island to the South American continent and its supposed Atlantean civilizations.[74] Thor Heyerdahl's mistaken belief that Easter Island had been comprehensively colonized from South America gave further impetus to the sunken-continent connection, as did his various, generally baseless, ruminations about the origins of Pacific peoples in general.[75]

Another focus of pseudoscience attention in the Pacific Islands has been on the monumental settlement named Nan Madol constructed by local people on ninety-two artificial

FIGURE 7.2. (*Facing page*) Some of the best-known places in the Pacific claimed, wrongly, to be evidence for a huge, now-submerged continent. **A.** One of the roads connecting the center of Enderbury Atoll in the central Pacific with the coast. These roads are pathways built by the island's earlier settlers, probably visiting temporarily for the purposes of obtaining food, to facilitate walking across the otherwise difficult terrain. (Photo by Anne Di Piazza and Erik Pearthree, used with permission.) **B.** Two of the carved stone *moai* on Easter Island in the Southeast Pacific Ocean. Note the scrub grassland on the hillside, evidence of the island's current degraded landscape, a contrast to that at the time the *moai* were produced. (Photo by Claire and Michael Poore, used with permission.) **C.** View of Nan Dauas islet, reserved for chiefly burials, with walls built from blocks of columnar basalt. This is part of the megalithic site of Nan Madol, just off the island of Pohnpei in Micronesia. (Photo by J. Stephen Athens, used with permission.) **D.** The slate forming the Sanninudai cliffs above the underwater Yonaguni structures has been fashioned solely by natural processes. Yonaguni is one of the southernmost islands in Japan, comparatively small and remote.

islands off the southeast coast of Pohnpei Island in Micronesia (Figure 7.2C). Despite its uniqueness, particularly because of its huge size and its stonework, Nan Madol is probably less than 1,500 years old,[76] certainly not an ancient navy base, nor ever visited in pre-European times by Egyptian explorers or anyone who had fought with Alexander the Great, as some best-selling new-age and pseudoscience writers would have us believe.[77] The slight submergence of Nan Madol is not attributable to the sinking of a vast Pacific continent but simply to the process of island subsidence that has probably been affecting Pohnpei for thousands of years combined with sea-level rise over the past 200 years or so.[78]

One of the more recent discoveries in which there is much pseudoscience interest, mostly on the Internet, is the supposedly submerged city off the south coast of Yonaguni Island in southernmost Japan. The principal structure, the Yonaguni Underwater Pyramid, has been studied extensively by geophysicist Masaaki Kimura, who reckons that it is a natural feature modified extensively by humans before it was submerged 10,000 years ago by rising sea level.[79] Geologist Robert Schoch agrees that the Yonaguni Underwater Pyramid is a natural structure but admits far less (if any) human modification than Kimura.[80] The Yonaguni Underwater Pyramid and associated structures found below water off southern Yonaguni continue above, where I have examined them extensively, and there seems no reason to suppose that they are artificial (Figure 7.2D). This has done little to discourage speculation that the Yonaguni Underwater Pyramid is part of a megalithic civilization that covered much of the western part of Mu (or some other lost continent).

Conveniently Hidden Stores of Cultural Origins: Submerged Continents and Diffusionism

Many of the people who believe in submerged continents have done so because it fits well with their diffusionist beliefs. This idea holds that modern human cultures all originated in one place (or only a few) before spreading out in all directions (diffusing) to other places where, despite acquiring distinctive cultural overlays, their original form can still be recognized. Diffusionism is implicit in many modern pseudoscience accounts of human origins, from those that regard Mu in the Pacific as the place where the human race originated to those that believe Egypt to be where an advanced civilization once existed long before that of pharaonic times. The problem for many diffusionists is that the place(s) of origin is not clearly identifiable in today's world, so that to say that it exists no more, perhaps because it was submerged or became covered with ice, is an easy and practically indisputable way out: a true deus ex machina.

Diffusionism grew out of a sense of disenchantment amongst anthropologists with the tenets of evolutionary anthropology. One of the earliest diffusionists, William Rivers, took great satisfaction from his discovery in Melanesia (western tropical Pacific) of a technique of human mummification similar to that once used in Egypt,[81] a discovery that led him to propose that Melanesians had come from Egypt. This idea, long since disproved to the satisfaction of scientists (see chapter 5), has come to be one of the touchstones for new-age interpretations of Pacific Islander cultural origins, ranging from the fevered idea that ocean-going Egyptians plied the waters of the Pacific more than 3,000 years ago[82] to the fanciful (yet attractive to some) notion that Fijians may be one of the lost tribes of Israel.[83] The truth is that mortuary practices, like symbolism, like words, like agriculture, like countless other cultural traits, have evolved independently in different parts of the world at different times. There are only so many ways to boil an egg.

That said, no scientist would ever deny that some diffusion of ideas has taken place as people have moved from place to place. The Lapita occupation of the western tropical Pacific Islands, discussed in chapter 5, is a fine example because it can be readily verified from the corpus of available information. The settlement of the Americas, discussed in the same chapter, is less certain but there are several competing scenarios, none of which is entirely implausible.[84]

Diffusionism is central to new-age and pseudoscience arguments that survivors of submerged continents/islands fortuitously escaped to nearby lands where they were able both to partly enlighten the people they encountered and also (more frequently) to hoard the secret knowledge that made them superior. Thus a whole range of theorists, from Jacolliot and Blavatsky to Churchward, and dozens of modern channelers claim to have been in touch with these survivors. Even Ignatius Donnelly, a prolific writer about Atlantis (and one-time United States congressman) who was not readily given to unsupportable notions about cultural origins, believed that sunken Atlantis had been the original seat of the Aryan family of nations, that a few people escaped and carried with them to east and west (diffused) the story of the catastrophe that remains in various cultures as a deluge story (so the deluge stories are themselves diffused).[85]

The basic thesis of the popular 1995 book *Fingerprints of the Gods* by Graham Hancock is that a lost continent must have once existed because the "fingerprints" of the people who inhabited it are found everywhere in the world. If you are a cultural diffusionist, as Hancock probably is, then every similar cultural trait you encounter anywhere in the world is of course evidence that those cultures came from the same root culture that, chances are, was located elsewhere.[86] Unlike most pseudoscience writers on Atlantis, Hancock faced up to the geophysical impossibility of a sunken continent existing anywhere in the modern ocean basins and so decided that Atlantis once existed where Antarctica is today. Just like locating Atlantis beneath the sea, this is a convenient conclusion for a diffusionist to reach because there is no evidence to show whether or not people lived on Antarctica before it became covered with ice,[87] so the cultural traits of the original Atlanteans will remain solely a matter for conjecture in the foreseeable future!

The Antarctica-as-Atlantis theory impinges on the Pacific, which is the principal focus of this book, so it is worth looking briefly at the idea in more detail. Hancock contends that "the simple facts about Antarctica are really strange and difficult to explain without invoking some notion of sudden, catastrophic and geologically recent change."[88] This is not true. The geological history of Antarctica from its incorporation into Gondwana around 220 million years ago, when it had a tropical climate, followed by its slow poleward progress is well known.[89] When the Antarctic continent finally separated from South America and Australasia, the Circum-Antarctic Current was established that insulated the continent from warm water and warm air, and terrestrial ice buildup began about thirty million years ago (see chapter 2). Simply no mystery there, but a credible scenario of environmental changes well supported by scientific observations.

Of course there are uncertainties, particularly whether or not the West Antarctic ice sheet periodically surges,[90] but to suggest that Antarctica was a continent supporting an advanced civilization at any time in the past that abruptly lurched poleward causing that civilization to become buried under ice is an explanation that clearly sells books but is not more real than any other science-fiction account.

8 Vanishing Islands

PROCESSES OF ISLAND DISAPPEARANCE
WITNESSED BY HUMANS

It could be argued that humans have a natural tendency to imagine uncharted lands. The reasons for this lie deep within the psyche of individuals of every ethnicity; imagined worlds appear to be a universal archetype.[1] In 1687, in the empty ocean more than 3,000 kilometers west of Chile, Edward Davis imagined that he saw islands within an area 15 kilometers across in the isolated Southeast Pacific [additions in brackets mine].

> about two hours before day we fell in with a small low, sandy island and heard a great roaring noise, like that of the sea beating upon the shore, right ahead of the ship. Whereupon the sailors, fearing to fall foul upon the shore before day, desired the captain to put the ship about, and to stand off until the day appeared; to which the captain gave his consent. So we plied off till day and then stood in again with the land, which proved to be a small flat island [later named Davis' Land], without any guard of rocks. We stood in within a quarter of a mile of the shore and could see it plainly, for it was a clear morning, not fogy or hazy. To the westward about 12 leagues [~67 kilometers], by judgment, we saw a range of high land, which we took to be islands, for there were several partitions in the prospect.
>
> This land seemed to reach about 14 or 16 leagues [78–90 kilometers] in a range, and there came great flocks of fowls. I and many more of our men would have made this land and have gone ashore on it, but the captain would not permit us."[2]

As Davis' Land, the sighting of these islands reignited the search by Europeans for *Terra Australis* (see chapter 7) and remains today a potent piece of evidence in the armory of those writers who imagine, incorrectly, that Davis' Land was/is the peak of a submerged continent or, more prosaically, the last traces of a sinking continent.[3] It is unfortunate that neither Davis nor any member of his crew landed on Davis' Land. More likely than a bedrock island, Davis' Land, if it was ever more than a mirage or a school of breaching fish, could have been a series of pumice mats produced by a shallow underwater eruption from the East Pacific Rise, that great range of volcanic mountains that runs beneath the surface of the Pacific from the Gulf of California to the fringe of Antarctica (shown in Figure 2.3).

The purpose of discussing this story in such detail here is to underscore how difficult it sometimes is to interpret historic accounts. The debate about Davis' Land has raged for decades and is destined to remain unresolved simply because there is insufficient information to enable any final judgment about it. The same can be said for many historical accounts of natural phenomena, particularly those from oral traditions that were written down a long time after the events described occurred, as is the case with many of those described in this chapter and the next.

There is no evidence that any continents have disappeared or become otherwise hidden anywhere in the Pacific (or elsewhere) during the time of the region's human occupation or that they will become so in the foreseeable future, so this chapter, together with chapters 9 and 10, considers only islands.

Islands That Disappeared as a Result of Sea-Level Rise

On the scale of a human life span we may not be aware of long-term sea-level changes, yet there is abundant evidence to show that these have taken place throughout the time that people have occupied the Pacific region (see chapter 2). More than that, sea-level changes have impacted human societies by changing the extent of the land available for occupation and altering the magnitude and nature of the food resource base, with sometimes-profound consequences.[4]

The native Australians of Arnhem Land in the north of the country trace their origins to an island to the northeast named Baralku, generally regarded, on the grounds that it cannot be located today, as a mythical island. In his 1998 tour de force about the effects of postglacial sea-level rise on the inhabitants of Southeast Asia, *Eden in the East,* Stephen Oppenheimer drew attention to similarities between that tradition and those of people in India and island Southeast Asia, regarding them all as descendants of people displaced by postglacial flooding of the Sunda Shelf.[5]

Chronologies of human occupation of islands off the coast of Australia have been linked to postglacial sea-level changes in ways that are likely to have been similar throughout the Pacific Basin, particularly in the case of other such near-continent islands.[6] For example, large (4,400 square kilometers) Kangaroo Island 14.5 kilometers off the south coast of Australia was unoccupied and unutilized by native Australians at the time of European arrival, but this was not the case much earlier on. About 7500 BC, when there were evidently many people living there, Kangaroo Island became separated from the Australian mainland by rising sea level. Human occupation continued until at least 2350 BC but then apparently ceased; the island was abandoned, probably owing to a decrease in wild foods arising from the reduction in land area associated with continuing sea-level rise.[7] Similar arguments have been presented for the end of human occupation of Flinders Island and King Island, in the Bass Strait between Tasmania and the Australian mainland, respectively about 4800 BC and 5700 BC.[8]

Although there is good, albeit inconclusive, evidence suggesting that bursts of rapid post-glacial sea-level rise may have initiated the migration of people out of lowland areas of East Asia and into the Pacific (see Figure 5.2), there are no unequivocal records of this. Distant echoes of such enforced migrations, referring particularly to the postglacial drowning of Southeast Asia, may be preserved in various myths and myth-motifs, but these are at best equivocal. Yet myths are an important source of information about periodic encroachments of the ocean across the land, be they short-lived such events or more enduring periods of island submergence.

Many creation myths recorded in Pacific Island traditions regard the world before humans appeared as having been covered by water. For example, the cosmogonic chant of Tiwai Paraone[9] from the Hauraki district of New Zealand's North Island told of Io, the supreme being.

I noho a Io i roto i te aha o tea o	Io dwelt in universal space
He pouri te ao, he wai katoa.	The universe was in darkness, all was water.
Kaore he ao, he marama, he maramatanga	Day was not, nor moon, nor light
He pouri kau, he wai katoa	Darkness alone was, all was water

Flood myths from the Pacific Islands are similar to flood and deluge myths from other parts of the world.[10] It is clear that such myths existed in the Pacific Islands before the arrival of Christian missionaries, who both introduced biblical elements into existing myths and even added to the corpus of supposedly indigenous mythology.[11] Yet it is also argued that such indigenous myths were not always created from observations of what actually happened in those islands where they were recorded but diffused there from elsewhere, perhaps the western rim of the Pacific, particularly from Southeast Asia, Papua New Guinea, and nearby parts of Australia. In this region, the submergence of shallow shelves during the postglacial sea-level rise transformed its geography and displaced large numbers of people. The descendants of some of these people carried the memory of the great floods with them into the Pacific Islands, where they were adapted to local circumstances and environments.[12]

Within the past 200 years or so, the trend of sea-level rise has been upward, with perhaps a net 20–30 centimeters of rise within this period in the Pacific.[13] This has undoubtedly been a cause of island disappearance, and, should it continue (as seems likely), many more islands may disappear as a consequence (see chapter 10). Yet it is simplistic to think of sea-level rise alone as the cause of island disappearance; were it so, then probably many more islands would have disappeared within the past 200 years.

No persistent islands in the Pacific 200 years ago rose less than 30 centimeters above sea level; it is just not possible for an island to endure at such a low level, largely because of the day-to-day variability of the sea surface and the effects of periodic storm waves. Those islands that have disappeared in the past 200 years in the Pacific were higher (typically 2–3 meters above sea level) but made mostly from unconsolidated material. Most such islands drew the sediments that supplied their beaches from associated coral reefs and from nearby islands of similar character (atoll *motu*). As sea level rose, so the system by which such bio-genic sediment was produced and moved around the areas offshore these islands changed, and, largely because of these changes, some island beaches became starved of sediment. This resulted in shoreline erosion and, eventually, the erasure of the island from the reef on which it was built. This sequence of events probably explains the disappearance of islands like Bikeman and Tebua on Tarawa Atoll in Kiribati described in chapter 9.

When considering the disappearance of low islands made from unconsolidated mate-rial, it is difficult to separate the effects of slow, long-term sea-level rise, such as occurred within the past 200 years, from the effects of rapid short-term sea-level rises such as occur during storms (tsunamis are considered in the section on giant waves later in this chapter). Slow sea-level rise can destabilize such an island, rendering it far more vulnerable to later erasure by a storm surge, for instance, than it might otherwise be. Most of the islands that disappear as a result of storm surges in the Pacific are classified as cays, which, by defini-tion, are simply accumulations of unconsolidated sediment on a reef flat that are so young that they have not had time to develop any of the natural defenses against erosion (such as beachrock) that more persistent *motu* have.[14] Given their transitory nature, accounts of cays that have vanished are not knowingly included in the discussions in chapter 9; examples of such islands disappearing are known from the larger atoll reefs in the Cook Islands, Kiribati, and parts of Micronesia.[15]

Island-Flank Collapses Witnessed by Humans

In chapter 3, an account of island-flank collapses that occurred before humans had signifi-cantly populated the Pacific Basin is given. This section describes some of the largest flank collapses in the Pacific to have been witnessed and to have affected humans. The importance of discussing this topic is that not only does it demonstrate the potential for entire islands to vanish during flank collapses (as in Figure 3.6), but it also shows how some island-flank collapses witnessed by humans were so memorable that it is likely they informed myths about vanished islands.

The Hawaiian Islands have a long history of flank collapses. Numerous subaerial flank collapses have occurred around the island of Hawai'i (the Big Island) within the islands' human history,[16] the most recent being focused on the Hilina Slump on the southeastern flank of active Kīlauea Volcano. A series of slides in 1868 was triggered by the magnitude

8.0 Ka'ū earthquake and generated a large tsunami. The earthquake and its effects were described in graphic terms by the adventuress Isabella Lucy Bird [added material in brackets mine].

> The earthquakes became nearly continuous . . . they were vertical, rotary, lateral, and undulating. . . . Late in the afternoon of a lovely day, April 2, the climax came. The crust of the earth rose and sank like the sea in a storm. Rocks were rent, mountains fell, buildings and their contents were shattered, trees swayed like reeds, animals were scared, and ran around demented. . . . The people of the valleys fled to the mountains, which themselves were splitting in all directions . . . looking towards the shore, they saw it sink, and at the same moment, a wave, whose height was estimated at from forty to sixty feet [12–18 meters], hurled itself upon the coast, and receded five times . . . the whole south-east shore of Hawaii [Hawai'i Island] sank from four to six feet [1.2–1.8 meters], which involved the destruction of several hamlets and the beautiful fringe of cocoa-nut trees.[17]

More recently, on November 29, 1975, during the Kalapana earthquake, a 60-kilometer stretch of Hawai'i Island's south coast sank 3.5 meters and moved seaward some 8 meters causing a 10-meter-high tsunami (Figure 8.1), perhaps presaging a major collapse of this flank.[18] And then in November 2000 over a thirty-six-hour period, in what is probably the most scientific observation of a large flank failure, an array of Global Positioning System (GPS) receivers set up on the southeastern slope of Kīlauea, the most active volcano on Hawai'i Island, detected a 20-kilometer-long and 10-kilometer-wide piece of the volcano moving seaward at a rate of 5 centimeters per day. The movement was aseismic (no earthquake was involved) and probably triggered by heavy rain over the preceding nine days. Whether this event was significant or not in terms of a future catastrophic flank collapse is uncertain.[19] The most recent field survey has concluded, reassuringly for many people following the debate, that the Hilina Slump may, by virtue of having been moved southwest, have actually become "comparatively stable."[20]

Other well-studied large-scale flank collapses of single Pacific Islands that occurred during historic time include the 1741 event on Oshima-Oshima Island in Japan and the 1888 event on Ritter Island in Papua New Guinea.[21]

Located some 50 kilometers southwest of Hokkaido Island, the 10-square-kilometer Oshima-Oshima Island volcano erupts only infrequently, but the eruption on August 29, 1741, was memorable because it was accompanied by a 15-meter-high tsunami that killed 2,000 people as it ran across adjoining coasts. The eruption is known to have been accompanied by a collapse of part of the emergent island, having an estimated volume of material of 0.2 cubic kilometers, but this was long regarded as insufficient to have produced such a large tsunami. Sea-floor mapping subsequently revealed that the 1741 eruption and subaerial collapse were also accompanied by a massive collapse of the island's submerged flanks,

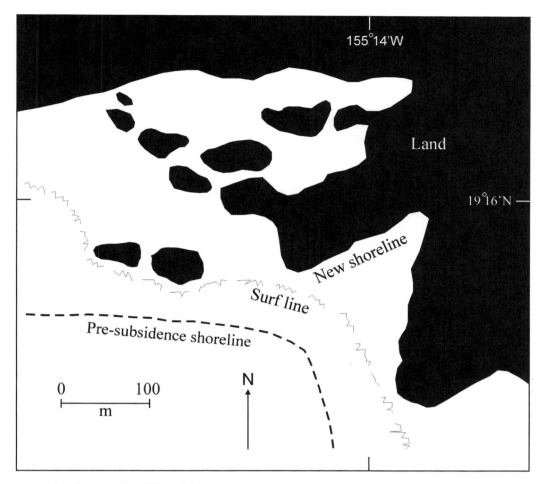

FIGURE 8.1. The coastline of Hawai'i Island around Keauhou Bay showing the new shoreline formed after rapid subsidence during an earthquake on November 29, 1975. (Adapted and redrawn from Nunn [2001b] after Henderson [1980].)

involving the displacement of a massive 2.4 cubic kilometers of material. There seems little doubt that it was this submarine collapse and the associated debris slide that were the main causes of the destructive tsunami.

The Ritter Island volcano (see Figure 9.1 for location) is more lively than Oshima-Oshima: five eruptions have been recorded since 1700. Of these, the 1888 event was by far the largest, destroying most of the island (Figure 8.2). The 1888 eruption involved the displacement of an estimated 4 cubic kilometers of material from the island's flanks that moved as a debris slide into deep water, producing a 12- to 15-meter tsunami that killed an estimated 3,000 people. Ritter is one of a line of island-arc volcanoes that includes Long Island and Hankow Reef, believed to be the remains of Yomba Island that blew itself to pieces and vanished around AD 750–850 (see chapter 9).

There are other examples of island-flank collapse known from the last few hundred years

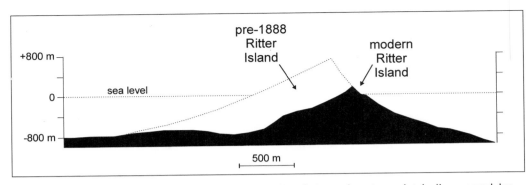

FIGURE 8.2. Ritter Island, Bismarck Archipelago, Papua New Guinea, almost completely disappeared during the eruption and submarine landslide in 1888. The modern island is shown in black, the pre-1888 island in outline. This example demonstrates the potential for islands to disappear catastrophically as shown in Figure 3.6. (Adapted and redrawn from Johnson [1987].)

in the Pacific in which the association between collapse and the generation of a tsunami is clear. In these cases, even a small flank collapse can generate a quite-sizeable tsunami, an observation that has implications for understanding and managing hazards in the Pacific.[22] For example, the large collapse that affected the island of Fatu Huku in the Marquesas Group of French Polynesia about AD 1800 saw the entire northwest quadrant of the volcano composing the island collapse, probably quite rapidly.[23] The coincidence between this collapse and the date when huge blocks of reef rock, the largest weighing 1,500–2,000 tons with a volume of 1,000 cubic meters, were thrown up onto the reef platform that surrounds Rangiroa Atoll, 1,250 kilometers away, led to the suggestion that a tsunami produced by the 1800 Fatu Huku collapse deposited these.[24]

Island-flank collapse is not an exclusively Pacific concern. In the year 1692, part of the waterfront of Port Royal on Jamaica in the Caribbean was removed as a result of an earthquake-induced submarine landslide on the island's flanks. The resulting tsunami destroyed most of the town and killed 2,000 of its inhabitants. Given that Port Royal was a town notorious at the time as a refuge for "pirates, cutthroats, whores and some of the vilest persons in the whole of the world," most commentators at the time considered its fate entirely fitting.[25]

The volcanic islands of the Canary Group in the eastern Atlantic are steep sided and have a long history of flank collapse.[26] The western flank of Cumbre Vieja Volcano on La Palma Island is currently a source of concern; during its 1949 eruption a 200-cubic-kilometer section dropped about 4 meters, and a major flank collapse may occur soon.[27]

The examples just quoted are all of comparatively large-volume collapses, but it is worth recalling that interspersed with these are smaller collapses, manifestly less important in human terms, but contributing to the progressive reduction in size of island edifices. Today many of these smaller slides are more than mere minor irritations, particularly when they result in the severing of ocean-floor cables.[28]

Collapses that caused, through the removal of its crestal parts, an entire island to disappear below sea level also appear to have occurred within the period of human occupation of the Pacific; examples include Teonimanu and Vanua Mamata, discussed in chapter 9. As with all types of flank collapse, it is likely that those that cause an entire island to disappear will occur more frequently in those parts of the Pacific where there are more potential triggers, typically volcanic or seismic, that might fast-track gravity collapse. But it is also important to note that gravity alone is sufficient to cause flank collapse of a steep-sided oceanic island. By applying these rules to the disappearance of entire islands, it is clear that most islands that have disappeared were in places where there is comparatively frequent large-magnitude volcanic and earthquake activity. But there are also a number of islands that have disappeared in other places simply as a consequence of gravity-induced flank collapse. The importance of this point will become clear in chapter 9 when explanations for the likely disappearance of islands like Burotu, Tuanaki, and Los Jardines are discussed.

Although just outside the Pacific Basin, the Sunda Arc in Southeast Asia marks an uncommonly active convergent plate boundary,[29] and several myths from the area speak of fabulous sea creatures destroying parts of the islands. Such accounts are likely to be recollections of island flank collapses during the earthquakes that often occur there. One example refers to the island named Luang off the eastern end of Timor, which was once part of larger Luondona-Wietrili Island before it became the victim of the fury of a large sailfish.[30] In the Pacific a comparable story from the steep-sided islands of the Marquesas recalls how a shark lashed the submerged pillar that held up the island of Fatu Huku, causing it to topple into the sea;[31] the story is discussed in more detail in chapter 9.

Throughout the Pacific Islands it was commonly believed that Māui (or another godlike being) held up the land to stop it sinking. Earthquakes were thought to result from Māui nodding while falling asleep, so when one occurred the islanders made as much noise as they could to awaken him and prevent the entire island sinking beneath the ocean. It could be argued that such a comparatively sophisticated response had derived from a memory of a real island-flank collapse (or island disappearance) that had reached most people through oral traditions.

Although there is obviously no possibility that the giant flank collapse of Tahiti Island shown in Figure 3.3B was witnessed by humans, the people of that island and others nearby interpreted island landscapes as fashioned by the spirits. One such myth from Tahiti states that an isolated coastal mountain, once "united" with those of the island's interior, was "removed" one night by the spirits.[32] If this refers to a flank collapse of steep-sided Tahiti, then it could have been one of the many that have occurred within the giant collapse scar.[33] A myth from the Marquesas Islands represents the peaks of the high islands as warriors; the headlands around the islands' coasts are "the bodies of warrior peaks who had been slain and had fallen headlong into the sea," also a possible recollection of flank collapses.[34] One version of the early history of Easter Island refers to a god named Uoke who traveled around the Pacific with a giant lever (crowbar) that he used to pry entire islands up from their foundations and toss them into the sea; when he reached Easter Island, he went around the coast breaking off parts with his lever until it broke on the hard rocks at Puko Puhipuhi.[35] A comparable myth comes from Aniwa Island in southern Vanuatu, in which once the "richer" part of the island "split off" and "sailed" to nearby Aneityum Island,[36] an unobstructed 100 kilometers away. This is plausibly the record of a flank collapse of Aniwa and of the associated tsunami wave striking Aneityum.

Yet the possibility that such myths are metaphorical, referring not to islands but to groups of people, should always be borne in mind. The metaphor is readily apprehended when the islands said to have separated are so far apart that the same collapse could not possibly have affected them both. A good example is provided by the myth that talks about the island of Tahiti (actually Tahiti Nui) breaking off of and swimming away from the island of Raʻiatea, some 200 kilometers away. This is almost certainly a metaphor for a group of people ousted from Raʻiatea who subsequently settled in Tahiti.[37]

Giant Waves

Giant waves can be generated by island-flank collapses: sometimes by displacement of the ocean floor in an earthquake, sometimes for reasons that are unclear. When comparing giant waves with other hazards that might cause an island to disappear, wholly or partly, three points are worth pondering.

- The size of the collapse is not necessarily proportional to the size of the wave generated: much depends on the water depth and the speed of the landslide.[38] For example, in 1792, a comparatively small landslide from the collapse of the Mayuyama lava dome on Mount Unzen Volcano on Kyushu Island in Japan moved into Ariake Bay, generating giant waves that resulted in the deaths of 14,524 people.[39]
- Neither are giant waves confined to the areas of their generation. They also affect coasts far away. A recent example of this was the Indian Ocean tsunami, which occurred on December 26, 2004. This tsunami series was generated by a sudden vertical displacement of the ocean floor off Sumatra and killed more than a quarter of a million people along Indian Ocean coasts. It was still large enough to kill hundreds when it swept ashore along a 650-kilometer stretch of the coast of Somalia, at the other extremity of the Indian Ocean.
- Giant waves can also apparently come out of nowhere, generated by gravity collapses of island flanks deep under the ocean surface that go undetected by people above. The 1985 wave that hit the islands of Kosrae in the Federated States of Micronesia is believed to have been generated by such a submarine landslide.[40]

Giant waves are commonly generated along convergent plate boundaries, triggered by an abrupt movement of an associated fault line. So it is no surprise to find that the mythical record of these waves is also linked with islands in such locations. One good example is from the mythology of the Ami, an indigenous group in Taiwan, who recall that long ago the sea rose high across the land, sometimes boiling, following an earthquake.[41]

The wave that led to *te mate wolo* (the great death) on Pukapuka Atoll in the northern Cook Islands early in the seventeenth century is also believed to have been a tsunami,[42] possibly one generated from a landslide on the steeper side of the ocean trench off the Pacific coast of South America. The oral traditions give few clues, describing the nighttime impact of the wave thus: "waters raged on the reefs, the sea was constantly rising, the tree tops were bending low. On the next day all the island was broken and everything destroyed." Only two women and seventeen men "with remnants of their families" survived to repopulate Pukapuka.[43] A comparable event recorded in Maori tradition as *Te Taraitanga* (the shaving-off) is held responsible for the disappearance of Taporapora Island off the North Island of New Zealand; Taporapora had been an important landmark for long-distance ocean voyagers, so its absence was much commented on.[44] Finally, it was reported that Beveridge Reef (location shown in Figure 2.4C) near Niue Island in the central South Pacific was once "a fine isle, with many coconut palms growing thereon, but that it was swept bare by a fierce hurricane, which carried away both trees and soil, leaving nothing but the bare rock."[45] Rather than solely wind, a large wave is more likely to have been responsible for the disappearance of this island.

Lituya Bay along the Pacific coast of Alaska was renowned among early voyagers in the area as a safe haven. French naval officer Jean-François de La Pérouse thought it "perhaps the most extraordinary place in the world" and planned to make it a French base. But this optimism was misplaced because the Fairweather Fault that runs across the inland side of Lituya Bay, parallel to the coast, periodically ruptures, causing giant waves to sweep seaward along the bay. Such waves were generated in the 1850s, about 1874, in 1936, and, most recently, on July 8, 1958. On that day, an earthquake affected the Fairweather Fault, causing a rockslide that crashed into the bay, producing a surge of water that reached the incredible height of 524 meters followed by a tsunami with a maximum height of at least 30 meters that swept seaward along the bay. Figure 8.3 is a photograph of Lituya Bay taken soon after the 1958 event.

Large waves have been considered responsible for the destruction of coastal settlements at various times in the history of the Pacific Islands. Some of these are recorded in myth, such as those from New Zealand, and others have been discovered by excavation, such as that on Huahine Island in French Polynesia.[46]

It is salutary at this point to reflect how mistranslations of myths from vernacular languages can sometimes lead to significant misinterpretations of what those myths actually recall. The fifteenth-century Hao-whenua event that affected the Wellington region of New Zealand has been linked by geologists to a coseismic uplift of the land in this unstable part of the world. The subject of a large body of oral traditions, the Hao-whenua event was first

FIGURE 8.3. View across Lituya Bay, Alaska. The active Fairweather Fault forms the cliffs at the head of the bay and at least four times since 1853 has ruptured abruptly, causing giant waves to rush seaward down the bay. Caused by an earthquake-induced rockslide, the most recent wave (July 8, 1958) surged to 524 meters in height, washing right over Cenotaph Island in the center of the bay. The nonforested (lighter-colored) areas on the sides of the bay were inundated by this wave. (United States Geological Survey, original photo by Byron Hale.)

interpreted in English as an event that "swallowed up a land surface," a contradictory explanation to one involving coseismic uplift. A more-appropriate translation of the Maori description of the event is "to sweep the land clean," and the myth has been reinterpreted as recalling the effects of a tsunami across the Wellington coastline at that time.[47]

Islands That Regularly Appear and Disappear

In 1796, the "voice of God" spoke loudly to the inhabitants of the Aleutian Islands in the northernmost Pacific. In the Aleut language, the voice of God is *bogoslof* and was applied to the roar accompanying the underwater volcanic eruption that led to the first known appearance above the ocean surface of the island that is today known as Bogoslof. This island has successively appeared and disappeared through time: shallow underwater eruptions sometimes result in an island being built (largely from rock fragments) above sea level, and then waves erode the island when the eruption ends, removing it completely.

This type of behavior is not uncommon elsewhere in the Pacific, where it is markedly more common than in the other oceans of the world. And in fact, over long periods of time, some shallow underwater volcanoes may persist in behaving this way until such time as a

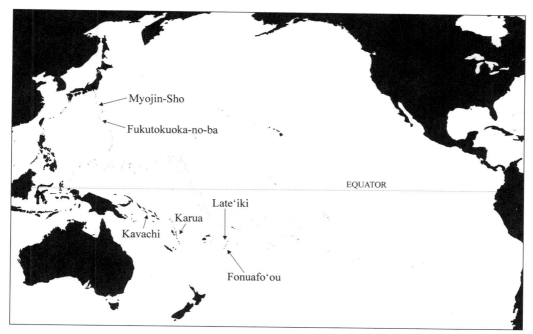

FIGURE 8.4. Map of the Pacific showing the locations of the "jack-in-the-box" islands mentioned in the text.

sufficiently voluminous and prolonged eruption allows lava rather than just fragmented volcanic rock (which results when magma encounters seawater) to be produced.[48]

It has been argued that the type of volcanic island that regularly appears and disappears has influenced Pacific Island origin myths,[49] yet the disappearances involved are not those that are a principal focus of interest in this book. In most cases, the islands remain emergent for perhaps a few weeks, although parts of Bogoslof have persisted since 1796, with other parts being successively added and then eroded.

This section describes some of the better-known such "jack-in-the-box" volcanic islands from the Pacific (Figure 8.4), the activities of which sometimes feature in the mythologies of people on nearby islands. Some such myths have been classified as employing the land-raiser myth-motif, which is found in various forms throughout the Pacific Rim as well as the Pacific Islands and which has been argued as having ultimately been derived from the experiences of people in Southeast Asia when that area was drowned by postglacial sea-level rise.[50] Others of these myths are more plausibly interpreted as largely autochthonous creations derived from repeated observations of shallow-water submarine eruptions.

Among the underwater volcanoes that rise from the Izu-Ogasawara Ridge, an active volcanic island arc in Japan, are Myojin-Sho and Fukutokuoka-no-ba (Shin-io-jima).

The earliest-known island-building eruption at Myojin-Sho occurred in 1896, but there have been more than ten appearances and disappearances of the associated island since then. The 1952–1953 period stands out because of repeated eruptions and island building/

destruction episodes.[51] Since 1953, no other island-building eruptions have occurred at Myo-jin-Sho, and its current activity level is low.[52] Eruptions continue at the nearby Bayonnaise Rocks that, like Myojin-Sho, rise from the rim of the huge submerged caldera named Myojin Knoll, discussed in chapter 3.[53]

In contrast to the lack of recent activity at Myojin-Sho, the Fukutokuoka-no-ba Volcano farther south has been conspicuously active in the past few decades. Although only four eruptions have been observed here in the past 100 years, since 1950 discoloration of the ocean surface, suggesting an underwater eruption, has been reported on more than thirty occasions. In January 1986, an eruption of Fukutokuoka-no-ba produced a 10-meter-high is-land measuring 600 meters by 200 meters that disappeared within three months.[54] Its most recent eruption in July 2005 produced a column of steam more than 1 kilometer high.

Probably the most active shallow-water submarine volcano anywhere in the world over the past 100 years has been Kavachi in Solomon Islands, which has erupted sixteen times since 1939, building an island on at least eight occasions. There is a conspicuous absence of long-enduring myth associated with Kavachi, which may be because its activity only became visible at the ocean surface in 1939. It may also be that, because of an abundance of active volcanoes in Solomon Islands, the occasional activity of Kavachi was considered insignificant by the mythmakers. Explosive eruptions of shallow underwater volcanoes have often been characterized in Pacific Islander myths as large fish being hauled up from the depths (see later in this section). A good example from Kavachi is shown in Figure 8.5.

Karua is the name given to an island that alternately appears and then disappears above the huge submerged Kuwae Caldera, formed in the massive eruption about AD 1453 when the entire island of Kuwae blew itself to pieces (see chapter 9). Since 1897, Karua is known to have erupted at least ten times, building an island on five occasions (see Figure 9.4C). The 1948–1949 eruptions, for example, built an island with a diameter of 1.6 kilometers that reached 100 meters in height but had been washed away within a year of its appearance.[55] A huge amount of oral tradition grew around the 1453 eruption, its precursors, and its after-math but relatively little about the comparatively benign activities of Karua subsequently.

In the Pacific, the greatest concentration of shallow-water active volcanoes is found in the Kingdom of Tonga, where two, Fonuafo'ou (Falcon Island) and Late'iki (Metis Shoal), have been particularly active within the last century.[56] In both 1885 and 1927 Fonuafo'ou erupted, building on each occasion islands more than 6 kilometers long that survived only a short time, but it is not known to have been active since 1936. There have been at least nine erup-tions of Late'iki since 1851, six of which are known to have formed short-lived islands; the 1995 eruption produced an island that was 280 meters in diameter and 43 meters high.[57]

A typical eruption of Fonuafo'ou was described in 1886 as follows by the Reverend Shirley Baker.[58] The eruption was sighted and heard from Tongatapu Island, 90 kilometers distant, and the vessel *Sandfly* was sent to investigate.

FIGURE 8.5. The October 10, 2002, eruption of underwater Kavachi Volcano in Solomon Islands. Note the resemblance of the eruption plume to a huge fish, a significant detail in many island-origin myths in the Pacific. (Photo by Corey Howell, The Wilderness Lodge, Solomon Islands, used with permission.)

As the *Sandfly* neared the spot the scene was most magnificent, great volumes of steam, of carbonic and sulphurous gas . . . being shot forth from many jets out of the sea, in a direct line of over two miles . . . to the height of 1,000 feet [305 meters] and more, then expanding themselves in all directions, in clouds of dazzling whiteness, and assuming the most fantastic shapes; sometimes presenting themselves as a mountain of wool, the tips of which were fringed with gold, caused by the rays of the setting sun, then again occasionally forming into a large cauliflower head of snowy whiteness, backed by clouds of intense darkness formed of dust and ashes mixed with watery vapour, which the wind was carrying down for miles on the distant horizon. As the heavier matter kept continually falling, it gradually raised in height the new-made island."[59]

In terms of Tongan myths, there are three elements in this account that are worthy of comment.

The first element is the huge shapes formed of steam that rise from the ocean above the eruptive center. Many of these resemble a huge fish (see also Figure 8.5), and because the memories of such eruptions were being told to audiences who were routinely acquainted with fishing, the common use of the metaphor of the emerging island as a giant fish being hauled in is unsurprising.

The second element is the raising in height of the new-made land, which occurs in many

Tongan myths. In the following extract from the jottings of the Reverend J. B. Stair, the god Tangaloa is asked by his son Tuli (a species of plover) for a resting place in the ocean. So,

> Tangaloa fished up a large stone from the bottom of the sea with a fish-hook. Having raised the stone to the surface, he gave it to his son for a dwelling place. On going thither to take possession of his new home, however, Tuli found that every wave or swell of the ocean partially overflowed it, which compelled him to hop from one part to another of the stone to prevent his feet being wetted by each succeeding wave. Annoyed at this, he returned to the skies to complain to his father, who, by a second application of the mighty fish-hook, raised the land to the desired height.[60]

Third, in these myths there even appears to be some recollection of the island disappearing shortly after it was formed. The idea of island disappearance was probably a popular component of Tongan oral traditions and diffused with Tongan migrants throughout the Pacific. The myth from Tahuata Island in the Marquesas that involves Māui fishing up a land called Tonaeva[61] and then letting it go may contain Tongan elements of this kind.

It is also worth pointing out that such shallow-water eruptions are inevitably accompanied by great noise, the voice of God for eighteenth-century Aleuts, and considerable disturbance of the surrounding ocean surface. This disturbance naturally lent itself to the metaphor of the island being fished up: myths often talk of a fish-island "squirming" while being hauled up "gurgling foam and bubbles."[62] The metaphor is arguably best developed and most widespread in Tongan tradition, suggesting that this was the part of the Pacific where shallow-water submarine eruptions witnessed by people have been most common. Consider this account by the Reverend Lorimer Fison about when the demigod Māui went fishing with his two sons: suddenly

> they were aware of something very heavy that the hook had caught. . . . The waters rose bubbling and foaming around the canoe, and smoke came from them with a thunderous rumble and roar, and the god [Māui] cried out in deadly fear. . . . From the midst of the waters rose a land, mountain after mountain, till there were seven mountains in all, with valleys between, and flat lands lying at their feet. . . . The sons thought the mountains too high so Maui, leaping ashore, sprang to the top of the highest mountain, and stamped upon it with his feet. And as he stamped, the earth shook, and the mountain crumbled away beneath his feet, and rolled down into the valleys below, till they were filled up to the level on which he stood.[63]

Note that this extract also contains examples of all three elements (see earlier in this section) that are argued as characterizing myths recalling shallow-water submarine volcanic eruptions. These are the representation of the eruptive column/cloud as a fish, the raising in height of land, and finally the disappearance of (in this case) part of the land. Many other island-origin myths in Tonga tell that Māui hauled up a whole ridge (or a land larger than modern Tonga), but then his line snapped and part of the ridge was submerged, leaving only its peaks emergent.[64]

Islands That Blew Themselves to Pieces

Preliterate Pacific Islanders were able to distinguish a range of the manifestations of volcanic activity. One good example comes from the islands of northern Tonga, where some myths are believed to recall the caldera-forming eruption on Niuafoʻou Island. The myths refer to the central part of Niuafoʻou, today a caldera lake about the same size as the land area of the island, being removed by mischievous "imps" who carried it away, dropping it to create the island of Tafahi, almost 200 kilometers to the east. In reality, the central part of Niuafoʻou would probably have collapsed (not been blown out), but this event would, by analogy with better-observed events elsewhere, have probably been accompanied by the ejection of large amounts of ash into the atmosphere, which would likely have been blown east, some settling on Tafahi.[65]

One of the least ambiguous eruptive phenomena is when a volcano blows itself to pieces. It may be exactly the sudden, highly explosive event that the phrase suggests, although that need not imply that it occurred without warning. There are numerous examples of preliterate people demonstrating a profound knowledge of the precursors of volcanic eruptions (and other geologic hazards) that enable these people to remove themselves from the danger area well in advance of the eruption occurring.[66] In the modern world, most governments are in a position to obtain advice about an imminent volcanic eruption and evacuate the people likely to be affected.

Probably the best-known example of an island that blew itself to pieces is that of Krakatau on August 27, 1883, part of the Sunda Arc in Southeast Asia.[67] This phenomenal eruption, which was audible across 57 percent of the earth's surface, led to the destruction of Krakatau Island. Another island-destroying eruption, one that may have fuelled the Atlantis myth, was that of Santorini in the Mediterranean Sea 1650–1620 BC.[68] Yet both these eruptions were dwarfed in magnitude by the 1453 eruption of Kuwae in Vanuatu, discussed in chapter 9 along with Yomba Island in Papua New Guinea, as examples of islands in the Pacific that blew themselves to pieces — and disappeared.

It may seem unnecessary to spell out just how the large eruption of an island volcano can lead to its destruction. Yet in most cases we cannot be sure of the processes involved because much of the evidence is dispersed and unrecognizable. That said, these kinds of island-destroying eruptions almost always appear to be associated with the formation of a caldera, implying the collapse of a volcanic edifice into an underlying void, created by the emptying (through eruption) of a magma chamber. Although the level of the sea around such a volcano is immaterial in terms of the outcome of such an eruption, if the void is well below sea level and large enough to accommodate most of the material forming the volcano above sea level, then its collapse will naturally result in the disappearance of the island. Caldera formation, especially in areas of steep slopes, may also be accompanied by landsliding that may also contribute significantly to island disappearance.

9

Recently Vanished Islands in the Pacific

The vanished islands described in this chapter are adjudged authentic (see Table 6.1). The test of authentication is largely based on both the details of the oral traditions (particularly whether the same tradition was obtained independently from different groups in the same area) and its credibility given the geological context of the island(s) alleged to have disappeared. Other supporting information, such as mention of the vanished island in written accounts not based on oral tradition and the presence today of some indicator of a sunken island in the place where it is reputed to have disappeared, is clearly helpful in judging the authenticity of a particular island's alleged disappearance.

Much of the information in this chapter is new, acquired from indigenous informants as part of my research projects between 2002 and 2005. This information is privileged in the sense that it was given for a particular purpose and should not be used elsewhere without appropriate acknowledgment of those informants and their affiliations.

Recently Vanished Islands in Papua New Guinea

The rich body of oral tradition in the islands of Papua New Guinea contains much detail about volcanic eruptions. The 1982 book *The Time of Darkness* by Russell Blong is a superb illustration of this, showing how people throughout the New Guinea Highlands recalled through oral tradition the deposition of the andesitic Tibuto Tephra associated with an eruption of the Long (Arop) Island volcano about 1660. While also showing that variations in the details of stories about the *yuu kwia* (the time of darkness) were manifold, Blong demonstrated the validity of using oral traditions from preliterate societies to reconstruct the extent of volcanic phenomena such as ashfall.

Long Island is off the northeastern coast of the main island of New Guinea (Figure 9.1), and what made the 1660 eruption exceptional was that it was powerful enough to eject

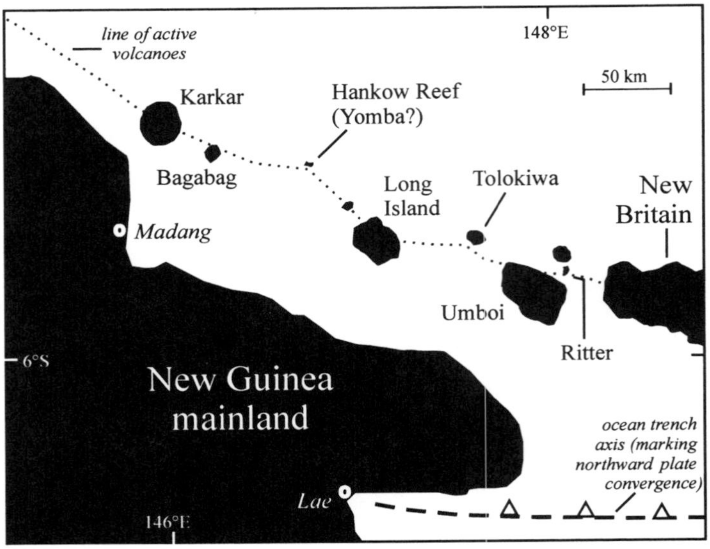

FIGURE 9.1. The modern geography of the southwestern Bismarck Archipelago, showing the line of active volcanoes that may have once included Yomba. The activity of these volcanoes is attributable to northward plate convergence along the ocean trench shown to the south.

tephra across the New Guinea Highlands, not to mention plunging them into darkness for several days. Most Long Island eruptions are not so powerful.

Some 60 kilometers northwest of Long Island lies a small coral reef, Hankow Reef, that is believed to mark the site of another island volcano named Yomba, now disappeared, about which oral traditions were extensively collected and analyzed by Mary Mennis.[1] Many people in the Madang lowlands of New Guinea trace their ancestry back to Yomba, which apparently disappeared following a volcanic eruption preceding the 1660 eruption of Long Island. Most people questioned by Mennis insisted on this; for example, the people of Kranket Island (just off Madang on the New Guinea mainland), who settled there only after Yomba blew up, have a clear recollection in their traditions of their ancestors being on Kranket during the 1660 ashfall from Long Island. Consider what some of them had to say about Yomba (material in brackets mine).

Yomba Island was as big as Karkar [Island: approximately 425 square kilometers]. . . . It had two mountains—one was a large mountain with smoke inside. When it erupted the island sank into the sea. . . . When the island sank it caused a large tidal wave.[2]

The island of Yomba was bigger than the islands near Madang. It had a mountain on it. . . . When Yomba fired up it erupted on its own. Karkar did not erupt, nor did Arop. Arop is there, and Karkar and Bagabag, but Yomba has gone. It used to be in line with the other islands.[3]

Yomba went down a long time before Arop [Long Island] erupted [in 1660]. There was a loud noise and the mountain on Yomba fired up and the island sank. The men from Yomba swam from the big island to Bilibil. . . . The island of Yomba was as big as

Karkar. The people there made pots. When they got the warning that the island was about to blow up, some left and went to Mindiri, others went to Kranket, Yabob, and Bilibil.[4]

People who are dubious about the possibility that Yomba actually blew itself to bits would do well to study the 1888 eruption of the volcano on nearby Ritter Island. This eruption, which involved the displacement of a massive 4 cubic kilometers of material from the island's flanks, saw the disappearance of all but a tiny part of the island (see Figure 8.2).

Dating the event in which Yomba appears to have blown itself up is problematical because the only clear evidence comes from genealogies. The figure of eight to ten generations is plausible from the oral traditions, but this gives a date *after* the 1660 eruption of Long Island, not before it, which is what informants insist was the order of events. It is possible that the eruption in which Yomba blew itself up and disappeared was the same that caused the deposition of the Olgaboli Tephra, which, throughout the New Guinea Highlands, consistently underlies the Tibito Tephra and has been dated to AD 750–850.[5] Yet this scenario is beyond the likely range of oral traditions and contradicts the figure of eight to ten generations given by informants.[6] It may therefore be that Yomba disappeared during an eruption shortly before the 1660 eruption of Long Island, although such an event has not been identified geologically.

The northeast-facing coasts of the island of New Guinea adjoin an active plate margin and are affected regularly by tsunamis generated by landslides along the offshore ocean trench (Figure 9.2A). The July 17, 1998, tsunami that swept onshore at Aitape killed more than 2,000 people but was not associated with island disappearance. Yet tsunamis are but one expression of plate convergence in this area, as the incident on the night of December 15–16, 1907, at nearby Warapu where Sissano Lagoon is now located confirms. In this an entire island reportedly sank.

Understandably, compared to the 1998 Aitape tsunami, little is known about the 1907 event, which occurred in the earliest period of European contact and has therefore been formally recorded largely by the subsequent activities of anthropologists.[7] Yet about eight months after the incident, Dr. Richard Neuhauss visited the area, approaching the site of the sunken island by canoe through a border zone of partly submerged coconut palms 1.0–1.5 kilometers wide. Here "the dead and leafless palm stems are a melancholy spectacle . . . one has to wind in and out among the dead trunks with the utmost care, for the slightest bump would be sufficient to bring rotting stems crashing down." Neuhauss reported that the island that sank was formerly home to 2,000 Warapu people, almost all of whom survived the tragedy by taking to their boats.[8]

There are many similarities between the 1907 and 1998 events, and ample reason to suppose that they were generated by similar processes and had much the same manifestations —the emphasis on the tsunami in the more recent event is due to the horrific loss of life aris-

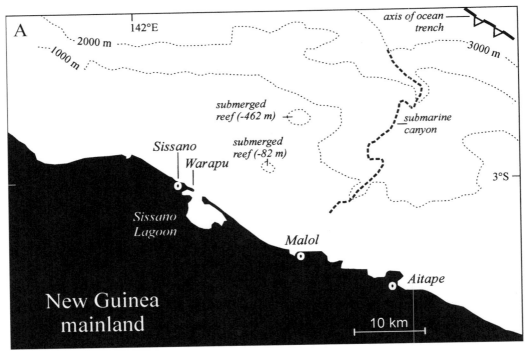

FIGURE 9.2. **A.** Map of the coastal area of northeastern New Guinea affected by the 1907 and 1998 events. Submarine landslides on the sides of the offshore trench often produce tsunamis that are funneled into the submarine canyon that leads onshore. This explains why such destructive tsunamis are regularly experienced in the Aitape-Sissano area. During the 1998 Aitape tsunami, some subsidence was noted at Warapu (see **B**). During the 1907 event, Sissano Lagoon was created by the subsidence of an island that was the site of Warapu Village at the time. The presence of submerged reefs offshore suggest that islands may have disappeared here at earlier times. (Adapted and redrawn after McSaveney et al. [2000], Davies et al. [2003], and Matsumoto et al. [2003].) **B.** Subsidence at the mouth of the Sissano Lagoon both during and after the 1998 event led to some coconut palms being partly submerged, a similar situation to that described by Neuhauss in 1908 (see text), and is permitting erosion of the Warapu Peninsula where pre-tsunami Warapu Village was located. (Photo by Hugh Davies, used with permission.)

ing from the higher population density (compared to 1907) in the worst affected areas. Yet it also seems that coseismic and postseismic subsidence in the area were more marked during the 1907 event, claimed by Neuhauss to have led to the formation of the Sissano Lagoon, but subsidence also occurred in the 1998 event. Figure 9.2B shows the mouth of Sissano Lagoon in the area where Warapu Village was eventually reestablished after the 1907 event. The partly submerged coconut palms testify to subsidence of this area in the 1998 event.

Recently Vanished Islands in Solomon Islands

As in Papua New Guinea, oral traditions are well preserved today in most parts of Solomon Islands. Considerable research into the stories of vanished islands has been carried out in that part of the central Solomon Islands shown in Figure 9.3A, notably on the islands Makira, Malaita, and Ulawa. The stories recall islands that vanished in three main areas: between Ulawa and Olu Malau (Three Sisters), from which stories about a vanished island called Teonimanu are ubiquitous; off the south coast of Maramasike (small Malaita); and off the northeast coast of Malaita and Maramasike.

This area of Solomon Islands is one of the most geologically active in the region, providing plenty of potential reasons why islands might indeed have abruptly sunk.[9] Malaita and Ulawa (but not Makira) and nearby smaller islands are part of the Malaita Accretionary Prism, an elongate crustal body composed of material scraped off the top of the Ontong Java Plateau while it has been subducting beneath the Solomons Arc for the past ten million years or so (see Figure 4.2). The rocks of the Malaita Accretionary Prism are intensely folded, a process that is continuing as the Ontong Java Plateau continues to push southwestward and downward.

The southeastern end of the Malaita Accretionary Prism, known as the Ulawa Structural Domain, is the area where there are traditions of islands vanishing. The Ulawa Structural Domain is that part of the Malaita Accretionary Prism beneath which the Ontong Java Plateau dips most steeply (7 degrees) and where, at 5,700 meters below the sea surface, the ocean trench marking the plate boundary is deepest. These two facts combine to make this domain uncommonly unstable, something that might account for stories about islands vanishing, not only through abrupt (coseismic) subsidence and gravity but also as a result of being overrun by tsunamis produced, perhaps, by landslides along the steeper side of a nearby ocean trench.

Figure 9.3B shows the outline of the area's structure. A ridge, marking the crest of the Malaita Accretionary Prism, runs parallel to the northeast coast of Malaita and joins Ulawa and Olu Malau. As can be seen in the cross sections in Figure 9.3C, the outer (northeast and east) sides of the ridge in both these areas is downfaulted and shows, by the accumulation of sediments at the foot of the underwater slopes, signs of a long history of downslope

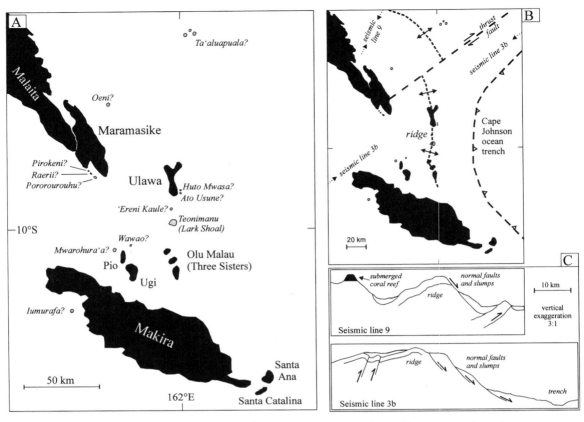

FIGURE 9.3. Part of the central Solomon Islands showing locations and possible interpretations of vanished islands recorded in oral traditions. **A.** Geography of the central Solomon Islands showing the locations of islands (shaded black) where informants related oral traditions about vanished islands (island names in italics) and the reported/likeliest locations of these (shaded light gray). **B.** Outline of the geology of the area shown in **A.** Note the presence of the Ulawa–Teonimanu–Olu Malau ridge, the eastern flank of which dips steeply down to the Cape Johnson Trench. (Adapted and redrawn from Phinney et al. [2004] and Taira et al. [2004].) **C.** Simplified interpretations of part of the two seismic lines (in B) run during the 1995 cruise of RV *Maurice Ewing*. Note in seismic line 9 the presence of a submerged coral reef (island?) within 10 kilometers of the northeast coast of Malaita and also the evidence of a long history of slumping along the northeast side of the ridge farther north. Seismic line 3b crosses the Ulawa ridge and note the evidence for slumping along its eastern flanks. (Adapted and redrawn from Phinney et al. [2004].)

movement of material. The Cape Johnson Trench east of Ulawa reaches depths greater than 5,000 meters.

Teonimanu

In the oral traditions of some of the inhabitants of the central Solomon Islands there is a persistent myth concerning a sunken homeland. The first written report of this myth, in 1925, named this homeland as an island called Teonimanu;[10] the author noted that survivors from the island's sinking escaped to Makira (formerly San Cristóbal) Island, to Ulawa Island, and to the Sa'a district at the southern tip of Maramasike (Figure 9.3A).[11] A later collection of oral traditions from Santa Ana Island, off the eastern tip of Makira, revealed that the Pagewa clan living there also claims descent from the people of this vanished island.[12] The earlier set of traditions claimed that this island sank because of sorcery; the later set claimed it was because of a natural disaster. The traditions from Santa Ana were sufficiently precise to be interpreted as meaning that Teonimanu lay south of Ulawa where the reef named Lark Shoal lies today,[13] a detail that agrees with information from traditions obtained more recently.

This disappearance of Teonimanu was accepted widely in this part of Solomon Islands in the early part of the nineteenth century, as recounted in the 1927 memoir of W. G. Ivens. Not only is Teonimanu often identified as Lark Shoal, but its inhabitants are credited with a number of distinctions, such as being acclaimed ocean voyagers and having a comparatively elaborate material culture. Among their artifacts was a type of canoe with a mast called *iola sara'a* (branched canoe), one of the last of which was apparently sighted in 1769,[14] and the bowl "anciently used at feasts" called *nima sara'a* (branched bowl) that had frigate birds carved on its sides.[15]

In 2004, I coordinated the collection of oral traditions in this part of Solomon Islands, targeting the four places where survivors of the sinking of Teonimanu are alleged to have settled.

Among the people on Makira Island, and its satellite islands Santa Ana and Santa Catalina, it is widely acknowledged that today the Aiga Tatari clan, being the direct descendants of the people who once lived on Teonimanu, are the only ones who are entitled to speak about its disappearance.[16] The Aiga Tatari say that they came from two islands that disappeared, Teonimanu and Teoniramarama (known collectively as Finuasi), located between the Olu Malau Islands (commonly known as the Three Sisters) and Ulawa Island. Following the disappearance of the islands, a considerable time ago,[17] the Aiga Tatari "floated" to the Wainoni area of northern Makira, then slowly migrated along the coast, settling finally on Santa Ana and then Santa Catalina. Uniquely among the survivors, the Aiga Tatari still have a strong recollection of Finuasi, recalling its place names, the people who lived there, their cultural practices and beliefs.

The myth recalling the disappearance of Finuasi begins when a man called Rapu'anate took a wife (he already had several) from Teonimanu back to his home on Ali'ite, the northernmost of the Olu Malau Group. Soon afterward, when Rapu'anate was elsewhere, the woman's brother went and brought her back to Teonimanu. Rapu'anate tried everything to get her back but to no avail, so, bent on revenge, he paid "an old custom lady" on Dai, a small island off the north coast of Malaita, to call the sea to sink the island(s). She gave Rapu'anate three taro plants, one he was to plant on Teonimanu, one on Teoniramarama, and the other that he was to keep; when it sprouted new leaves, that was the sign that the islands were about to sink.

Rapu'anate did as he was bid, and, when the new leaves started to appear on the taro plant he had kept, the islands "became salty" (probably a sign of seawater infiltrating the groundwater, a sign of submergence). As the sea rose further some people left the islands because it was becoming too difficult to survive. Then the winds picked up strength and waves crashed into the islands; Teoniramarama sank quickly but Teonimanu was "slow to disappear." Some people had already made rafts on which to escape, but others had to cling to coconut fronds or banana trees.[18]

Traditions collected recently from Ulawa Island also state that Teonimanu (or Hanua Asi [submerged land]) was located where Lark Shoal lies today (see Figure 9.3A). But the Ulawan traditions[19] also recall that four islands in total vanished in the area between their island and the Olu Malau group: Teonimanu, ʻEreni Kaule, Ato Usune, and Huto Mwasa.[20] In addition, at least two islands, Mwarohuraʻa and Wawao, are reputed to have disappeared north of nearby Ugi and Pio islands. Of all these islands, Teonimanu was the largest and highest, possibly the only one that had (many) people living on it. The people of Teonimanu were known as industrious people, renowned for their interisland voyaging achievements, but their island was destroyed by a series of eight tsunamis. These waves were described as *luelue* in Ulawan, a word reserved for tsunamis and not applied to large waves generated by other means. The cause of the tsunamis involves a myth that is essentially the same as that collected from Santa Catalina; an original rendition of the Ulawan myth is given in Appendix 1. Most informants on Ulawa regarded the islands that had disappeared in the area as having done so in a single event. One informant recalls being told that they existed when Mendaña came to the area in 1568 but had disappeared by the time that James Cook was close by.[21]

Teonimanu is more than just a memory; it is sometimes invoked in practical situations. For example, in the memoir by Ivens mentioned earlier, there is an incantation deriving from Ulawa that can be used by one in a canoe caught in a storm. In translation:

A paddle rent the waves
A fish rent the waves
All the fish of Him who drifted to Ulawa
It rent the waves, all the bottoms of the waves
From off my little canoe
All my little canoes shall be as the young of the swallows.

Upon reciting this, a man from Teonimanu in the incarnation of a shark appears and leads the canoe to safety.[22]

Apparent confirmation of the 1925 report that some of those who survived the sinking of Teonimanu (named as Maroʻrouhu in this tradition) Island settled in the Saʻa district of southern Maramasike was recorded by Sereana Usuramo during her 2004 study of the meanings of the island's place-names.[23] She collected a tradition from Oroha Village that explained that it was established by three brothers from Maro'rouhu (now a reef) after that island sank; the name Oroha means "swimming across."

Together with Ulawa and the islands of Olu Malau (Figure 9.3A, B), the former island Teonimanu lies along the ridge marking the crest of the Malaita Accretionary Prism in the area near the south of the Ulawa Structural Domain. Although, as shown in the interpretation of seismic line 3b in Figure 9.3C, the western flanks of this ridge are gently sloping and upthrust in parts, the eastern flanks that extend down to the floor of the Cape Johnson Trench are steep and cut by at least three normal faults. From the amount of sediment accumulated at the foot of this slope, it is reasonable to conclude that it has experienced numerous slumps (landslides) and that Teonimanu slipped downward and vanished during one or more of these.

A similar explanation is likely for the disappearance(s) of the islands 'Ereni Kaule, Ato Usune, and Huto Mwasa shown in Figure 9.3A, all of which lie on the same ridge crest as Ulawa and the islands of Olu Malau. It is also noted that the east coast of Ulawa resembles the crescentic head scar of a large landslide, which adds weight to the idea that the eastern flank of this ridge is inherently unstable.

Other Recently Vanished Islands in Solomon Islands

Another tradition from the Oroha area of southern Maramasike refers to the disappearance of three islands that are distinct from Teonimanu although some of the traditions may have become mixed.[24] The three islands (Pororourouhu, Raerii, and Pirokeni) were all low "coral islands" located close to the south coast of Maramasike between Cape Hartig and Cape Zelée (see Figure 9.3A). The largest of the islands, Pororourouhu, was settled by a man named Roraimanu Paina and his people after he had fought with his two brothers on the Olu Malau Islands, some 70 kilometers southeast. Yet even though Roraimanu Paina settled peaceably on Pororourouhu, the enmity of his brothers persisted until one day they sent sharks to make "tidal waves" that destroyed all three islands, forcing their inhabitants to the main island where they established Oroha Village.

Tangible evidence of the disappearance of these islands includes a "forest of tree trunks," washed away during Cyclone (Hurricane) Namu in 1986, and three elongate reef platforms, now 4–5 meters below mean sea level, corresponding to the three islands.[25] The disappearance of these islands appears to have been within the past 50–100 years.

The most parsimonious explanation for the disappearance of Pororourouhu, Raerii, and Pirokeni is that given in the myth, namely that the sand covers of three submerged reefs were washed away during a large wave. Both tsunamis, perhaps generated by slumps along the steeper eastern wall of the Cape Johnson Trench (see Figure 9.3B), and large waves associated with cyclonic storms are common in this part of Solomon Islands.

But there is also some reason for suspecting that the disappearance of these islands may have been helped by subsidence. The observation that the surfaces of the stripped reef platforms are now well below sea level is curious although not unprecedented (see various examples later in this chapter). Most modern sand cays and *motu* are built on reef platforms lying at or above low-tide level. Tidal range in this area is no more than 1.1 meters, meaning that platforms that are 4–5 meters below sea level are well below low-tide level and could not support a pile of unconsolidated sand and gravel for long. This suggests that about the same time as the destructive tsunamis were generated these platforms may have sunk, perhaps coseismically, which implies that the wave deemed responsible for the removal of their sand cover was earthquake-associated. In support of this idea, it seems clear that the south coasts of Malaita (including Maramasike) are unstable and that coseismic events do occur; during an earthquake in 1930, there was a rapid 1-meter uplift at Rohinari in southwestern Malaita.[26]

There are oral traditions of other allegedly vanished islands in the central part of Solomon Islands: Oeni, the Ta'aluapuala Group, and Iumurafa.

One tradition, collected by Sereana Usuramo from Manieli Village in northeastern Maramasike, talks of a "submerged reef" named Oeni that was once an island; its possible location is shown in Figure 9.3A. The people who survived the sinking of the island came to Manieli and were given land near the sea on which to settle.[27] The area off the northeastern coast of Malaita-Maramasike, where Oeni was located, is also part of the Malaita Accretionary Prism (see earlier in this chapter) that was crossed by seismic line 9, an interpretation of which is given in Figure 9.3C. This shows a submerged reef knoll, "a shallow water carbonate build-up" in geological terms,[28] that was almost certainly once an island above the ocean surface. Although this reef knoll is probably of Quaternary age and could conceivably have sunk within recent human memory and therefore be Oeni, it probably is not; it would be a coincidence if a single seismic line happened to pick this up. Yet what is important about its presence here is that it shows that there is likely to be a long history of islands sinking in the area.

Another tradition was obtained in 2004 by Bronwyn Oloni from nearby Tawanikeni Village on Maramasike and recalls a group of two to three low inhabited islands,[29] collectively named Ta'aluapuala, that apparently disappeared before European arrival in Solomon Islands in 1568. The Ta'aluapuala Islands were located a considerable distance offshore — it is said it would take four to six hours to reach them in a boat with an outboard motor from Tawanikeni[30] — and after the catastrophe the people were able to swim to the mainland only with the aid of the shark god they worshipped. There had been 200–400 people living on the Ta'aluapuala Islands, but today all that can be seen is a submerged reef, more than 10 meters deep. It is possible that a few of Oloni's informants were confusing Ta'aluapuala with Sikaiana, a group of low atoll islands still very much emergent and inhabited that lie some 212 kilometers northeast of Malaita.

It is also possible that these vanished islands referred to artificial islands once constructed offshore in this area. Artificial islands in the lagoons fringing the coast of Malaita are a distinctive feature of settlement in this part of the Pacific.[31]

The oral traditions about Ta'aluapuala also implicate "tidal waves" in the disappearance of these islands, although all informants stressed the incompleteness of their knowledge. The Ta'aluapuala Islands, perhaps also Oeni (see earlier in this section), all appear to have been sand cays (motu) that most probably disappeared as a result of the impact of a large wave, either associated with a storm or an earthquake. The deepest part of the North Solomons Trench (-5,700 meters) lies just north of this area and is a typical location for sizable tsunami generation.

Yet, as argued in the case of Pororourouhu, Raerii, and Pirokeni, coseismic sinking is also likely to have contributed to the disappearance of these islands given that the reef platforms marking their locations are today so deep. A likely site for the Ta'aluapuala Islands is shown in Figure 9.3A, on the outer (northeast) side of the ridge marking the crest of the Malaita Accretionary Prism in this area (Figure 9.3B). According to the interpretation of seismic line 9 (Figure 9.3C), this side of the ridge in this area is unstable, cut by normal faults and characterized by an accumulation of slump debris indicating a long history of probably largely abrupt subsidence. It is quite possible that the Ta'aluapuala Islands vanished as described in the oral traditions as a result of this process.

During interviews with the Aiga Tatari clan on Santa Catalina Island, mention was also made of an island named Iumurafa located off western Makira that disappeared quite recently. The island, which was low and uninhabited, was probably a sand cay, and its disappearance most likely a result of a tsunami. Informants reported that on a fine day, the outline of Iumurafa can still be seen from the sea surface.[32]

Recently Vanished Islands in Vanuatu

The Vanuatu Archipelago is one of the centers of seismic and volcanic activity in the Pacific, the site of some huge eruptions in the past, and a place where culture has developed in the context of almost unparalleled hazard.[33] Among the traditions of vanished islands in Vanuatu, four areas (shown in Figures 9.4A and 9.5A) stand out as those to which such traditions are probably autochthonous rather than introduced from elsewhere.

For reasons similar to the situation in Solomon Islands and Papua New Guinea, discussed earlier in this chapter, the islands of Vanuatu mark the convergent boundary between two plates. To the west, the Indo-Australian Plate is thrusting eastward, down beneath the plate to the east. The result has been the formation of three chains of islands, all running approximately north to south, parallel to the axis of the trench that marks the line of plate convergence. Only the Central Chain of islands, which includes the islands of Efate, Tongoa, Epi, Ambrym, and Ambae, is volcanically active today, but the Western Belt, which includes Malakula (Malekula) and Espiritu Santo islands, is being pushed upward and sideways as

FIGURE 9.4. (*Right and facing page*) Location and pictures of the Kuwae Caldera that formed after the disappearance of Kuwae Island during the 1453 eruption. **A.** Map of central Vanuatu showing the principal islands discussed together with the 1,000-meter isobath and the location of the trench axis. (Bathymetric detail and interpretation largely from sources in Greene and Wong [1988]; form and location of the Kuwae Caldera from Monzier et al. [1994].) **B.** Photo showing the shallow-water submarine eruption of Karua taking place on February 22, 1971, with the southeastern tip of Epi Island in the background. (Photo by Don Mallick, used with permission of Vanuatu Cultural Centre.) **C.** Photo showing the island of Karua that formed on February 23, 1971, and remained as a subaerial island until shortly after the eruption ended. (Photo by Don Mallick, used with permission of Vanuatu Cultural Centre.)

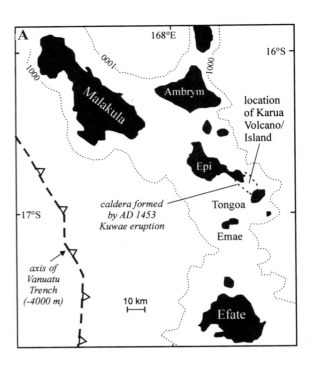

a result of the attempted subduction of the D'Entrecasteaux Ridge (see chapter 3). As elsewhere along the Ring of Fire in the western Pacific, these geological conditions provide ingredients for a range of natural hazards including, as argued here, the disappearance of entire islands. Although it is possible for islands to sink abruptly in Vanuatu, it is also possible for them to be washed away by large waves generated locally.

Early in 1453, people in the Shepherd Islands, including Emae and Tongoa in central Vanuatu (Figure 9.4A) were undoubtedly unsettled because the precursors of an impending volcanic eruption had been noticed for a long time. When it happened, it proved to be one of the largest eruptions anywhere on earth within the last 10,000 years, certainly the largest within the past 700 years.[34] It is not surprising that this event became a landmark in the history and societal development of central Vanuatu, but it also had visible and enduring effects as far afield as Europe. The eruption, which resulted in the disappearance of an island called Kuwae, provides an excellent example of how myth and geology complement one another in the reconstruction of ancient events of this kind.

The earliest written account of the incident, made by the Presbyterian missionary Oscar Michelsen more than 400 years after it occurred, stated that

> when all seemed to be peace and safety. . . . Suddenly there was an alarming subterranean report, accompanied by a violent earthquake. The shock was prolonged into an irregular vibration, and the explosive roar was continued day after day. . . . Slowly but surely large tracts of land sank into the sea, and other parts of the earth's crust were

raised several hundred feet. At three different places fountains of fire were opened up, and glowing lava sprang into the air to an appalling height.[35]

One of the mythical accounts of the incident recalls the increasing premonitions of impending eruption. A man named Tombuk organized a feast at which he killed six pigs, inflating their bladders and attaching them to a tree. Then he climbed the tree and, starting from the top, burst each bladder in succession. As each bladder burst, the earth shook more strongly, and when the sixth was burst, a volcano erupted from beneath the tree, destroying the island.[36]

Geological understanding of the nature and legacy of the Kuwae eruption has been significantly aided by oral traditions of the people on nearby islands, particularly Tongoa. On the northeastern coast of this island, facing the place where Kuwae once was, the people have names for several of the volcanic strata exposed in the cliffs; one layer named Nato Nantia is linked by them (and subsequently by geologists) to the Kuwae eruption.[37] There is also evidence from nearby islands for an associated tsunami, both from oral traditions and geological observations.[38]

The effects of the 1453 Kuwae eruption were experienced beyond the Pacific Basin. In what is now Turkey, the city of Constantinople (now Istanbul), almost the last remnant of the Byzantine Empire, was under siege at the time by the Ottomans. On May 22, a solar eclipse was made darker and more prolonged by the presence of Kuwae ash in the atmosphere, and the associated blood-red sunsets were thought premonitory and caused great apprehension amongst the city's defenders, who succumbed on May 29.[39]

But more than this, chronologies of last-millennium climate variations obtained by tree-ring density measurements show a marked cooling attributable to the Kuwae eruption that had severe consequences for agriculture across the Northern Hemisphere in 1453.[40] We can assume that similar effects were experienced elsewhere in the world, with evidence from ice cores for sulfate from Kuwae having been deposited across Antarctica.[41]

Kuwae blew itself to pieces and, although there is evidence that people recognized the precursors of the event and moved away from the area beforehand, nothing in their experience could have prepared them for the violence of the eruption. Many must have perished as a result. The Kuwae eruption was a caldera-forming eruption, similar to those described in chapter 3, that was an expression of activity along the convergent plate boundary that runs along the western side of the Vanuatu archipelago. There are other islands in Vanuatu that have apparently disappeared, probably not directly as a result of an eruption as with Kuwae, and these are described later in this section.

Today the site of Kuwae is covered by ocean, whose placidity is occasionally broken by the underwater eruption of the Karua Volcano that has grown on the submerged rim of the Kuwae Caldera. These eruptions sometimes produce short-lived islands (Figure 9.4B, C) similar to those described in chapter 8.

Turning to other vanished islands in central Vanuatu, there are a number off Malakula, a high limestone island thrust upward within the past few million years as a result of the convergence of two plates along the ocean trench 40 kilometers west. Malakula is also being bulldozed eastward, a result of collision between the Vanuatu Arc and the D'Entrecasteaux Ridge (chapter 3). With its long axis running parallel to the trench, the uplift of Malakula is far from uniform; recent rates of uplift have varied from 4.3 millimeters/year along the west coast, closest to the ocean trench, to 0.6 millimeters/year along the northeast coast.[42]

Off the northeast coast of Malakula, there are a number of smaller islands beyond which the ocean floor plunges steeply into the 2,000-meter-deep South Aoba Basin (Figure 9.5B). Bathymetric mapping suggests that some of the smaller islands that were once emergent

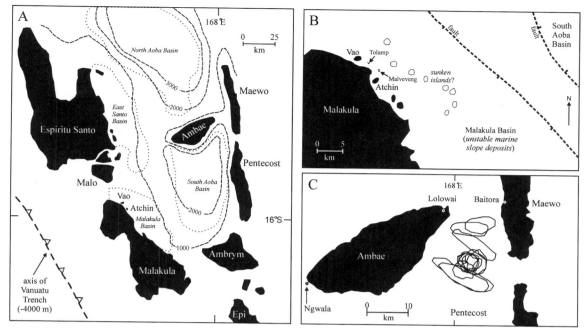

FIGURE 9.5. **A.** The islands of north-central Vanuatu, showing the principal islands discussed, together with the bathymetry of the Aoba Basin (isobaths at 1,000-meter intervals) and the location of the axis of the ocean trench west of Malakula. The three areas where vanished islands were found are off northeastern Malakula (between Atchin and Vao), off the western end of Ambae, and in the area between eastern Ambae, southern Maewo, and northern Pentecost. (Bathymetric detail and interpretation largely from sources in Greene and Wong [1988].) **B.** The northeast coast of Malakula Island and areas offshore showing the locations of the two vanished islands Malveveng and Tolamp, and other possible sunken islands. (Bathymetry and structure redrawn and adapted from Greene et al. [1988].) **C.** The area between the islands of Ambae, Maewo, and Pentecost for which numerous reports of one or more vanished islands were obtained. The locations and shapes of all islands as described and/or rendered by all informants in this area are shown.

here have slid down this slope, a suggestion supported by oral traditions relating to the disappearance of two such islands, Malveveng and Tolamp.[43]

Malveveng and Tolamp are said to have disappeared at the same time, and there is no mystery about their locations, for today they are marked by submerged coral-reef knolls, often visited for fishing by people from nearby villages. The surface of Malveveng is said to be 4–5 meters below water level; that of Tolamp is deeper, around 15 meters. When diving it is apparently possible to see a number of signs of the islands' former human occupation, such as stone-ringed meeting places (*nasara*), paths, and stone fences.[44] The islands disappeared sometime before around 1817 when a custom ceremony was held by (descendants of?) the Tolamp refugees on the Malakula mainland.

If one regards Malveveng and Tolamp as having been sand cays (or *motu*) then the simplest explanation for their disappearance is that a large wave washed them away, leaving bare their coral-reef foundations. In cyclone-prone and earthquake-prone Vanuatu, there is no shortage of mechanisms for producing such large waves. But this explanation ignores the likelihood that Malveveng and Tolamp are more likely to have been bedrock islands, akin to present-day Atchin and Vao that rise more than 20 meters above sea level. In support of this, numerous submerged islands are present on the underwater slopes leading to the floor of the South Aoba Basin (see Figure 9.5B). These are likely to have disappeared for the same reasons as Malveveng and Tolamp, probably long before humans reached these islands around 3,000 years ago. Another observation supporting this explanation is that, as for several vanished islands in Solomon Islands discussed earlier in this chapter, the surfaces of the reef knolls corresponding to Malveveng and Tolamp are deeper than would be expected had they simply had an unconsolidated covering of sand washed off them, suggesting that subsidence contributed to their disappearance.

The explanation favored is that Malveveng and Tolamp disappeared as a result of a structural failure off the northeast coast of Malakula that also, as is common in such situations, produced a large wave that swept back across the northeast Malakula coast and the islands offshore. An analogous event occurred on August 12, 1965, when a 7-meter-high wave, generated by a local earthquake, washed across this coast.[45] Even though no islands are known to have disappeared in this event, it demonstrates the area's vulnerability to seismic-wave impact.

A final example of a vanished island in Vanuatu is Vanua Mamata. Many of the modern inhabitants of Maewo Island, together with some in eastern Ambae (Aoba) and northern Pentecost, are familiar with the story of an island that disappeared in the area *bifo bifo yet* (very long ago). It is generally believed that the island, known by various names including Vanua Mamata and Lingland, sank and that its inhabitants (known as Tambweel) paddled across to southern Maewo and established the village of Baitora. All this happened *taem we Maewo igat wan vilei nomo long hem be nao ia igat 20 vilei oltugeta* (when Maewo [Island] had only one village on it, although now there are 20 villages altogether).[46]

Unusually for such Pacific Islander traditions, this particular story receives some support from an outside source. When the USS *Narragansett* was steaming through this part of central Vanuatu in 1872–1873, Commander Meade heard the story of the disappearance of Vanua Mamata, noting it in his Remark Book and commenting that "the survivors saved themselves with great difficulty."[47]

Within the Vanuatu Archipelago, traditions about Vanua Mamata are probably more widely known than stories about any other such island, an observation that suggests both that this island vanished quite recently and that it was comparatively large and populous — its surviving inhabitants were sufficiently numerous to have dispersed across a large area. Oral traditions specifically targeting Vanua Mamata were gathered in 2003–2004 from eastern Ambae, northern Pentecost, and southern and central Maewo by Mary Baniala.[48]

Informants were also asked to sketch the location and the outline of Vanua Mamata on maps, the results of which are shown in Figure 9.5C. But if this suggests concord, then consider that 67 percent of informants said the island was high, the rest low; 63 percent of informants who claimed to know stated that the island disappeared as a result of earthquakes, the rest cited large waves.

Whatever the precise location of Vanua Mamata, it certainly appears to have been at the northeast corner of the South Aoba Basin, an area characterized by steep slopes, crisscrossed by faults, and draped with mass-movement deposits, all indicative of an active location where seismic or volcanic triggers, or even just gravity, could plausibly account for the disappearance of an entire island. The bathymetry shows several possible locations for Vanua Mamata, but none is more compelling than the others. It is possible when looking at the map of the recalled locations of the island that there were actually two islands, a smaller rounder one and a more elongate one.

The 1872–1873 record of Commander Meade is invaluable in determining the time when Vanua Mamata supposedly disappeared. It is reasonable to assume that because Meade saw fit to report the event, it had been relayed to him sufficiently often and/or sufficiently forcefully for him to believe it, in turn implying that the incident was still newsworthy to his informants; in other words, it had occurred quite recently. And indeed, about 1870 a notable fumarolic eruption breached the caldera wall on nearby Ambae Island and caused a mudslide that destroyed a village in the southeast of the island,[49] the side facing Vanua Mamata. Although it might be considered odd that the island's disappearance was not mentioned in the same context by oral traditions on Ambae, it is entirely plausible to suppose that this eruption may also have triggered a submarine landslide that carried Vanua Mamata away, and that the memory of this event was sufficiently fresh in the minds of the area's inhabitants to have been one of the first things they told Meade.

If the disappearance of Vanua Mamata was indeed caused by a submarine landslide that was coincident with an eruption on Ambae, this would be an analogous situation to the 1741 eruption of Oshima-Oshima (Island) volcano in Japan (see chapter 8). Another possible analog from Vanuatu is the 1913 collapse of part of Ambrym Island shown in Figure 10.5.

Recently Vanished Islands in Fiji

Fiji's oral traditions are replete with stories about vanished islands, particularly that named Burotu (cognate with Pulotu in the islands to the east) to which people on many of the ninety or so inhabited islands in the group today trace their ancestry. Belief in Burotu has been very strong amongst Fijians. For example, Dr. Thomas St. Johnston reported in 1918 that stories about Burotu (material in brackets mine) so inflamed that adventurous spirit Ratu Mara [half brother of Ratu Seru Cakobau, the powerful nineteenth-century Fijian chief] that, like a knight-errant of old, he started forth and swore that he would find Burotu or perish in the attempt. As a matter of fact he did neither, but the story . . . shows how earnestly he believed in it.[50]

Throughout Fiji and western Polynesia, Burotu (or Pulotu) is alternately believed to be the abode of the gods, of the spirits of the dead, sometimes the homeland of the ancestors of modern Fijians, and sometimes (particularly in Fiji) an earthly paradise. All such reasons have led to its existence as a real place being doubted. This conclusion appears valid when the Pulotu myth as a whole is considered, as explained in chapter 6, yet there are certain linguistic and geographical pointers to the site of the original Burotu being near Matuku Island in southern Fiji.

The current geological condition of the Fiji Islands is generally more benign than those of the island groups to the west discussed in the preceding sections. A convergent plate boundary (the Hunter Fracture Zone or Kadavu Trench) runs south of the group but does not currently appear to be accommodating plate convergence.[51] There is no evidence of recent coseismic subsidence in Fiji, and this explanation is therefore not favored for the observed disappearance of islands. But there is no shortage of steep island flanks in the group, and the gravity failure of these is a significant potential hazard here.

Linguist Paul Geraghty has compiled much of the evidence about Burotu and remains the foremost authority on it. As part of my research projects, current oral traditions about Burotu were collected from neighboring islands by Elia Nakoro in 2004–2005.[52]

It was noted in chapter 6 that, although the homeland for people of eastern Polynesia represented by Hawaiki might have been the island Savaiʻi in Samoa, there was no agreed equivalent for the homeland of Pulotu, claimed as a homeland by the people of western Polynesia, particularly the chiefs of Tonga and Samoa. Yet almost all the stories about the location of Pulotu say that it lies west of western Polynesia, most likely in Fiji. This suggestion is supported by numerous other lines of evidence. For example, the ruler of Pulotu (Hikuleʻo) is often portrayed as half-human, half-eel, and the foremost pre-Christian deity of much of Fiji (Degei) is almost everywhere regarded as half-serpent.[53]

To many people in western Polynesia, Pulotu was the abode of the gods, the place where the spirits of the dead chiefs went, and the homeland in the west from which the chiefs had come. But in Fiji, Burotu is none of these. Instead it is paradise, a place where everything is perfect.[54] This misfit has been explained by assuming that the Burotu as Pulotu traditions are an extension westward of the western Polynesian traditions of Pulotu. The tradition of Burotu-as-paradise seems to have been built on an earlier pan-Pacific myth about a paradisical island, also known as Fanuakula (or Vanuakula) in this region.[55]

It has been convincingly argued that Burotu was located on or close to Matuku Island in southeastern Fiji (Figure 9.6A). The evidence comes largely from place-names and clan

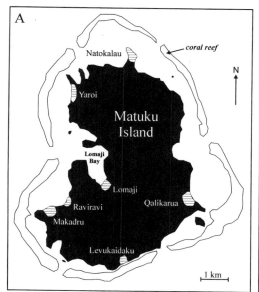
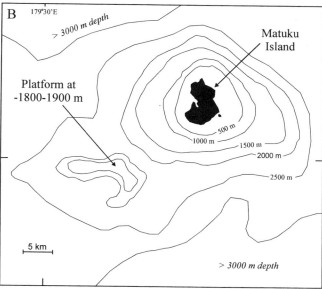

FIGURE 9.6. The island of Matuku, southeastern Fiji, and a possible location for the "vanished island" Burotu. **A.** The island of Matuku showing places mentioned in the text. Most traditions of Burotu (including its periodic resurfacings) are associated with the people of Levukaidaku and Makadru villages in southwestern Matuku. **B.** The bathymetry of the Matuku area showing the submarine platform that may be part of the main island that slid down its flanks. (Adapted and redrawn from Nunn et al. [2005].)

names, especially the names of the two halves of ancient Matuku — Burotu and Babajea (now Babasea) — and the name of the hill fort Togaviti, from which the Tongafiti invaders of Polynesian island groups elsewhere took their name.[56]

The island of Matuku is the surface expression of an isolated long-extinct volcanic edifice, steep-sided and with a history of Quaternary uplift, that shares the same origin as three other island-reefs in the area, Moala (or Muala), Navatu, and Totoya. Together with Matuku, these make up the group of islands known as the Yasayasa Moala.[57]

There is a definite possibility that Burotu exists still, as part of the island Matuku, its precise location and boundaries long since forgotten. But this explanation seemingly conflicts with the abundant stories recounting details of its disappearance, and its periodic reappearances (see later in this section), although it has been suggested that such details could be embellishments, "part of the inherited tradition of an island paradise."[58] Although not discounting this possibility, it seems clear that the representation of Burotu as a sunken land preceded the earliest time at which Europeans could have influenced indigenous stories[59] and that it is therefore an authentic detail.

It seems plausible to suppose that Burotu actually did sink, perhaps catastrophically and therefore more memorably, or that, as with Atlantis, the stories about Burotu incorporate details of another island that sank. No other candidate appears obvious, given our current state of knowledge, so it is assumed here that Burotu once existed as part of Matuku or nearby and subsequently sank.

By analogy with similar islands elsewhere, a coastal settlement on an exposed coast anywhere on Matuku Island could have been wiped out as a result of large-wave impact, either during a tropical cyclone (hurricane) or as a tsunami. The possibility that Lomaji Bay on Matuku (see Figure 9.6A) was the site of Burotu, suggested by Paul Geraghty,[60] is viable when the similarity with enclosed bays like Lituya Bay (see Figure 8.3) is considered. A large rockfall at the head of Lomaji Bay would be likely to displace a large amount of water that could wash over lowland bayside settlements with catastrophic effect.

Another possibility, illustrated in Figure 9.6B, was suggested recently.[61] Some 15 kilometers southwest of Matuku, on the underwater flanks of the island, is a 17-square-kilometer platform at a depth of 1,800–1,900 meters. By analogy with some of the large island-flank failures known (from Hawai'i and Johnston Atoll, for example, which involved as much as 1,650 meters and 700 meters of vertical movement, respectively),[62] it is possible that this platform was once emergent and slid down the sides of the parent volcano. This suggestion is a very long way from being proved, with no targeted survey of this platform ever having been carried out. Yet its location does concur with the observation that the Burotu associations of the people of Levukaidaku and Makadru villages in southwestern Matuku, the part closest to the submerged platform, are stronger than elsewhere on the island.

For example, it is the people of Levukaidaku who claim to be able to smell the scent of the flowers from Burotu when the wind blows from a certain direction. A stone causeway near this village at a place called Vatuvoleka is claimed to connect to Burotu. The women of the Burotu clan in Makadru Village claim to be the only ones who know the dances (*meke*) from Burotu. Finally, most of the sightings of reemergent Burotu are off this area of Matuku.[63]

In Fiji, Burotu goes by many names. There are adornments of the basic name, the most common being Burotukula, which means "red Burotu" in Fijian: red has all the connotations of golden in English as an indication of wealth and power. Some accounts of Burotu claim that everything on the island is red. A 1929 account stated that "if the women of Matuku go out at low tide, at dawn they find red reeds that come from Mburotu [Burotu]."[64] Informants during the 2003–2004 survey also emphasized the redness of everything that came from Burotu,[65] red of course being the color of the parrot feathers (*kula*) for the supply of which, it is inferred, the island of Pulotu became famed across the Pacific. In 1907, the missionary Lorimer Fison reported that the people of Matuku Island "say that sometimes burnt-out fishing torches of a strange make, with handles of shell, drift ashore on their land, and when they pick them up they say 'see the torches from Burotu!'"[66]

As with many oral traditions about vanished islands in the Pacific (and elsewhere), contemporary and recent traditions about Burotu are loud on the moral issues behind the sinking but almost silent on how it actually happened. The most common explanations of why

Burotu sank involve the people of Matuku giving unwitting offense to those of Burotu who responded by causing their own island to sink. One version of the story states that the chief of Makadru Village on Matuku, having been sent a white pig from Burotu, killed it rather than keeping it as was intended.[67] In a similar vein, a more recent story tells that one day the chief of Burotu (*Tui Burotu*) sent a special bird as a gift to Matuku, but the people killed it, thinking it might be a devil. And the next day, where Burotu had been, there was nothing but ocean.[68]

The circumstances of the submergence of Burotu are unnecessary to recall if you believe that it occurred magically, which may be why there are so few such details in the oral traditions. One tradition that mentions submergence following an earthquake may be European influenced.[69] Another tradition states that the island sank after it was trodden on by the ancestor hero Kubunavanua.[70]

The vagueness of details about the nature and the timing of the sinking of Burotu in accounts obtained around the time of European contact suggests that, unlike the disappearance of Vanua Mamata in Vanuatu, the memory of the disappearance of Burotu was far from fresh in the minds of informed Fijians in the mid-nineteenth century. This makes the suggestion that it disappeared gradually, starting in 1760,[71] improbable. A date considerably earlier, around AD 1000, close to the limit of oral-tradition survival, is more plausible.[72]

One distinctive aspect of the Burotu myths in Fiji generally, but most frequently on Matuku, is the alleged periodic reappearance of Burotu. Perhaps most commonly explained as a good-luck omen or, conversely, a portent of disaster, the reappearances of Burotu play an important role in keeping this particular tradition alive in the minds of many Fiji people.

The island of Burotu is said to have reappeared off Matuku at least seven times since about 1933, sometimes enduring only a few minutes, sometimes several hours, but always vanishing again. Sightings of Burotu on Matuku may not be routine, but neither are they considered universally newsworthy in the minds of the observers. Although it is impossible for a high island to appear one moment and disappear the next without considerable disturbance, some accounts of the periodic reappearances of Burotu are difficult to readily discredit. Consider this one by Emitai Vakacegu, the head teacher of Babasea Primary School in Levukaidaku Village, who, early on the morning of September 4, 2003, was exercising with his students, the other teachers, and some villagers in front of the village when the mystical island of Burotu — high, mountainous, well vegetated, with clouds around its summit — appeared. This is what Vakacegu wrote in his log book:

> at about 6 o'clock in the morning, the teachers, students and villagers witnessed a historical scene, the famous and mysterious island of Burotukula [Red Burotu] surfaced again after 10 years. The first reddish orange light of the rising sun on the horizon formed a beautiful background to the island, which floated alone on the calm sea. The strange island was slowly sinking as the sun rose over the horizon.

It is simply not possible for such an island to appear and disappear so fast, at least not without the accompaniment of some unmissable seismic or volcanic outburst, and yet such observations should not be dismissed simply as fantasy, particularly because they tend to be made by people not generally given to flights of fancy. It seems likeliest that such sightings represent the glimpse of some uncommon sight, such as distant dark clouds obscured by the glare of the rising sun, to which the observers transfer their deeply held beliefs about the periodic reappearances of Burotu. Alternatively, these alleged sightings could occur when a tsunami hit the coral reef encircling Matuku. When funneled into reef passages, locally generated tsunamis in Fiji have been observed to form "large brown bubbles" that could, depending on the observer, be seen as Burotu reappearing.[73]

It is worth noting that regular sightings of illusory islands in modern times are not confined to Burotu. Off the eastern seaboard of North America, an island covered in greenery is said to have periodically emerged near Boston, and there are apparently many instances of people trying to reach it by boat, but never succeeding.[74]

Besides Burotu, there are also traditions of a vanished island named Vuniivilevu in Fiji; some of these were related in chapter 1. Off the east coast of Fiji's largest island, Viti Levu, lies the island of Moturiki, to the south of which lies the site of Vuniivilevu, a sizeable island reputed to have disappeared (Figure 9.7A). This event remains prominent in the oral traditions of the people living in the surrounding area, to the extent that, when sailing across the area where Vuniivilevu once stood, people keep silent and bare their heads or face the wrath of the spirits below.[75] Although I have conducted research in the Moturiki area for many years and heard the oral traditions, most of those presented in this account were collected by Alifereti Nasila and Samuela Tukidia as part of my targeted research project in 2003–2004.[76]

According to the oral traditions collected from the area around Vuniivilevu, the island was a high island, larger than present-day Moturiki (11 square kilometers), that encompassed the modern sand cays of Caqalai, Leleuvia, and Nasautabu. The name Vuniivilevu means large *ivi* tree,[77] a typical way of naming places in Fiji. Combining the oral traditions with maps of the area's bathymetry, it proved possible to reconstruct the probable form and location of Vuniivilevu (Figure 9.7B). It seems likely that the island had some high parts, probably part of the same volcano that formed Moturiki Island, but also an area of lowland where the principal settlement was located. Some oral traditions suggest that Vuniivilevu disappeared in an earthquake and a "hurricane." If this island every truly existed, the most likely explanation for its disappearance seems, on geological grounds, to be that it was partly washed off the edge of the Viti Levu Island shelf and partly collapsed into the reef passage known as Davetalevu; tropical cyclones and earthquakes would have provided likely triggers.

The suggestion that the people of Vuniivilevu were punished for particular excesses is, as elsewhere, commonly cited as a reason for the island's disappearance. One writer attributed

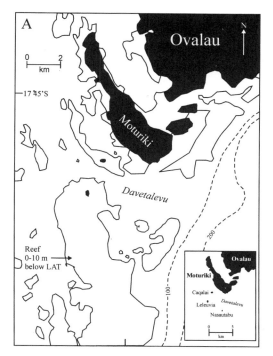
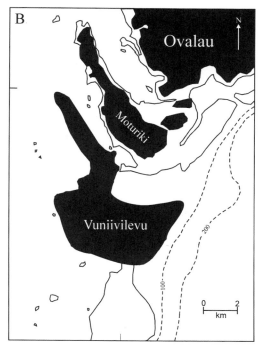

FIGURE 9.7. The geography of the area off the east coast of Viti Levu Island, Fiji, showing a plausible reconstruction of the "vanished island" Vuniivilevu. **A.** The modern geography of the Vuniivilevu area, showing the modern islands (shaded black [see also inset for locations]), together with shallow reef surfaces (less than 5 meters below Lowest Astronomical Tide [LAT] level) and deeper reef surfaces. The reef pass (*daveta* in Fijian) named Davetalevu is customarily associated with Vuniivilevu. **B.** A possible reconstruction of Vuniivilevu, shown in association with the other islands in the area today and the shallow reef surfaces. In this reconstruction, the central and northern parts of the island are considered higher, like Moturiki part of the large Ovalau (Lovoni) Volcano, and the southern and eastern parts are considered lower. (Adapted and redrawn from Nunn et al. [2005].)

it to the cannibalism practiced on the island, noting that the island's chief was one Komai Vuniivilevu with a particular predilection for consuming human heads, for which reason he was also known as Kamunagauluna, roughly translated as "he for whom a head is as valuable as a whale's tooth."[78]

The geological evidence for the existence of Vuniivilevu is not compelling by itself. Were it not for the extraordinary prominence of oral traditions concerning Vuniivilevu, it is unlikely that the story would be taken seriously. But various people from different parts of eastern Fiji, and beyond, trace their ancestry back to Vuniivilevu. These include the clan of the Tui Motuliki, a noble family in neighboring Tonga, that arrived there from Fiji about AD 1200, according to genealogies, and are descended from the son of the last chief of Vuniivilevu. It is possible that his arrival in Tonga was associated with the literal collapse of the chiefly seat on Vuniivilevu.

This is the only clear indication of when Vuniivilevu disappeared. It is supported by the inference that, because the story is not mentioned in any accounts from the period of European settlement of this part of Fiji in the middle and late nineteenth century, the disappearance of Vuniivilevu must have taken place long before. Were it otherwise, had it been fresh in the minds of the area's inhabitants, it is likely to have been the subject of discussion and written down.[79]

Recently Vanished Islands in Kiribati

Covering an area of more than 3 million square kilometers, Kiribati is the largest Pacific Island nation. Yet with a land area of around 800 square kilometers, Kiribati is also one of the smallest. Another extreme concerns island elevation. Were it not for high Banaba (Ocean Island), Kiribati would be composed exclusively of thirty-two atoll islands, none rising more than a few meters above sea level. The lowness of such islands (cays and *motu*) makes them extremely vulnerable to erosion, and this vulnerability is exacerbated by their composition, mostly unconsolidated sand and gravel derived from nearby coral reefs.

The midplate location of the islands of Kiribati, far away from plate boundaries and from intraplate hot spots, might be regarded as insulating them from the processes that appear to have caused islands elsewhere in the Pacific to disappear. Yet this is manifestly not the case. Islands have disappeared in the Kiribati island group in the past and are likely to do so again this century as sea level rises (see chapter 10). There are similarities with the traditions in the Tuamotus, another mainly atoll-island group, where the unusually high numbers of wandering and vanished island myths may be founded in observations of real islands (cays) that periodically form and then disappear from shallow submerged reef knolls.

The islands of Kiribati are very scattered, and most examples of islands that have disappeared come from the western (Gilbert) group, which is the most densely and longest populated. Some examples are well known, and the islands seem to have disappeared as a result of processes that are likewise well known. But other examples are known only from oral traditions, and the causes of their disappearance are more conjectural.

The disappearance of culturally important Bikeman and Tebua islands in the lagoon of Tarawa Atoll[80] was probably associated with changes in lagoon circulation arising from causeway construction. But it serves as a good analogy for what might happen in the future.

Bikeman was an island within the Tarawa Lagoon composed entirely of sand that had accumulated on a large patch reef. The island was notoriously changeable in shape, reconfiguring both intra-annually as a result of seasonal variations in wave direction and interannually as a result of the replacement of tradewinds by westerlies during El Niño events. But key to the continued survival of Bikeman as an entity was the movement of water through the gap in the reef to the south between Betio and Bairiki islands (both inhabited *motu* on the main atoll reef). In 1986–1987 the Nippon Causeway joining these two islands was completed and considerably reduced water and sediment exchange between the ocean and the lagoon through that gap. Bikeman disappeared shortly afterward, starved of the sediment it needed to counteract the effects of wave erosion. The disappearance of Bikeman is merely a casualty of twentieth-century development.

Also in the Tarawa Lagoon, the sand island of Tebua reputedly existed "since creation" but has now disappeared.[81] Lying off Naa Village at the northernmost tip of Tarawa, Tebua in the 1940s was a place to which the villagers used to wade to harvest the abundant coconuts and breadfruit and to fish. But in the 1950s, the trees on Tebua began dying, the sand that composed the island began visibly eroding, "the island began to shrink," and within a decade had vanished. Today Tebua is awash even at low tide. More so than Bikeman, it appears likely that the disappearance of Tebua was a result of twentieth-century sea-level rise some time before this was generally realized to be a serious issue,[82] but at the same time it is possible that other factors, of which we are currently unaware, also played a role.

There is evidence that another island vanished between Makin, the northernmost atoll of western Kiribati, and Mili, the southernmost in the Marshall Islands, some 280 kilometers away (Figure 9.8). The evidence is solely from oral traditions, most collected by Teena Kum Kee and Vicky Claude in 2004–2005.[83] This island, named Karawanimakin in some accounts, was evidently a sand cay (*motu*) on a separate reef 3–5 kilometers northeast of Makin. The erosion of this island was slow, allowing the few families living there ample time to relocate to neighboring Butaritari Atoll. Today a patch of sand, home only to a few nesting birds, sometimes marks the site of Karawanimakin. It is unclear when the island disappeared.

There are slight variations on this story, some naming it Te Bike[84] and placing it much farther from Makin, some making it formerly home to 40–50 people. If these details are correct, then it may be that there is a second vanished island in this area. This may be the same island that in 1908 the crew of a trading vessel landed on and found to be covered with salt bush and nesting birds. According to Sir Arthur Grimble, this island was later searched for systematically but was never found.[85] One of the informants, Tamuera Nakoro, questioned by Kum Kee and Claude stated that in 1984 he was on a vessel that sailed about 200 kilometers north from Makin to search for an "islet" close to Mili Atoll that was reportedly awash at low tide.[86] Soundings in the area revealed a knoll at a depth of 60–70 meters but no islet.

With reference to the map in Figure 9.8, Karawanimakin may have been an island developed on the shoal at location 1, northeast of Makin. Alternatively this may have been the site of Te Bike, although it is possible that some of these accounts of this, including that of Tamuera Nakoro, refer to Keats Reef, east of Mili.

Both islands described from this area are likely to have vanished as a result of wave erosion, perhaps a combination of the progressive erosion associated with recent sea-level rise[87] that allowed the occupants of Karawanimakin to leave at their leisure, and the rapid erosion that can erase whole sand cays like Te Bike from exposed reefs in a single large-wave event. This area is not volcanically or seismically active, although it is exposed, as are most coasts

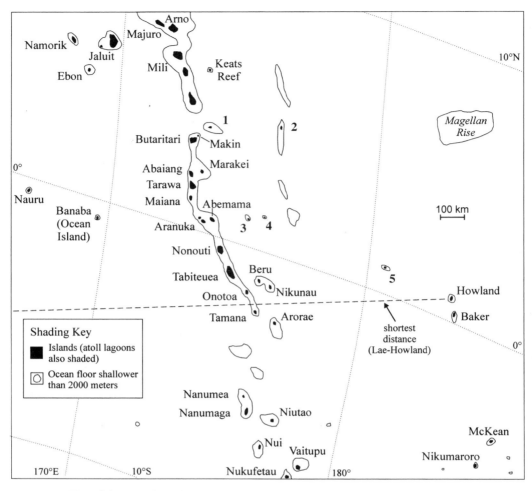

FIGURE 9.8. Map of the central Pacific showing the islands of Kiribati and others. Note that atoll reefs are filled to enhance the understanding of the map, but this makes island land areas appear far larger than they actually are. Numbers 1–5 refer to shoals that may have featured in reports of "vanished islands" in Kiribati, as explained in the text. (Outlines of bathymetry are mainly from the 1981 maps of the Circum-Pacific Council for Energy and Mineral Resources, Houston.)

in the central Pacific, to long-range tsunamis, particularly those generated along the Peru-Chile ocean trench.

The way in which vanished islands in this area of Kiribati have passed into local myth is quite distinct from that in the other areas described in this chapter. The culture of I-Kiribati (the people of Kiribati) is suffused with stories of the events of this world being entwined with those of other invisible worlds.[88] The physical disappearance of islands like Karawani-makin and Te Bike has transformed them into ghost islands (*abaia anti*) whose inhabitants, past and present, are half human and half spirit (*anti ma aomata*). Perhaps prompted by the same observations of islands that alternately appear and then disappear are stories about islands "wandering" across the ocean. One elderly informant, Tiata Koriri from Makin Island, explained that once all the islands in this northern group of Kiribati wandered "until one mighty ancestor held them down with an anchor made from strong wood."[89]

To be truly objective about oral traditions in Kiribati, it is necessary to include the decidedly nonindigenous traditions that have multiplied in recent decades about islands where aviator Amelia Earhart and her navigator Fred Noonan might have crashed on July 2, 1937. En route from Lae in Papua New Guinea as part of her pioneering aerial circumnavigation, Earhart was expected to land at Howland Island, a U.S. territory, but failed to do so. The relevant part of her planned route is shown as a dashed line in Figure 9.8. Since then, myths about her destination have extended from the initial idea that she crashed in the ocean, through the idea that she landed in the Marshall Islands or one of the inhabited islands in Kiribati.[90] Current intelligence, supported by some hard data, points to Nikumaroro (formerly Gardner) Atoll as Earhart's likely crash site.

Many myths surround the disappearances of Earhart and Noonan, but one unusual explanation relevant to this book is that they intentionally landed on what looked from the air like an island but that they found to be entirely underwater when they crash-landed on it. Whether or not this is true is immaterial, but the suggestion has been made with reference to an island known as Bikenikarakara that is alleged to have disappeared in Kiribati, close to the line that Earhart and Noonan were supposed to be following.

Bikenikarakara has not been located precisely, and the only evidence of its existence is from oral traditions. The island was first mentioned by Sir Arthur Grimble as being due east of the midpoint between Makin and Marakei atolls.[91] In 1985 a short article in *Pacific Islands Monthly* by Captain E. V. Ward told of some traditions he knew concerning Bikenikarakara. He reported that the old people of Marakei and Butaritari atolls in western Kiribati say that Bikenikarakara is a "lost land" lying almost 100 kilometers east of a line bisecting the two. Ward quoted a man named Beatau whom he met at Matang Village on Nonouti Island in the 1950s. Beatau said that Bikenikarakara "used to be there but has now gone," adding "That is where that American aeroplane woman landed . . . she thought she saw an island but it had sunk."[92]

Studies of the bathymetry of the area show no submerged reef knolls where the traditions say that Bikenikarakara was, but there are other possibilities. These are submerged island edifices that may have once been shallow enough to allow the development of a sand cay or, alternatively, have once been sufficiently clear from the air to have looked like an island. These are shown in Figure 9.8 as shoal 2, a little over 300 kilometers east-northeast of Makin; shoals 3 and 4, respectively about 145 and 210 kilometers east of Abemama; and shoal 5, almost 300 kilometers due west of Howland and closest to the route that Earhart was supposed to have taken.

Bikenikarakara is a generic term that can mean "growing island," which, in the context of Kiribati, most plausibly refers to the piling up of reef detritus on a reef platform to form a sand cay, particularly after a large storm when the geography of coral reef islands can change quite profoundly. It is quite possible that there is an isolated reef knoll in this region where the *bikenikarakara* discussed here periodically forms in this manner. But such islands are equally likely to vanish as a result of erosion by large waves during storms. The transient nature of these islands may have helped their characterization by I-Kiribati as *abaia anti* (ghost islands) and of course may be one reason why no unequivocal trace of Earhart and Noonan has yet been found.

Recently Vanished Islands in the Cook Islands

There are several myths concerning the origin of atolls in the northern Cook Islands. One of these myths involves Māui fishing up islands — in this case, Manihiki and Rakahanga — as he did in many other parts of the Pacific.[93] This is a familiar story, rendered unique here by the details of what led Māui (and his brothers) to fish up an island here in this "empty" area of the Pacific. The story begins with a man named Iku from Rarotonga Island, the largest in the Cook Islands, who, fishing a great distance away from his home, discovered a huge stone on the ocean floor that he thought was a sunken island. It was hearing of this find that led Māui and his brothers to sail some 970 kilometers northward from Rarotonga to fish up the "submerged island." When they fished up the island, there was a great struggle, as a result of which it broke into two pieces, the modern islands of Manihiki and Rakahanga.[94]

Other myths about submerged and floating islands refer to the southern Cook Islands, where islands are more numerous and generally much higher (volcanic) than in the northern group. A myth about the origin of Aitutaki Island refers to the island, when discovered, as "partly submerged."[95] Another myth identifies the "floating island" Nuku-tere as Rarotonga.[96] In this myth, the wandering island Nuku-tere was located by the god Tonga-'iti who stamped on it to stop it moving. His wife Ari then dived down to fix the island's foundations, after which the island was named Tumu-te-varovaro. Both these myths are likely to be purely etiological, having no bearing on island movements but rather recalling those of the various human groups to have occupied these islands.

But there are in the Cook Islands two oral traditions that are far more persuasive: those about Tuanaki in the southern group and Victoria Island in the northern group, discussed next.

The principal source of information about the vanished island of Tuanaki (also Tuanahe) in the southern Cook Islands is the remarkable narrative of the mission teacher Maretu, a native Cook Islander, the key parts of which are reproduced in Appendix 2. The gist of the relevant part of Maretu's narrative is that he received information in the early 1840s from a person called Soma who had visited Tuanaki two years earlier. Soma described the island and its people, stating that it took just one day to sail to Tuanaki south from Mangaia Island. With Maretu and the Reverend William Gill on board, the mission ship spent a week in this area looking for Tuanaki in 1844 but could not find it.[97] The same year the Reverend Platt went in search of Tuanaki but was also unsuccessful.[98] Notwithstanding these setbacks, William Gill was confident enough in 1856 to say of Tuanaki that "such an island exists, there seems to be no doubt."[99]

A little later, Dr. Wyatt Gill took an interest in Tuanaki, reporting a conversation in 1897 with "old Tamarua" of Ngatangiia on Rarotonga Island who asserted that there was once frequent communication between Rarotonga and Tuanaki. Tamarua knew of two ocean-going canoes, of the kind that crossed the 2,900 kilometers of ocean between Rarotonga and New Zealand, that had headed for Tuanaki, which he said lay two and a half days sail south of Rarotonga. Wyatt Gill also quoted a Judge Wilson and Lieutenant-Colonel Gudgeon as having known of Tuanaki. Writing in 1911, Wyatt Gill wrote of Tuanaki that "it is no doubt correct to say that the island has disappeared, due, probably, to some volcanic disturbance."[100] He also noted that there was a shoal that he called Heymet Reef (Haymet Rocks on most modern maps) marking the place where Tuanaki once existed.

Tuanaki was in a part of the Pacific where the underwater geography is still not well known.[101] Haymet Rocks was named for the captain of the vessel *Will Watch* that in 1863 came upon this shoal and, while trying to pass between two emerged rocks, damaged its false keel because the water was so shallow.[102] If Tuanaki became Haymet Rocks, then it appears to have been located, approximately, it must be admitted, at 26° S, 160° W. There are other candidates for Tuanaki in the area, including Fabert Seamount (24° 7' S, 158° 33' W), discovered in 1887, the summit of which lies about 120 meters below the ocean surface.

Some more information about Tuanaki was gathered from the Cook Islands by Kori Raumea in 2003–2004.[103] All his informants placed Tuanaki southwest of modern Mangaia Island, one even stating that the island's *papa* (fringing reef platform) could be seen from Mangaia on a clear day.[104] This seems unlikely; most informants agree with earlier published accounts that Tuanaki was up to 100 kilometers from Mangaia. One informant stated that Tuanaki was the same size as Takutea Island (1.2 square kilometers), also in the southern Cook Islands Group. Many of Raumea's informants told of how people long ago from the islands of Rarotonga and Mangaia once routinely visited Tuanaki to collect food and fish, and to interact with the island people, who were related to those of Mangaia and spoke a similar language. One informant said that Tuanaki was a "refuge for defeated warriors from Mangaia tribal wars."[105]

The idea that Tuanaki once existed and disappeared was accepted by that most critical of modern lost-island professionals, the oceanographer Henry Stommel, and by many others[106] who have noted that the island was known to whalers traversing the South Pacific and to French government officials in western French Polynesia in the nineteenth century. A similar vanished island from the northern Cook Islands is Victoria.

In 1946, at the age of 84, William Marsters II died. At the time, Marsters was the patriarch of remote Palmerston Island in the central Cook Islands (Figure 9.9) and the only person known to claim to have visited Victoria Island.[107] As a young man in the 1880s, Marsters

FIGURE 9.9. Photograph of the family of William Marsters II, taken on Palmerston Atoll in the 1930s. Marsters is the bearded figure. He claimed to have visited now-vanished Victoria Island when he was a young man. (Photograph courtesy of *The Fiji Times*.)

joined a gang of laborers from Suva in Fiji and a man named Levy on board a ship that took them to uninhabited Victoria Island where they were paid to plant coconuts, a response to the high demand for copra in the late nineteenth century. Marsters described Victoria as a fairly well-wooded volcanic island (not an atoll) with a barrier reef and lagoon. The men were left on Victoria with food for a few months, but in the interim the firm that employed them went bankrupt and they were not rescued for eighteen months, by which time they had eaten all the food on the island including the coconuts they had brought to plant. Victoria has not been seen since they left.

Victoria appeared on several maps in the first half of the twentieth century, but it was removed subsequently after it was reported missing. I have a copy of the 1932 edition of Philip's *Handy-Volume Atlas of the World* in which Victoria Island is prominently named at an approximate location of 5° S, 164° W, some 800 kilometers north-northwest of Penrhyn (Tongareva) Island in the northern Cook Islands.

The fact that vessels actively searched for the missing Victoria Island in the first half of the twentieth century may not, in retrospect, be considered excellent evidence that it ever existed, but it is evidence of a kind. The earliest such report appears to have been that of Captain Andy Thomson of the trading schooner *Tagua* in 1921.[108] Then in 1923, Robert Dean Frisbie set out toward the equator from Penrhyn in the yawl *Motuovini* to look for Victoria but failed to find it;[109] it was, he believed, about 320 kilometers north of Penrhyn.[110] An informant from Palmerston Atoll, William Powell, interviewed in 2004, was adamant that an island north of Penrhyn had existed in the 1900s but had "perished" about 1930.[111]

There is no shortage of candidates for Victoria Island; at least six alleged islands were shown in the Penrhyn-Fanning area by Henry Stommel, none of which he regards as having truly existed, and numerous "dangerous shoals" in the area around Fanning.[112] It is odd that, given the thoroughness of his compilation, Stommel made no mention of Victoria Island.

In this context it is perhaps worth looking more closely at one version of the myth recalling the fishing up of Manihiki and Rakahanga outlined earlier. This version, recorded from Rarotonga in 1899 by S. Percy Smith, states that the island was broken up into three parts (not two): Manihiki, Rakahanga, and Tu-kao, the latter a name that Smith could not identify.[113] Possibly Tu-kao refers to Victoria or another island in this area that subsequently disappeared.

Less prosaic is the idea that Victoria was Beveridge Reef, shown in Figure 2.4c and located 620 kilometers west of Palmerston, which was "once a fine isle" but "swept bare by a fierce hurricane," perhaps sometime around the start of the twentieth century.[114] Yet on balance Beveridge Reef is unlikely to have been Victoria Island, not just because it is in the wrong direction from Penrhyn, but also because there is no evidence that it was ever the high volcanic island that Marsters claimed Victoria to have been, although perhaps this detail was embellishment by an old man recounting an extraordinary adventure from his youth.

Fatu Huku: An Almost-Vanished Island in the Marquesas

In the Marquesas Group of northern French Polynesia, the island of Fatu Huku exists today, a small rock northeast of Hiva Oa Island (see Figure 6.1). Yet not too long ago, Fatu Huku was much larger. Maps of Fatu Huku made by Cook in 1774 and Hergest in 1792 show it to be a far larger island than it is today. A map by Porter (~1820) shows Fatu Huku as it is today. Backed up by the oral traditions, the cartographic evidence has led to the suggestion that Fatu Huku did experience a catastrophic collapse between 1792 and 1820 that led to the disappearance of almost the entire island.[115] Again there have been no definitive geophysical studies, but the location of Fatu Huku, like Clark Bank, on the slope rather than the crest of the Marquesas volcanic ridge does support the inference that most of it disappeared in this way.

There are oral traditions concerning the disappearance of Fatu Huku in which the catastrophe is explained by the shark guardian of the island lashing the submerged pillar of rock on which the island supposedly balanced, causing it to topple over into the sea, so that only a small part remained emergent.[116] Another version of this story recalls that "the land was overturned and sank down in the depths, and the people perished in the sea. No one was left alive. Only a little piece of land was left."[117]

Los Jardines: An Island That Vanished
in the Northwest Pacific

No island exists today at the location 21° 40' N, 151° 35' E. The nearest land is Minami-tori Shima (Marcus Island) 340 kilometers northeast, itself a remote island in a region of the Pacific that is almost void of land (Figure 9.10). Yet a succession of navigators reported an island at that location, some of whom might be forgiven for being inaccurate, given the problems of determining longitude until the middle of the eighteenth century, but others of whom are unlikely to have made significant errors. This has led to the suggestion that the island, named Los Jardines, once really existed but has since vanished.[118]

The first report of two islands at this location was by the Spanish navigator Alvaro de Saavedra, who named them Los Buenos Jardines. Saavedra spent several days in the midst of these islands in 1529 and reported that the islands' inhabitants were very friendly. He concluded

> that the people had come originally from China, but had, by long residence, degener-
> ated into lawless savages, using no labour or industry. They wear a species of white
> cloth, made of grass, and are quite ignorant of fire, which put them in great terror.
> Instead of bread they eat cocoas [coconuts], which they pull unripe, burying them for
> some days in the sand, and then laying them in the sun, which causes them to open.
> They eat fish also, which they catch from a kind of boat called *parao*, or *proa*, which
> they construct with tools made of shells, from pine wood that is drifted at certain
> times to their islands, from some unknown regions.[119]

Some fourteen years later Spaniard Ruy Lopez de Villalobos reported sighting a small group of islands in the same area (21–22° N, 153° E).[120] In 1788 Captain William Marshall sighted two islands at 21° 40' N, 151° 35' E. Marshall was a competent navigator, having charted much of the Marshall Islands archipelago, which bears his name, and unlikely to have made a significant mistake in his positioning of these islands. There was a subsequent sighting by a whaling vessel of the islands at 20° 50' N, 151° 40' E, but since then they have not been seen. Figure 9.10 shows the area in which Los Jardines may have been located; its location as determined by Marshall is marked with a cross.

The question of whether Los Jardines was a misidentification of Minami-tori Shima (Island) is of course pertinent to this discussion. Minami-tori appears to have been discovered in 1694 by Andres de Arriola, at which time it was uninhabited.[121] This detail clearly conflicts with Saavedra's description of Los Jardines, although it is also possible that he was referring to one of the Marshall Islands, perhaps Enewetak Atoll. Yet Villalobos, Marshall, and the whaler all have Los Jardines in approximately the same position. If they had mistaken it for Minami-tori or Enewetak, then it is highly unlikely that their errors would all have been the same distance and in the same direction.

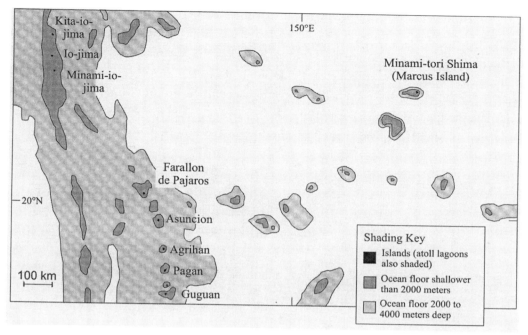

FIGURE 9.10. Map of part of the Northwest Pacific showing the reported location of Los Jardines (marked by a cross) relative to Minami-tori Shima and other nearby islands. (Outlines of bathymetry are mainly from the 1981 maps of the Circum-Pacific Council for Energy and Mineral Resources, Houston.)

The question of whether or not Los Jardines existed caught the interest of Captain G. S. Bryan of the United States Navy, who in 1940 wrote the most comprehensive account of the island(s). He reported that a number of searches were made for Los Jardines during the early twentieth century, all unsuccessful, except for that of the USS *Ramapo* that found an underwater mountain close to the alleged position of the islands.[122] The highest peak of this mountain is 2,050 meters below the ocean surface, yet Bryan was left in no doubt.

> The fact that the submarine mountain is located in the exact position as given by the most reliable of the discoverers will undoubtedly convince many that we have a clear case of the disappearance of an island beneath the ocean surface.[123]

Aside from Minami-tori, the nearest islands are the northern Marianas (Guguan to Farallon de Pajaros) and the Volcano Islands (Minami-io to Kita-io), all a considerable distance away.

Bryan also added that geologists with whom he discussed the question of Los Jardines remain unconvinced that a peak now more than 2,000 meters below sea level could have projected above it "within modern times."[124] Therein lies the problem, for it certainly does seem a great amount for an island to subside. Conversely, what we now understand about the great magnitudes of some past landslides along the flanks of islands and continents might have allayed some of Bryan's informants' skepticism.

For the moment, the last word about Los Jardines belongs to a group of Japanese scientists who have been mapping the seafloor around Minami-tori and have found a number of seamounts rising to within a few hundred meters of the ocean surface.[125] Unfortunately their surveys have not yet extended to the location of Los Jardines as reported by Marshall, but the presence of apparently shallower submerged edifices around Minami-tori, hitherto unknown, might be taken to suggest that there is better reason for believing this area to be the site of an island that disappeared within the past few hundred years.

Recently Vanished Islands off the Continental Margins of the Pacific

Oceanic islands throughout the Pacific also disappeared in the period 17,000–4,000 years ago as a result of postglacial sea-level rise, but there is no evidence that any of these islands were inhabited at that time.[126] Yet islands adjoining the contiguous continental margins of the Pacific, many of which were inhabited, were also being submerged at that time for the same reason.

A good example of islands that disappeared during their human occupation as a result of sea-level rise comes from the area of the Pacific off southern California where today are found the Channel Islands (Figure 9.11). At the coldest time of the last ice age (the Last Glacial Maximum), these islands were more numerous and more extensive. The islands shown in Figure 9.11 were all part of a single island, the largest in the group, which has been named Santarosae. Some 12,000 years ago the area of Santarosae was nearly two and a half times greater than that of the current islands combined. Humans almost certainly occupied Santarosae at that time, perhaps much earlier, exploiting the area's rich marine food resources. Late in prehistory, the Chumash Indians continued these traditions and maintained a vigorous trade with the mainland, some 25 kilometers distant.[127]

Much attention has focused on the effects of the changing geography of the Channel Islands resulting from postglacial sea-level rise.[128] About ten of the twenty-six islands that existed here during the Last Glacial Maximum (about 22,000 to 18,000 years ago) have since been wholly submerged, and others like Santarosae significantly reduced in size and dismembered. In 2002, a hitherto unknown island that was once emergent, named Isla Calafia, was recognized from newly made sonar images of the Santa Barbara Channel.[129] During the Last Glacial Maximum, Calafia rose almost 10 meters above sea level and is thought to have provided an important stepping-stone for various animals between the California mainland and Santarosae. Too small to support a permanent human population, Calafia was finally submerged about 16,000 years ago.

Numerous islands also drowned during the most recent period of postglacial sea-level rise in the Southwest Pacific.[130] Although some of these islands may have had human inhabitants, most of whom probably settled on nearby higher islands by the time it became

FIGURE 9.11. Westward view across the northern Channel Islands: in the foreground are the Anacapa Islands, behind are the larger islands Santa Cruz and Santa Rosa, with San Miguel in the background. Together these islands were once all part of the large island Santarosae that was dismembered and largely drowned by postglacial sea-level rise. (Photo by Frans Lanting, used with permission.)

obvious that the island was about to be submerged, there have been no formal scientific investigations to confirm this.

The most promising candidate for study might be the 22,800-square-kilometer Bellona Platform (including the Îles Chesterfield, the Avon Islands, and Bampton Reefs) that probably formed a single landmass about 20,000 years ago (as shown in Figure 2.7A). The Bellona Platform lies between the Queensland coast of northeastern Australia and the large island of New Caledonia (La Grande Terre) and may have been an important stepping-stone for humans and other biota that moved between Australia and the Southwest Pacific Islands.

10 Vanished Islands of the Future

Any student of environmental change quickly learns that many popular predictions of the effects of future climate change and sea-level rise in the Pacific (and elsewhere) are highly, and unhelpfully, exaggerated. Such natural changes are also far from unprecedented, although the predicted rates of twenty-first-century change may indeed be so. The specter of future sea-level rise is one that has loomed large over the Pacific countless times before. The principal difference with the situation today is that, for the first time in human history, we can predict with a high degree of certainty that temperature and sea level will rise and by how much they will do so. So the specter is no longer ethereal; it can be discerned and its effects anticipated. The other important difference is that the human societies of the Pacific are far more vulnerable today than they were once to the effects of future climate change and sea-level rise. This is both because people are more settled in particular places — they cannot move so readily — and because there are far more people living today in the lower-lying parts of the Pacific than there have ever been. Together these two considerations are understandably causing a great deal of anxiety.

The Intergovernmental Panel on Climate Change, the body established by the United Nations in 1988 to look specifically at the nature and effects of future climate and sea-level change, estimates that average earth-surface temperature will rise by 1.4–5.8°C between 1990 and 2100, and sea level will rise 9–88 cm in the same period. Given that in the period 1900–2000, temperatures in the Pacific rose by around 0.6°C and sea level by around 15 cm, it looks like there will be a significant acceleration in both temperature and sea-level rise this century. Specifically, the rate of temperature rise will be at least 2.3 times faster this century, perhaps more than nine times faster. The rate of sea-level rise may be nearly six times faster. Much of the variation in these estimates depends on how earth systems respond to the global warming that is thought to be driving these changes.[1]

During the twentieth century, temperature rise and sea-level rise caused significant changes to the Pacific that have impacted on its people and their livelihoods. Principal among these has been the loss of lowlands by seawater inundation, erosion, and groundwater salinization. Also, as discussed in the next section, recent sea-level rise has probably been responsible for the erasure of some surficial islands off the map of the Pacific. If sea level rises as projected during the twenty-first century, more will undoubtedly vanish. One of the main reasons for writing this book is to raise awareness about particular geohazards in the Pacific. To this end, in the final section of this chapter there is a discussion of what we can learn about geohazards from the study of vanished islands.

Short-Term Changes in Pacific Geography

The ocean surface (sea level) is never still. Even were it perfectly flat for as far as the eye could see from a particular coast, it still would rise and fall every day. There are also changes that occur regularly every month and within the course of a single year. Then there are the less-regular, and less easily predictable, intra-annual changes, such as those associated with El Niño events. Then finally there are longer-term changes, barely discernible within a human life span, that range from those taking place over centuries to those taking place over millennia.

Currently the world is experiencing a period of sea-level rise. As far as we can tell, this began about AD 1800 at the end of the Little Ice Age during the 400 years of which sea level had been consistently low compared with more recent times. About AD 1800, for reasons we do not completely understand, sea level began rising and, despite some variation,[2] has been doing so ever since. As noted earlier, the best-possible estimate of future sea-level changes (to AD 2100) is that sea level will continue to rise, almost certainly accelerating within the next few decades (Figure 10.1).

Just as it is clear that sea-level rise over the past hundred years brought about significant environmental and societal changes in the Pacific Basin,[3] so it is inevitable that future sea-level rise will bring change, probably more widespread and having greater impacts on humans than those of the twentieth century. The popular media have made much out of the possibility that entire islands will vanish as a result of sea-level rise. This is regrettably alarmist, but it does express the probability that the geography of the Pacific will be radically altered in the next hundred years or so. Of particular concern are the low islands made from largely unconsolidated sediment (*motu*) that are found on many Pacific atolls and that are home to considerable numbers of people: notably citizens of Kiribati, Marshall Islands, Tokelau, and Tuvalu.[4]

Most such islands only formed and/or became inhabitable a few thousand years ago as a result of sea-level fall, so in a geological sense they are transitory. Although most *motu* have an emerged core of resistant fossil coral reef, it has been shown that once sea level rises above the top of this then the sediment that covers it (what the island itself is largely made from)

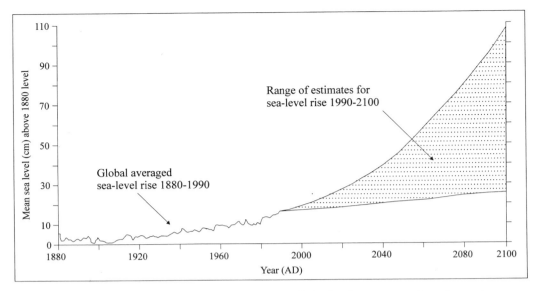

FIGURE 10.1. Sea-level changes, AD 1880 to AD 2100. Upper limit of estimates comes from the maximum sea-level rise under the A1F1 scenario of the Intergovernmental Panel on Climate Change (IPCC). Lower limit comes from minimum sea-level rise under the B1 scenario. (Global sea-level change 1880–1990 is simplified from Barnett [1984] and Warrick et al. [1993]. Estimates of 1990–2100 sea-level rise from IPCC WGI [2001] employing SRES [sea-level rise scenarios] from Nakicenovic et al. [2000].)

will be rapidly eroded and the island may disappear.[5] If this is unconvincing, then a simpler equation is to consider that the maximum height to which a typical *motu* rises above mean sea level is 2–3 meters. If sea level rises even half a meter in the next few decades, then the island will inevitably become smaller and the freshwater lens within it will shrink and rise, which may cause permanent ponding of the lowest areas, making human life significantly more difficult to sustain in such places.[6]

The issue is not simply one for *motu* and other low-lying islands in the Pacific. In the past, a lack of available coastal lowland was one of the main reasons why some people in parts of Solomon Islands built artificial islands offshore. Many of these remarkable structures are hundreds of years old, built entirely by hand. If sea level rises as predicted in the future, so people throughout the Pacific region may consider similar adaptation options.

The future threats to Pacific Island coastal environments do not come solely from accelerated sea-level rise.[7] Among the other natural threats are increasingly frequent and intense tropical cyclones (hurricanes) that are affecting an increasing area of the tropical Pacific.[8] Rising temperatures will stress many ecosystems in the future, notably coral reefs on which many Pacific Island peoples depend for sustenance.[9] Unsustainable environmental practices, ranging from logging to coral-reef degradation, threaten the inherent abilities of many smaller-island environments to sustain human life, at least at current levels. Little wonder the prognoses for people on the most vulnerable Pacific Islands are bleak.[10]

There have been various models (simulations) of future sea-level rise during the twenty-first century (and sometimes beyond) that depend on particular scenarios involving more or less reduction in emissions of greenhouse gases and variations in other human influences on global warming, which is assumed to be the principal cause of recent and future sea-level rise.[11] Pacific Islands are uncommonly vulnerable in this regard, with thousands of people likely to be displaced from their lower areas within the next few decades as a result of sea-level rise.[12]

Figure 10.2 shows the island of Tongatapu, the principal one in the Kingdom of Tonga (South Pacific), which rises to heights of more than 70 meters above sea level, particularly in its south-central part. The island's geography is shown in Map A. Most people live along the island's low (north-facing) coasts, not only because of access to marine resources but also because Tonga's singular land-tenure system means that most of the freely available land on which to settle is located below the high-water mark.[13] Map B shows how the habitable area of the island might be transformed were the high-water mark to move upward about 4 meters. This would happen with less than that amount of sea-level rise (perhaps just 1–2 meters are needed) as a result of shoreline erosion, lowland waterlogging, and, most important, overtopping of the resistant core of fossil reef that underlies the island's coastal lowlands. Such a change would inundate nearly one-quarter of the land area of Tongatapu, perhaps forcing more than 63 percent of its inhabitants from the places they currently occupy. The more probable scenario for the end of the twenty-first century (sea level rising by 1 meter) would see 14 percent of the current population displaced.[14]

Obviously Tongatapu Island is not predicted to disappear in an absolute sense in the fore-seeable future, but it is highly likely to become smaller and for a significant proportion of its current population to be displaced as a result. But many other islands may truly disappear if twenty-first-century sea level rises as projected. Most vulnerable are the atoll islands, some of which have already disappeared, others of which may do so shortly.

Most land subsidence in the Pacific Basin is so slow that it can be discounted on short time scales where processes such as sea-level change dominate. Yet recent subsidence in some places has amplified the effects of sea-level rise. A good example is provided by the monitoring of recent sea-level changes on the Hawaiian islands of Oʻahu (at Honolulu) and Hawaiʻi (at Hilo). The observed rate at Hilo is more than twice that at Honolulu because of the effects of the subsidence of Hawaiʻi Island.[15]

In the context of islands that may shortly disappear (in whole or in part) as an apparent result of sea-level rise, it is therefore critical to know something of their tectonics (land-level movements). Fortunately most atolls, by definition, are subsiding only very slowly, so there is little possibility that tectonic changes will influence the rates of submergence. The eastern group of islands in Fiji, the Lau Islands, is highly variable in a tectonic sense, both in space and time.[16] Many islands have had a long history of short-term bursts of uplift separated by longer-duration subsidence. Compared with most of the surrounding islands, the Vanua

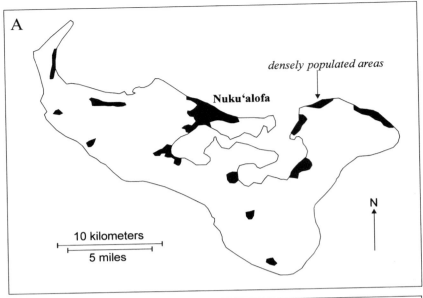

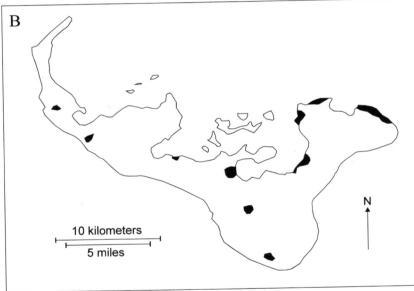

FIGURE 10.2. Maps of Tongatapu, the main island in the Kingdom of Tonga, South Pacific. Map **A** shows the current geography of the island, with densely populated areas (including the capital Nuku'alofa) shaded black. Map **B** shows the geography of the island were the sea level to rise an effective 4 meters above its current level. Most of the currently densely populated areas would be submerged and the Nuku'alofa peninsula converted to a series of smaller islands. (Maps based on my unpublished research, Nunn and Waddell [1992], and Mimura and Pelesikoti [1997].)

FIGURE 10.3. Shoreline erosion along the lagoon coast of southernmost Vanua Balavu Island in the northern Lau Group of eastern Fiji is faster than on adjacent islands because of the subsidence of Vanua Balavu. There is no chance of Vanua Balavu disappearing in the next few decades, but it is likely that significant areas of coastal lowland will become flooded, forcing the people who currently occupy them to move elsewhere.

Balavu island group in northern Lau appears to have been subsiding over the past few thousand years, which has combined with sea-level rise to produce conspicuously faster rates of shoreline erosion here than on nearby islands (Figure 10.3).

There are faster types of land subsidence, typically associated with large-magnitude earthquakes, that affect those parts of the Pacific Basin where plates are converging. Such coseismic subsidence can involve 1–2 meters of downward movement in just a few seconds, and clearly it is important for coastal management purposes to understand where and how often such processes occur. Fortunately, they do not appear to be widespread in the Pacific and do not seem to manifest long-term trends in most places where they are known to have occurred. Yet, as discussed in the next section, the locations and recurrence times of coseismic-subsidence events have implications for understanding the potential for the catastrophic disappearance of islands.

In the context of future subsidence, mention must also be made of some of the areas of the Pacific Rim, commonly highly urbanized, where the outpumping of groundwater has produced some horrifying examples of subsidence in the last few decades. For example, parts of Osaka in Japan sank more than 2.8 meters between 1935 and 1972, and parts of Shanghai in China sank 1.64 meters between 1921 and 1992. In both places, the rate of subsidence has today been reduced by the artificial recharge of groundwater aquifers, but this has not been enough to reverse the downward movement altogether.[17]

In the Atlantic Ocean, threats exist to low-lying coastal areas from giant waves generated by sudden and catastrophic island-flank collapse. Unfortunately, this is not the stuff of some pseudoscience rant but of numerous credible scientific investigations, particularly in the Canary Islands.[18] The wave likely to be generated by a future flank collapse of Cumbre Vieja Volcano on La Palma Island in the Canary Islands has been modeled as having heights of 13–18 meters when, six hours after the collapse, it reaches the eastern seaboard of the Americas.[19]

The question as to whether comparable collapses have occurred in the history of the Pacific is an unequivocal "yes," and such collapses will occur in the future. The important questions are then "where" are these collapses likely to occur and "when." It is debatable whether our geological knowledge of the Pacific is currently sufficient to be able to answer such questions with pinpoint accuracy. Most surveys, including those employing Pacific Islander oral traditions, point to (parts of) four island groups as the most likely to contain islands liable to collapse: the southeastern Hawaiian Islands, the Marquesas (French Polynesia), central-northern Vanuatu, and the islands off the northeastern coast of New Guinea (Papua New Guinea).[20]

A large collapse in any of these island groups could lead to a large wave being generated, but its propagation across the Pacific and onto densely inhabited continental coasts (such as those in North America or East Asia) depends on there being few obstructions in the form of other islands in its path. In this regard, it is likely that a large wave generated within the vulnerable areas of Vanuatu and Papua New Guinea would largely be contained within the immediate areas, but waves produced in either the southeastern Hawaiian Islands or the Marquesas would have a relatively unobstructed run across to various parts of the continental rim of the Pacific.

But the focus of this book is not on large-wave hazards, although these are manifestly important to the future of humankind in the Pacific, but on islands that are liable to disappear. The four areas identified are also those where future island disappearance appears most likely, and they are discussed in detail in the following section.

Geohazard Implications of Vanished-Island Studies

Attempting to validate stories about vanished islands is fulfilling enough by itself, but, over and above this, there are excellent reasons for wanting to understand broader questions, such as

- how often do islands vanish, and how often do island-flank collapses occur
- where in the Pacific (and elsewhere) are the danger areas for island disappearance and island-flank collapse, and
- who are the people most threatened by island disappearance and island-flank collapse.

In all this it is essential to maintain perspective. As Stephen Ward put it, "to raise a proper degree of awareness, without needless alarm, the potential and frequency of landslide tsunami [from oceanic islands] must be assessed quantitatively."[21]

The latest estimate of how often large flank collapses of oceanic islands occur is once every 10,000 years.[22] This is not a helpful yardstick for estimating either the frequency of likely island disappearances or the incidence of large waves associated with island-flank failures. Island disappearances may be associated with large flank collapses, but they may also, depending on the nature of the island and the geometry of the flank collapse, be associated with quite small events. Similarly, large island-flank collapses undoubtedly produce large waves, but these can also be generated by comparatively small collapses; much depends on the rapidity of the collapse and the geometry of the water body into which it falls.

Myths can provide important insights into questions about the frequency of island-flank collapses. For example, within the past 2,000 years in central Papua New Guinea there appear to have been two significant failures: Ritter and Yomba. Perhaps three island disappearances attributable to island-flank failure have occurred within the same period in part of the central Solomon Islands: Teonimanu, Oeni, and the Taʻaluapuala Group. A similar conclusion applies to Vanuatu. Although crude, it is plausible to suppose that once every hundred years or so, a large island-flank failure capable of generating a sizable wave occurs in the Pacific.

In the longer term, it is also worth noting that the frequency of the kinds of island-flank failure that can cause islands to disappear (and a host of other phenomena) varies through time depending on the relative height and rate of change of sea level. Over the past few million years, in response to regular variations in the amount of solar radiation received at the earth's surface, earth-surface temperatures have oscillated between cool and warm every 70,000 years or so. These variations have in turn led to corresponding oscillations of sea level, with low sea levels characterizing cool periods and high sea levels warm periods.

When the sea level is high, the ocean exerts hydrostatic pressure on the submerged flanks of a steep-sided landmass (island or continent) that is akin to wearing a belt tightly around your waist. When sea level is low, then this buttressing effect is gone, and your waist is free to expand. And indeed there is evidence that large-scale collapses occur more frequently along continental margins when the sea level is relatively low, as it was during the ice ages (glaciations) of the past few million years.[23] One effect of more-frequent continental-flank collapse at such times has been to release vast quantities of methane (hydrates) stored in marine sediments into the earth's atmosphere, causing its rapid short-term warming — the so-called Clathrate Gun Hypothesis.[24]

Yet the converse appears true for oceanic islands, where flank collapses appear to have been more frequent during times when sea level was relatively high (as it has been for the past few thousand years). The main supporting study comes from the Hawaiian Islands and has been explained by the wetter conditions that generally prevailed in the Pacific during such interglaciations. Comparatively high levels of water retention in volcanic islands at such times increase the likelihood of water interacting with magma (liquid rock). In turn this will increase the incidence of explosive (phreatomagmatic) eruptions, which are more likely to trigger flank failure than nonexplosive eruptions.[25]

Yet more than simply sea-level oscillations, it is sometimes the rate of sea-level change (as sea level is rising or falling) that can trigger volcanism or flank collapse. Sea level falls when temperatures are cooling at the start of an ice age, and normally this fall is slow and prolonged compared with the sea-level rises that occur at the end of an ice age, when temperature commonly increases rapidly. Within these periods of sea-level rise there are often rapid bursts of rise associated with the breaching of dams holding back vast meltwater lakes on the continents. The most recent such catastrophic rise event in the Pacific (CRE-3) occurred 7,600 years ago and involved 6.5 meters of sea-level rise in less than 140 years. It has been shown that both rapid sea-level rise and sea-level fall is liable, albeit for different reasons, to trigger volcanic activity in volcanoes that are suitably primed.[26] A similar inference for the triggering of flank collapses, at least during rapid sea-level falls, appears reasonable.

Figure 10.4 shows a map of the Pacific marked with the four areas where some future island collapse is considered most likely (see earlier in this chapter). The circles of broken lines show the probable extent of any associated wave a few hours after collapse and the directions in which the wave would move. The locations of the Aleutian Islands and the atolls of Moruroa and Fangataufa in French Polynesia are also shown. This map emphasizes the point that island-flank collapse is a concern not only for people living in the Pacific Islands but also for those living, particularly in low-lying areas, along the continental rim of the Pacific Basin.

Any hazard manager in the Southwest Pacific concerned with the generation of large waves from island-flank collapse should pay special attention to the islands of Papua New Guinea. Perhaps the best example is Ritter Island, which almost completely disappeared during the 1888 event (see Figure 8.2), but also those described in chapter 9. In terms of large-wave generation and potential island disappearance, it is the islands of the Bismarck Archipelago that appear most hazardous. In addition to Ritter, the smaller-island volcanoes of Bam, Kadovar, and Tolokiwa also show signs of partial collapse, as do some of the volcanoes along the Willaumez Peninsula on New Britain Island.[27]

In terms of hazard management in Vanuatu, Ambae Island (see Figure 9.5) is clearly worth watching closely.[28] Not only does it have a history of large, violent eruptions, but these are likely to have caused islands such as those like Vanua Mamata to disappear. In a situation similar to that of Ambae is the island farther south called Ambrym, which has a demonstrated potential for flank collapse associated with volcanic eruptions (Figure 10.5). For Tanna Island in the south of Vanuatu, there is a warning of "a considerable threat to the inhabitants of Tanna" from a variety of possible manifestations, including flank collapse, of the "continued accumulation of magmatic materials at shallow levels in the crust."[29] The

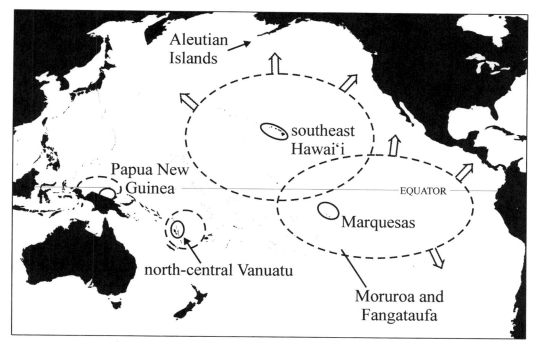

FIGURE 10.4. Map showing island groups in which an island might, by analogy with past incidences and by consideration of modern geological processes, disappear abruptly as a result of flank/crest collapse sometime in the next 500 years. Also shown are the likely directions and areas of impact of associated large waves along the Pacific Rim.

steep-sided nature of many islands in the Vanuatu Group and their active tectonism make a potent mix, and many islands in the group should, given our current state of knowledge, be considered liable to partial collapse.

As discussed in chapter 3, the islands of the Marquesas are falling to pieces, a process dominated by large-scale slumping of island flanks. The location of the Marquesas Archipelago in the central equatorial Pacific means that a catastrophic flank failure there could certainly send destructive ripples across the Pacific Ocean. Yet some reassurance comes from recent research that shows that, unlike the Hawaiian Islands (see next paragraphs), the history of island flank failure in the Marquesas appears to have been dominated by slumping, a process that is generally slow and is only rarely punctuated by catastrophic events.[30] Most recent collapses appear to have occurred along the edges rather than the crest of the Marquesas volcanic ridge, illustrated in Figure 6.1, and this is probably where they will occur in the future.

Although the past manifestations of flank failure in the Marquesas may have been comparatively benign, the same is not true for the Hawaiian Islands, where debris avalanches are considered to have been the dominant form of island (and island-ridge) denudation.[31] Debris avalanches are forms of denudation (or mass wasting) that occur rapidly and usually

FIGURE 10.5. One of the first achievements of the missionaries posted to Ambrym Island in central Vanuatu (for location, see Figure 9.4) was to construct a hospital (pictured in **A**). In December 1913, Ambrym "danced" under the influence of volcanic eruptions and earthquakes, and the hospital and mission station disappeared; the site became a deep-water harbor (pictured in **B**). "The configuration of the surrounding country had been entirely changed by the titanic upheaval . . . behind the hospital, a range of hills 500 feet [152 meters] high has been raised. So complete was the destruction of the hospital and the mission buildings that not even a match was left. Thousands of acres of fertile lands have been changed to barren wastes, forests were blasted, and large numbers of lives were lost." (Photos from Frater [1922]; quotation from Frater [1922:28].)

in a single event. As the study of past such events in Hawai'i shows, the associated hazard potential for Pacific coasts is great indeed.

The Big Island (Hawai'i) in the Hawaiian Archipelago is currently the location of what is probably the most active volcano on earth: Kīlauea. The 1975 Kalapana earthquake, which saw part of the island's flank subside and slip seaward some 8 meters, and the 2000 aseismic event, triggered by heavy rain, were potent reminders of just how unstable such edifices are[32] and will continue to be for the foreseeable future. The movements of the coast during the Kalapana earthquake were associated with rupturing along a series of normal faults that define the upper parts of the Hilina Slump, an active landslide 3–5 kilometers deep that is cut into the southeast flank of Kīlauea.[33] This slump appears to be an incipient form of a giant flank collapse of the kind that has occurred on numerous occasions throughout the history of the Hawaiian Islands (see Figure 3.3A).

There are two views about the future instability of the Hilina Slump. Most scientists who have worked most closely with the field data are cautious. For although they accept that recent observed slips, seismic and aseismic, are part of the process of flank collapse, they do not see any signs of an imminent catastrophic event.[34] But there is a case for arguing that such caution is excessive, and that hazard managers across the Pacific need to understand the existence of a real threat of flank collapse and the possibility that this will generate a large wave, perhaps as high as 30 meters when it hits the western seaboard of North America.[35]

There seems little prospect of an entire island disappearing through flank collapse in the Hawaii Group. Most of the islands are large shield volcanoes that have gentler slopes than many other oceanic islands in the Pacific. In support of this, there are no oral traditions that are unequivocally autochthonous to Hawai'i that speak of islands sinking.

The low proportion of island disappearances outside the four areas highlighted in Figure 10.4 is unsurprising but nevertheless a reminder that the development and collapse of oceanic-island flanks is a normal and expected part of such an island's life cycle, whatever its location. Collapse may be fast-tracked where there are regular seismic and volcanic triggers, but it continues elsewhere in the absence of these.

On June 24, 1949, on La Palma Island in the Canary Islands (East Atlantic Ocean) there was an eruption of Cumbre Vieja Volcano. Eventually a large part of the island volcano (as much as 200 cubic kilometers) dropped 4 meters, then stopped. This has been widely interpreted as an aborted collapse — one that may be reactivated during some future eruption, causing a megatsunami across the Atlantic.[36] The modeling of this, particularly the detail that a wave 13–18 meters high might slam into the eastern seaboard of North America, has understandably been the subject of much media attention, although there is a body of scientific opinion that cautions against alarmism. Irrespective of how it is perceived, the threat from Cumbre Vieja brought considerable attention to bear on the instability of oceanic islands and how this threatens coastal populations.

There can be no doubt that oceanic islands are inherently prone to flank collapse and that many observed collapses, far smaller than the largest of the ancient collapses, have caused tsunamis. The 1888 Ritter Island collapse (4 cubic kilometers) caused a tsunami that was 8 meters high when it reached Hatzfeldhafen 500 kilometers away.[37] Yet some flank collapses appear to have involved up to 5,000 cubic kilometers of material.[38]

Flank collapses of Pacific Islands have the potential to cause large waves that might impact any Pacific coast, and there is clearly an imperative to identify those islands where flank failures appear imminent and understand their specific likely impacts. Obviously the populations most at risk are those occupying the islands that disappear, although it is reassuring that many examples of island disappearance (such as those of Teonimanu and Vanua Mamata discussed in chapter 9) recall that there were numerous survivors, who settled elsewhere, suggesting that the process of island disappearance was neither sudden nor unheralded. Of course, contrasted with that are the stories of inhabited islands (such as Tuanaki) sinking with no known survivors, which suggest that some such events were indeed sudden and unexpected.[39]

It is clearly pointless for any practical purpose to focus only on hazards involving islands that are liable to disappear (or experience flank collapse) in the future rather than on those that would be impacted by large waves produced by such events. In this sense, the higher-risk archipelagoes identified earlier, the Bismarck Archipelago of Papua New Guinea, the Vanuatu Islands, the Marquesas Islands of French Polynesia, and the Hawaiian Islands, appear also to be those most vulnerable to impacts by large waves generated by island (flank) collapse and washed across an adjacent island. There is variation within those archipelagoes arising from their geography. Some islands are sheltered by others, making direct large-wave impact less likely.

The danger in identifying high-risk areas with any poorly understood or difficult-to-predict hazard is that those living outside those areas may therefore downgrade their awareness of such hazards and their alertness to their possible effects. With island (flank) collapse, this is a mistake for two reasons. First, because islands have been shown to have disappeared outside these higher-risk areas (examples include those in the Cook Islands and Kiribati discussed in chapter 9), there is a real possibility that these include islands that are similarly disposed to disappear/fail in the next few centuries. Second, the waves generated by island disappearance/failure within (and beyond) these higher-risk areas may reach their greatest heights locally but may also have significant long-distance impacts.

11 Vanished Islands and Hidden Continents in the World's Oceans

LAST THOUGHTS

Scientists are trained, at least in their professional lives, to think conservatively, to deduce only the deductible, and to express only that which is expressible given the information available. For such reasons the language of science often appears dry and detached to outsiders, who may therefore not wish to dig deep to find something that is personally relevant or enlightening. In contrast, for many people, the accessible and exciting writing of many pseudoscientists readily provides this, which is why their books sometimes sell in the millions. By selecting only those facts that appear to support their manifestly unsupportable models, by dealing cavalierly with topics that appear beyond the pale to many scientists, and by investing those topics with import far beyond rational judgment, they push the right buttons in many minds. Science cannot compete.

Nor should science try to compete. Yet it should periodically respond. This book has sought to inform, analyze, and enlighten. It tackles well-known accounts of vanished islands and hidden continents but also presents information that is likely to be unfamiliar to many readers. It is clear that the Pacific (as well as the Atlantic and the Indian Ocean) has many stories of its own about vanished islands and hidden continents. These include not only those written in the language of science, and those gratuitously imposed on the people of the Pacific by pseudoscience and new-age theorists, but also those stories of Pacific peoples themselves.

This last chapter describes some of the reasons why people are interested in vanished islands and hidden continents and then, as a coda, acknowledges the obvious dangers in writing a book such as this.

Reasons for Learning about Vanished
Islands and Hidden Continents

There is no shortage of good reasons for learning about vanished islands in the Pacific, or indeed elsewhere in the world. That islands have disappeared, both in whole and in part, in the past is beyond doubt; the collapse of unstable island flanks in the future poses a range of threats to coastal peoples throughout the region, of which hazard planners and others need to be better aware. That islands and fragments of continents once existed in the Pacific, long before people arrived there, and were instrumental in the dispersal of various biotas is likewise indisputable.

But this entire field of scientific inquiry has been overwhelmed by a tidal wave of pseudoscience theorizing. This has become institutionalized, in stark defiance of a huge body of empirical data, into a series of belief systems ranging from theosophy to the new age. There are dangers in allowing antiscientific beliefs to dominate any field of legitimate scientific inquiry, and one of the principal goals of this book is to counter such beliefs as they apply to vanished islands and hidden continents.

But again the argument is not quite so simple, or so neatly polarized. Anathemic as it may appear to many a conventional scientist, there are reasons why stories of vanished lands fulfill a range of human needs, discussed in the next paragraphs. In addition, for many continental dwellers, in particular, the image of islands is often quite different from that held by islanders, which is why, as also explained later in this section, islands are often the favored location for apparent mysterious happenings, uncommon cultural practices, and even parables for human futures.

In the past 200 years or so, the process of lifestyle change for many people has been so rapid that much that has been lost is being sought anew. When it proves unobtainable, it may be substituted with something else, something invented. For some, this explains the rise of the popularity of fiction, but it also explains the rise of interest in other worlds, including vanished islands and hidden continents. Michael Crichton, himself the creator of many such lost worlds, said it well.

> Direct experience is the most valuable experience I can have. Western man is so surrounded by ideas, so bombarded with opinions, concepts, and information structures of all sorts, that it becomes difficult to experience anything without the intervening filter of these structures. And the natural world — our traditional source of direct insights — is rapidly disappearing. Modern city-dwellers cannot even see the stars at night.[1]

In a similar vein, according to Sprague de Camp, the story of Atlantis, that famous lost island-continent, strikes a responsive chord by its sense of the melancholy loss of a beautiful thing, a happy perfection once possessed by mankind. Thus it appeals to that hope that most of us carry around in our unconscious, a hope so often raised and as often disappointed, for assurance that somewhere, some time, there can exist a land of peace and plenty, of beauty and justice, where we, poor creatures that we are, could be happy. In this sense Atlantis . . . will always be with us.[2]

Interest in lost worlds is an innocent pastime. Yet, when those lost worlds are claimed as tangible or when invented lost worlds are represented as having been real places, then guile begins to replace innocence. Many sincere people have been swept along in the tide of new-age thinking on the subject of lost worlds, and millions read the accounts of pseudoscience writers unaware of their charlatanry.

Lyon Sprague de Camp was perhaps the first person to understand the essential ambivalence of the new age—what he, writing in 1954, called the occult—toward science. On the one hand, new-age and pseudoscience writers realize that without reference to science their arguments will lack popular credibility; in de Camp's phrase "they would like to cut in on the enormous prestige of science."[3] To achieve this goal, such writers commonly choose those scientific arguments (or those parts of them) that support their goals and ignore (or misinterpret) those that do not.[4] Often they will cite work that is either well out of date (by the yardsticks of conventional science) or insufficiently conclusive to support their arguments. Sometimes people who are far from deserving of the epithet are referred to as "learned."

On the other hand, new-age and pseudoscience thinking is generally dismissive of the scientific method, given its roots in empiricism and its emphasis on deductive reasoning and verification, preferring information obtained by decidedly nonscientific reasoning: from automatic writing through inspiration to channeling. In de Camp's critical judgment, "material evidence . . . is worthless; truth is to be dredged out of the inner consciousness by mystical introspection."[5] If material evidence is needed, then it is usually claimed to have been seen, perhaps in a now-lost book or inscribed on pillars in India or Tibet, perhaps in secret libraries underground that can be accessed only by the chosen few. It must surely amaze the objective observer of such pseudoscience that anyone can take seriously such a selective and improbable mishmash of supposed fact and received wisdom. Yet many people do, and understanding why is the key to appreciating why those stories about vanished islands and hidden continents in the Pacific (and elsewhere), which have no basis in fact, continue to be so widely believed.

What is so special about islands that makes them such a convenient vehicle for parading utopian systems or, whether or not in the guise of continents, as places where polities of enviable order and achievement once existed and may reappear?[6] It is difficult to second-guess the intentions of everyone who has used islands in this way, but it is plausible to suppose that the characteristic of boundedness is paramount; "boundedness makes islands graspable, able to be held in the mind's eye and imagined as places of possibility and promise."[7] And, for continental dwellers whose conceptions of islands may owe most to imagination and least to experience, islands are often, on account of their boundedness and their effective unattainability, places of refuge, inviolate places, where life is at once simpler and safer.

Societies in which oral traditions are the principal vehicle by which a people's history is transmitted from one generation to the next may not appear as the most dispassionate observers of nature. Their myths portray the full arsenal of dramatic tools that obscure the details of the account but make it far more interesting, and, most important, more memorable, for the audience. Many myths also carry a message that often serves as a lesson for the current generation from their ancestors. The theme of divine retribution is a common element, an archetype in the sense of Jung, of many vanished-island stories. One possible purpose, at least soon after the catastrophe, would have been to ensure that future generations lived appropriately upright lives and did not therefore tempt the deity to such a drastic course of action ever again. But there is one example, conveniently involving a remote Pacific Island, that has been hailed by many modern commentators as an unimpeachable parable for modern times.

The Easter Island (Rapa Nui) of the past has long vanished. Famous for its numerous huge stone statues (*moai*), once erected in lines along the island's coast (see Figure 7.2B), the culture of Easter Island has attracted huge amounts of speculation and scientific interest.[8] Although the island is physically intact, its environment was merely a shade of its past glory when Roggeveen arrived there, the first European to do so, in 1722 and described it as having a wasted appearance.

Easter Island, Earth Island is the title of a book by Paul Bahn and John Flenley that caused a stir when first published in 1992. It is an accomplished volume, that details the human history of Easter Island through the times of plenty, when the *moai* were carved and erected, and the subsequent times of crisis and warfare. It draws the disturbing conclusion that the collapse of Easter Island society was caused by the profligate (unsustainable) use of the island's natural resources by its inhabitants, and therefore that Easter Island in microcosm represents a cautionary tale for the modern world; if we do not use our planet's natural resources sustainably then one day we shall pay a dear price for this.[9] The message was timely and no right-thinking person would doubt its importance.[10]

Easter Island has not vanished, but its ancient and evidently high-achieving culture has done so — and the message is clear. More pertinent to the theme of this book is the common idea that particular islands vanished because people failed to follow traditional protocols. Examples abound in the Pacific, from the story that Burotu sank because the people of neighboring Matuku Island failed to appreciate gifts sent to them, to the Micronesian tradition that the chief of Sipin Island caused it to sink to preserve its people's traditions.

It is worth noting that the same theme is popular with those who describe Atlantis. Many commentators aver that Atlantis vanished because its people had turned from uprightness to debauchery. In 1942, Lewis Spence took this one step further, arguing in his book *Will Europe follow Atlantis* that Atlantis was sunk by God for its sins and the same would happen to Europe unless it reformed itself.

The Danger of It All

Many people are less critical thinkers than others. To be less critical is not always synonymous with a lack of formal education but often something cultural. In the past many Pacific people were schooled in a highly disciplined system where questioning teachers and adults was not merely discouraged but punished, and grew up accepting everything that they read in print simply because it was printed.

For more than twenty years, I have worked at the University of the South Pacific, an international university serving twelve Pacific Island nations. At the main teaching campus in Suva, Fiji, is the university's largest library. Many of the pseudoscience books that describe hidden continents in the Pacific are included in the library's collections and, more worrisome, appear to be among some of the most popular recreational reading for students. This exemplifies the real danger of such work, namely that its representation as truth understandably deceives people who, through no fault of their own, know no better. And, unless scientists stand up and speak out against such misinformation, then future generations may know no better.

A lot of the information in this book could be misrepresented and used in ways that are far from appropriate or intended. So, in conclusion, I ask unwary readers to read not just the details of the examples in this book but also the numerous caveats. There is no sunken continent in the Pacific, the Atlantis of Plato never existed, Nan Madol and the Easter Island *moai* were simply created by the ancestors of the modern islanders, and so on. The geological history of the Pacific Basin is well known. The origins of the people who inhabit it are equally well known. There are lacunae of uncertainty and voids of information in both time and space but, truly, no real mysteries.

Appendix 1

The Story of Teonimenu,[1] Central Solomon Islands, and How It Vanished, from an Unpublished Manuscript by Zacchariah Haununumaesihaaʻa

The original manuscript was written in Ulawan by the late Zacchariah Haununumaesihaaʻa and incorporates an oral narrative by the late Owen Haununuʻoloaningau. The English translation is by His Excellency Sir Nathaniel Rahumaea Waenaporopaine and Tony Ahikau Heorake. Permission for publication of this story has been granted by Polycarp Haununu and Raphael Taloniweieu (sons of Zacchariah Haununumaesihaaʻa), Fr. Simon Ouou (son of Owen Haununuʻoloaningau), and Sir Nathaniel Rahumaea Waenaporopaine (brother of Zacchariah Haununumaesihaaʻa).

The Matrilineal line of Kaliitaʻalu

Saudjorangaasi was a woman from Pehuaraouou's clan of Arona (on Ulawa Island). She married Ouousinairaa of the Sulieuwo clan (also of Ulawa). They had two children, one of whom was a son named Poroirohaʻa whose only female child was named Saumoleasi. Saumoleasi was married to Pororihuanimae of Teonimenu Island. They had a son and two daughters. Their son was named Kalimatawarepa and their daughters Tekudjorangaʻasi and Huʻepepeʻanawe. Huʻepepeʻanawe married Porongarasimae of Teonimenu and they had a daughter named Sauweteʻau, who was married to Roraimenu (a Malaitan) of Aliʻite Island.

In old age, Porongarasimae became seriously ill so Sauweteʻau and Roraimenu moved to Teonimenu (from Aliʻite) to care for him in his illness. They stayed with him till his death. After this, both Roraimenu and Sauweteʻau returned to Aliʻite and, as is customary, Sauweteʻau fasted for eight years by not eating any food from the sea.

Kaliitaʻalu in Aliʻite

In the ninth year, Kaliitaʻalu and his brothers, the sons of Tekudjorangaʻasi, arranged to visit Sauweteʻau and Roraimenu on Aliʻite with their uncle Kalimatawarepa. They agreed to fish and take their catch with other foodstuffs to Sauweteʻau to mark the end of her fasting. The brothers with their uncle caught about 100 tuna in their fishing canoe (taʻeolu) named *Aamoloiseutangalau*. The women of the village collected shellfish and prepared cooked and baked food items such as yam, taro, and vegetables to take to Aliʻite.

When the feast was ready, the brothers with their uncle and the other relatives gathered the food items, put them in their canoes, and sailed across to Aliʻite. The brothers and their uncle traveled on a bigger canoe (*luusuinima*) named *Iiʻepwarilopona*. The entire fleet (of men, women, and food items) was known as the *Aalatangalau*. The fleet of canoes journeyed to Aliʻite and upon arrival they were greeted by Roraimenu, Sauwete'au, and the villagers on the beach. They pulled their canoes ashore and off-loaded the cooked foodstuff and fish, and transported it all to the village.

As the villagers were preparing for the feast, Roraimenu was busy chanting custom tunes (*teatea*) with the village elders in the custom house (*toohi*). He was also organizing the womenfolk of Ali'ite to prepare a performance of traditional custom dancing and had appointed his wife — Sauwete'au — to lead the women in this since she usually fulfilled this role.

Meanwhile at the beach in the canoe house (*taoha*), Kalimatawarepa was performing magical rituals on his nephew Kaliita'alu with the aim of attracting the attention of a young lady as a potential wife. He chewed some betel nut and kava leaves with lime (made of calcium carbonate from burnt coral limestone); he then sprinkled, puffed, and rubbed Kaliita'alu's shoulders, arms, and forehead with some powdered lime. He advised Kaliita'alu to wait in the canoe house until all the women and girls had started dancing; then both of them would go to the village. When the dancing commenced and many people had congregated around the perimeter of the dancing arena, Kaliita'alu and his uncle joined the chanting men as the main soloists. As Kaliita'alu was leading the chanting he got to a part where the dancing women and girls needed to respond to his chant and, as it happened, Sauwete'au was the first one to make the response. She sang "*oi poroineu 'a Kaliita'alu*" ("oh my husband Kaliita'alu"). Now, the ritual preparation of Kaliita'alu had been intended to determine that whoever (among the female performers) would respond first to the dancing chorus chanted by Kaliita'alu would become his bride. Ironically, yet unknowingly, Sauwete'au was the first person to respond.

The whole dancing ceremony abruptly stopped and was abandoned in embarrassment. Even the feasting was disrupted because of that particular incident. In the evening, the village chief called for a meeting and, after the gathering in the custom house (*toohi*), the elderly men of both Ali'ite and Teoni- menu continued chanting traditional tunes and chants long into the night. But because of the earlier incident, Kaliita'alu and his brothers with their uncle and most other people from Teonimenu decided to return home prematurely the next morning and did not participate in the evening's activities.

On the next day as they were preparing to leave, Sauwete'au approached them on the beach and requested that she be allowed to return with them to Teonimenu. Initially Kaliita'alu and his brothers were reluctant and debated amongst themselves whether Sauwete'au should travel with them since their canoes were ritually consecrated to fish for tuna and normally women were forbidden to travel on them. As they were arguing amongst themselves, Roraimenu intervened and assured them that Sauwete'au could go to Teonimenu in one of their canoes. Kalimatawarepa also offered counsel and eventually a consensus was reached and she was permitted to travel on the canoe of her cousin Sulupwau'edja. So, all the visitors returned to Teonimenu that day including Sauwete'au, but Roraimenu (her husband) remained on Ali'ite.

Some three months later, Kalimatawarepa went to Ali'ite on one of his usual across-the-sea voyages to visit some relatives. During this trip he met Roraimenu on the beach and, while they were convers- ing in the canoe house (*taoha*), Roraimenu questioned him persistently about Sauwete'au's well-being. After some hesitation Kalimatawarepa informed Roraimenu that Sauwete'au was now cohabiting with Kaliita'alu. Now that information really disgusted Roraimenu. He became infuriated and vowed to take his revenge on Sauwete'au and particularly Kaliita'alu for bringing him such shame.

Roraimenu's Plan of Revenge

So Roraimenu hatched a plan whereby he would travel to Ndai Island off the northern tip of Malaita and purchase with shell money a wave curse (from someone who had that powerful magic) to destroy Teonimenu, including Kaliita'alu and Sauwete'au. On the appointed day he set out on his journey to

Ndai in his canoe. He made several brief stops en route and oftentimes slept in some villages along the way. It so happened that whenever people from the villages he passed would ask about his purpose for traveling he would not disclose anything. When he arrived on Ndai he was also questioned by the people there and he told them that he was searching for someone from whom he could buy a wave curse that would cause an entire island to disappear. He said he wished that particular island to be sunk so completely that he would never even see any foliage from its tallest trees protruding from the sea. After spending a night on Ndai, his request was eventually granted. The wave curse would involve eight waves.

Roraimenu returned with the wave curse to Ali'ite—four waves at the front and four waves at the back of his canoe. He was also given two plants to take with him to be used in association with the "wave curse" on Teonimenu. He placed a *ngoli*[2] at the front and a *ngoli* and a taro leaf at the rear of his canoe. On his return journey to Ali'ite, he also stopped at the same villages he had visited on his way to Ndai and the villagers asked him fearfully "which island are you going to destroy" (*"hanua i hei oto a nai warea"*), but he did not tell them.

As Roraimenu approached Teonimenu from the eastern part of Ulawa, the sun was slowly setting on the distant horizon. When it was almost dusk, he pulled his canoe ashore at a secluded spot on the beach and went inland. He buried the *ngoli* from the rear of the canoe at the foot of a small hill and planted the taro leaf and the *ngoli* from the front of the canoe at the top of the hill in the center of the island, then he hurriedly returned to the beach, pulled his canoe out to sea, and paddled across to Ali'ite. When he arrived at Ali'ite, he went to the interior of the island and climbed onto a high rock and watched with grim satisfaction as Teonimenu was gradually swamped and crushed by the waves.

The Eight Waves That Destroyed Teonimenu

As Roraimenu was watching from afar, the waves began to swamp Teonimenu. The first wave hit the island and went up as far as the midportion of the coastal beach ridge. The second wave lifted off the roof of the canoe house (*taoha*) and smashed it among some coconut trees at the back margin of the beach ridge. One of the canoe house posts hit Poropwaheimatawa (Kaliita'alu's father) in the chest, killing him instantly. The third wave reached the center of the village and woke the people who were asleep at that time. The fourth wave went through the village and into the people's cultivated food gardens. The fifth wave reached the foot of the small hill where the *ngoli* from the rear of the canoe had been placed. The sixth wave reached the top of the hill where the *ngoli* from the front of the canoe and the taro leaf were buried. The seventh wave went over this hill and onto the eastern side of the island. The eighth wave completely submerged the island. Following the passing of the eight waves, the sea continued to swell, churning, twisting and rolling any floating debris as the island was finally submerging.

Very early on the next day Roraimenu climbed onto the high rock named *haunikeni* and observed the continuing submergence of Teonimenu. Smoke and haze rose from the waves, and rumbling noises still filled the air. Roraimenu then called out to the other villagers to come and witness the aftermath of the eight waves washing over Teonimenu. His voice full of sarcasm, he said *"molu lae kau mai losie dunga a Kaliita'alu e a hahi poo ana na e a lolo hahia maelani waieu"* ("come all of you and witness Kaliita'alu's fire, used for baking pig and tuna").

The eight waves ran across the three islands in the Teonimenu group until they were totally submerged. The entire population of the islands was swept away in the raging waves and scattered all over the ocean. Kaliita'alu managed to survive by climbing onto the top of the canoe house's (*taoha*) central post. He stayed up there for about eight days subsisting mainly from a bunch of ripe bananas he had been able to scavenge from among the floating debris. Others of the people who survived this terrifying ordeal were driven in various directions by the ocean currents and were washed ashore in different places in Solomon Islands. For example, some eventually landed in certain locations along the coast of the islands of Makira (San Cristóbal) and perhaps Santa Ana, and others eventually landed on Guadalcanal, probably in the Longgu region. A few survivors also landed on the southern tip of Ulawa (at Arona), and others landed at Su'u-i-Ooha in western Ulawa. Others probably drifted and landed on the southern end of Maramasike (Small Malaita). An unknown number of people perished as a result of this merciless event.

The names of three persons who landed at the Arona coast on Ulawa Island were Kaliita'alu, Tekuite'o, and Hu'e'anawe (a male and two females, respectively). As they were drifting toward the shore, clinging to banana stumps, the shark guardians (ancestral spirits) of that coastal region would not allow them to go ashore, so Kaliita'alu had to sacrifice the life of Hu'e'anawe to appease them and ensure a safe landing for himself and Tekuite'o, who were able to go ashore safely at Sulieuwo.

After some time had elapsed, Kaliita'alu and Tekuite'o were able, with the approval of Chief Poroiroha'a of Sulieuwo, to prepare and hold a ceremonial feast (*maeta*) to honor the spirits of their relatives who perished as a result of the submergence of Teonimenu. For this purpose, they cultivated three large gardens of yam and pana and named them as *papanaho, ho'opapanite,* and *i'epwarilopona.* Although Kaliita'alu's fate remains shrouded in mystery, he is thought to have remained on Ulawa for some time after Teonimenu was destroyed, eventually leaving for an unknown location by drifting on a banana raft out into the open ocean in the same manner that he came to Ulawa.

The three islands of Teonimenu that vanished between Ulawa and Ali'ite (northernmost island of the Three Sisters Group) are known to this present day as Hanua Asi (our home) by people throughout the Makira Ulawa Province (which includes the islands of San Cristóbal, Santa Ana, Santa Catalina, Ugi, and Ulawa). The term in the Ulawan language given to the kind of waves that destroyed Teonimenu is *luelue,* which is analogous to a tsunami. The traditional shell money (*te'ete'eni-roa*) that Roraimenu used to purchase the wave curse is believed to be in an overseas museum.

Appendix 2

Extracts from Maretu's Narrative of Cook Islands History Pertaining to the Vanished Island Tuanaki (from Crocombe 1974).

These two extracts were written by Maretu, one of the earliest Christian missionaries of Cook Islander ethnicity. The first extract reports the questioning of a man named Soma on Aitutaki Island who had visited Tuanaki. The second extract reports the circumstances of the missionaries' unsuccessful search for Tuanaki in 1844. All page numbers refer to Crocombe's translation, not Maretu's original manuscript. Extracts reproduced courtesy of Marjorie Tua'inekore Crocombe.

The Lost Island of Tuanaki (Pages 163–165)

Katuke and Ngatae were chosen to come with us to search for Tuanaki. "I want to go [as a missionary] to Tuanaki," I said to Pitman and Buzacott [European missionaries].

"No. You go straight to Mangaia [Island]," Buzacott answered. "After you disembark there, the ship can go on and search for Tuanaki."

The ship [the *Samuel and Mary*] left for Aitutaki [Island] and there we met a man named Soma who had stayed ashore from a ship for three months. He told us he had seen Tuanaki. The two missionaries and the captain [of the *Samuel and Mary*] met Soma, who told this story:

"It is two years since I saw that island. We had come from Rurutu [Island]. When we reached the island the captain sought the harbor and a boat was lowered. The captain and six others of us got into it and went ashore." No one was about on the shore so the captain said to Soma, "You go inland and see if there are any people about. When you see someone then come back." The captain gave him a sword. He headed inland and there he saw a house full of people. It was the *ariki*'s house [the high chief's house].

The *ariki* called out, "Where are you from? Are you from Araura?"[1]

"Yes," Soma replied.

"Come inside."

So Soma entered the house. Only men were in it; there were no women because the men and women had separate houses. They were sheltering from the sun; that was why they were all inside this house. Soma sat down and was asked again, "Are you from Araura?"

"Yes," he told them, "I am from Araura" (Araura was their name for Aitutaki).

"Where is the captain of your ship?"

"He's in the boat," Soma replied.

"Why doesn't he come inland?"

"The captain is afraid you might kill him."

"We don't fight; we only know how to dance. We don't know how to fight."

Soma ran back to the captain, who asked, "What are they doing?"

"They're there inside the house."

"Why are they in the house?"

"I don't know," he told the captain.

So the captain went ashore, taking with him *aukute,*[2] axe, and hat.

They entered the house and the captain presented the gifts to the *ariki* and asked him his name.

The *ariki* answered, "Maeva Rua; my name from Rarotonga [Island] is Te Tuikura."[3]

The captain and I slept ashore. The boat went back to the ship laden with food: chicken, pigs, yams, bananas, taro, and green coconuts. We stayed ashore for six days.

Gill [the missionary] asked Soma, "What are the people like?"

"They are just like us," he said. "They are subject to the authority of the *ariki* and have to render food tribute. They speak the Mangaian dialect and they wear the Mangaian *tiputa* or poncho. They use fans just like those of Mangaia [Island]."

Gill then said, "Let us all go [and look for Tuanaki]. We'll pay you thirty *moni* [unspecified units of currency] for the trip."

Soma said, "It takes one night[4] to travel from Mangaia to Tuanaki. Take your bearings from Tau.[5] I have just come back. I am staying here. My sister is dying. Another died [while I was away]. This one will die and I won't see her."

On our arrival at Aitutaki an epidemic of dysentery was raging and thirty people had died.

The Unsuccessful Search for Tuanaki (Pages 171–172)

Gill replied [to Maretu], "We'll have to go together [to Tuanaki], but there'll be trouble if we go tomorrow.[6] You know Soma's story about the [Tuanaki] language being like Mangaian and that they have fans and clothes like them too. Ask the Mangaians for some fans and clothing, then go aboard secretly with them. It won't be any use taking your wife with us. Leave her here."

Some children managed to collect ten fans and twenty ponchos. I told Parima that I was going on board the ship immediately. "Don't worry," I warned him. "There'll only be trouble if the others know about it. My wife is remaining here."

"Okay. But what a couple of liars you are," they said, "You'll never see Tuanaki. You'll come back with nothing. If you weren't leaving Mrs. Maretu behind you'd find yourself soaking wet with seawater by and by.[7] It's all right. You two can go, but we are not going to fix that house. We'll wait till you return."

Soon the boat arrived, and I boarded the ship. Two deacons followed me to the ship to throw me overboard. "Come down here," they called, "so we'll get you soaked with seawater."

"Why have you followed me?" I asked. "Why aren't you consoling my wife? Why did you follow me?" Then they went back. Presently, Mr. and Mrs. Gill and the captain came on board and we set sail for Tuanaki. That evening the boom broke in half and it was put down in the hold. Then at midnight, one of the ship's masts broke, followed by a breakage at the stern. On that same night too, I suffered badly from a recurrence of the dysentery I had contracted in Atiu [Island]. On Saturday the two Europeans came to see me.

"You are getting worse," Gill said. "We are near Rarotonga [Island] and we've reached the limit of mileage [set aside for the search for Tuanaki]. Let's go back."

"What about the directions Soma gave us?" I asked. "Let's zigzag back."

On Sunday evening the missionary [Gill] and the captain came back and said that I was very sick and that the ship should return to Mangaia.

On Monday we turned back for Mangaia and reached there on Tuesday morning. A canoe carrying Pati and Ruirunga came out and they called out, "Did you find the island?"

"No, we didn't. Maretu is sick. Put him in your canoe." Instead, the canoe turned around and headed for the shore. The men on board the canoe were angry with us. Gill called out, but they still refused to come back. It was my canoe and the two men in it were from my own village. They were sent by my wife to see us. They were still angry with Gill for making me go with him.

Notes

Chapter 1: Introduction: A Personal Odyssey

1. Ramage (1978:42–43).

2. Although pottery is not manufactured in this area today, the early ceramic history of the Moturiki area identifies it as one of the first parts of the Fiji Archipelago to have been inhabited by pottery-making humans, perhaps around 1000 BC (Nunn et al. 2007b).

3. Snow's most famous work was *The Two Cultures and the Scientific Revolution* (1959), in which he argued that a lack of communication between the humanities and the sciences was preventing the understanding and solution of many of the world's problems.

Chapter 2: The Earth's Dynamic Third: The Pacific Basin

1. For a thoughtful account of the various perceptions of the Pacific Ocean and Islands, see Ward (1989). The collection of essays edited by Hau'ofa (1993) is also representative of the views of many people living in the Pacific Islands. Good accounts of Pacific Islander voyaging include Lewis (1994).

2. I am talking here of islands large enough to sustain a viable human population, rather than smaller chunks of rock.

3. This issue was explained in more detail, using numerous examples, in the first chapter of *Oceanic Islands* (Nunn 1994).

4. Beer (1990:271). Beer went on to explain that the construction of the word "island" is a kind of pun: isle meaning "watery" and having land appended to make "water-surrounded land."

5. Continents are composed of generally lighter (less dense) materials than the ocean floors. The age of the continents is commonly far greater, most continents having cores (cratons) that date back at least 1,000 million years. In contrast, most of the oldest ocean floor is around seventy million years in age.

6. The oldest known formation is the Acasta Gneisses of northwestern Canada, dated to around 4,000 million years ago (4,014 ± 25) by Sano et al. (1999).

7. A systematic account of earth history in the Pacific is given by Nunn (1999).

8. Formerly known as Gondwanaland.

9. Dietz (1961). The observation that magnetic stripes on the ocean floor are symmetrical on either side of the mid-ocean ridges proved critical.

10. This has given rise to the idea of a recurring supercontinent cycle (Nance et al. 1988).

11. This topic is reviewed in Nunn (1994).

12. A good exposition of the expanding-earth hypothesis is by King (1983).

13. The Ring of Fire is usually represented as being the line of active volcanoes associated with the convergent plate boundaries along the Pacific margins, but it is occasionally represented as including the East Pacific Rise, a line of underwater active volcanoes that marks a divergent plate boundary.

14. The use of the word "plate" may confuse the lay reader. In this sense, the best analog is not something from which you might eat your dinner but rather the hard carapace of a turtle that seems divided into rigid interlocking segments.

15. Although it is convenient to regard transform boundaries as characterized by no net convergence or divergence, there is actually often some. This is because these boundaries are not perfectly linear; often they have slight irregularities, which mean that the strike-slip (sliding) motion that occurs along them is expressed locally as convergence or divergence. The Fiji island named Cikobia is believed to have been uplifted for this reason because it occurs at a convergent kink in the Fiji Fracture Zone, a major transform boundary in the region (Nunn 1994).

16. The name Lōʻihi means long in Hawaiian, a reference to the elongate form of the volcano, which is stretched along a rift zone trending north-south.

17. See, for example, the discussion by Williams (2001:91).

18. The event was described fancifully: "Three quarters of the Earth's surface, to a depth of 35 miles, was carried away in a trailing mass of ruin. New Zealand itself was just saved to the Earth" (Pickering 1924:32).

19. J. M. Brown echoed these views when he wrote that "most geologists who study the whole surface and crust of the earth assume a hypothetical Pacific continent" (1924:47). The idea fell into scientific disrepute for want of evidence within a decade or so yet has been kept alive by less-scrupulous writers.

20. The continental crust is commonly termed "sial," after its main elements silica and aluminum, and has an average density of 2,700–2,800 kilograms per cubic meter, and the surrounding oceanic crust, termed "sima" after the silica and magnesium that dominate it, is heavier, with an average density of 2,800–3,300 kilograms per cubic meter.

21. Typical is the comment of M. R. Williams, following a forthright explanation of isostasy given to him by geophysicist Tanya Atwater of the University of California, ". . . so continents can't sink after all. . . . Oh well, geologists have been wrong before. I guess the debate will go on forever" (2001:307). No, it will not.

22. Magellan actually named it *Mar Pacifico*. Although born Portuguese, Magellan renounced his citizenship in 1514 to become Spanish (Joyner 1992).

23. Dalziel (1991).

24. Knoll (1991).

25. Knoll and Walter (1992).

26. Hallam (1986).

27. Casamiquela (1980).

28. A range of articles about this Great American Biotic Interchange is in the book edited by Stehli and Webb (1985).

29. Thomson (1977).

30. The original work on Pacifica was done by Nur and Ben-Avraham (1977).

31. The original work, lauded for its innovation, was by Alvarez et al. (1980). A description of the Chicxulub site was given by Hildebrand et al. (1991).

32. Including many other groups of reptiles, ammonites, and calcareous plankton.

33. There have been large meteorite (or bolide) impacts during the Cenozoic, some of which have been claimed as the cause of prolonged periods of stepwise cooling (Keller et al. 1987).

34. Keigwin (1980).

35. Collins et al. (1996).

36. Repenning and Ray (1977).

37. The process is recorded by changes in oxygen-isotope values of ocean water on either side of the modern isthmus (Savin and Douglas 1985).

38. The process of closure led to the accumulation of warm water in the equatorial western Pacific, This Solomon Islands Warm Pool, as it is known today, is the main area of the Pacific in which tropical cyclones (also known as hurricanes or typhoons) are generated.

39. The effective closure of the two equatorial water gaps in the Pacific Rim and the opening of two at high latitudes gave birth to the modern ocean circulation of the Pacific (Nunn 1999).

40. Rawling and Lister (1999).

41. Many people wonder how, if the land is rising continuously, it is possible for reefs to form at all, let alone be uplifted discretely. There are two answers. One is that, even if uplift is truly continuous, its rate may vary substantially so that noticeable uplift may occur only when there is a rapid burst of uplift. The other is that, in places where slow uplift occurs over long periods of time, the effects of sea-level changes mean that sometimes (when sea level is falling) emergence is rapid, and at other times (when sea level is rising at the same rate as the land) there is no vertical movement and coral reefs can grow out rapidly from the coast. These explanations have been employed in the interpretation of many emerged-reef staircases from around the Pacific, including Fiji (Nunn 1996), Papua New Guinea (Bloom et al. 1974), Peru (Ota et al. 1995), Taiwan (Huang et al. 1997), and Vanuatu (Lecolle et al. 1990).

42. This swell is the South Pacific Superswell, first described by McNutt and Fisher (1987).

43. Duncan and McDougall (1976).

44. Spencer et al. (1987).

45. Rates for the Tuamotus are from Crough (1984) and for Taiwan from Chen and Liu (2000).

46. Taylor et al. (1980).

47. Uplift rates along the Huon Peninsula were calculated by Ota et al. (1993). Uplift in southern Taiwan was measured by Chen and Liu (2000).

48. Nunn (1995a, 1998a).

49. Davis (1913).

50. Jones (1995).

51. The theory was first published in Darwin's 1842 book entitled *Structure and Distribution of Coral Reefs.*

52. The history of coral-reef upgrowth and outgrowth can therefore be determined by examining in a drill core through a reef the types of corals at particular depths. A concentration of branching corals signifies a period of (rapid) upward reef growth, and the presence of slower-growing species indicates a time when the reef was not being submerged. A pioneering study from the Pacific was of Tarawa Atoll in Kiribati (Marshall and Jacobson 1985).

53. The vocabulary of reef islands has an international flavor. The word "atoll" comes from the Maldives, and the word *"motu"* is a pan-Pacific word. The word "atoll" is sometimes used to refer to the ring reef, sometimes to the islands on it, and sometimes to the entire complex.

54. Menard (1984, 1986) and Nunn (1994) gave accounts of island types. Note that the terms "guyot" and "seamount" are not consistently applied; the latter are most common and often applied to guyots.

55. Campbell (1984).

56. The main area of drowned atolls in the Pacific is west of the islands of Samoa. There are many more drowned atolls in the Indian Ocean per unit area, an observation that has intrigued many scientists (Stoddart 1971).

57. "Prodigious" was the descriptor used by Moore et al. (1989).

58. It is also due to groundwater extraction. In the period 1957–1961 recorded subsidence in Shanghai

was as high as 287 millimeters/year. Despite being reduced by artificial recharge of aquifers, subsidence rates there are still high (Wang 1998).

59. Plafker (1972).

60. This point is spelled out to avoid the kind of confusion that dogs nonscientific accounts of how and why sea level changes. One of the forerunners of modern new-age writing believed that when the continent of Lemuria sank (which it could not have done even had it ever existed) "it caused the water of the Pacific Ocean to recede on all coastlines . . . leaving them higher and allowing many islands to appear that had not been visible before" (Cervé 1931:94). This is of course the opposite of what would have happened!

61. Nunn (1999).

62. A general review was given by Grossman et al. (1998). One of the most detailed studies was of the Fiji Islands, where sea level has fallen a net 1.5–2.1 meters in the past 4,200 years (Nunn and Peltier 2001).

63. Dickinson (2003).

64. The formation of *motu* in the Pacific was described by Nunn (1994).

Chapter 3: Islands That Vanished Long Ago

1. Readers dissatisfied with this might pause to consider how much more imprecise is the term "small island." It has been recommended that this term, bandied about with little thought for its pejorative connotations, particularly by United Nations bodies, should be abandoned in favor of simply "islands." After all, one never hears talk of large islands as a supposed universally understood definition, so why should anyone talk of small islands? If one wants to refer to the smaller individuals in the global family of islands, then one should use the expression smaller islands (Nunn 2004a).

2. The classic study of the Louisville Ridge and its effects on the form of the Tonga-Kermadec Arc was by Dupont and Herzer (1985). More recent work confirmed their model by showing that Louisville Ridge melting contributed to the volcanic rocks erupted two to three million years ago on the northern Tonga island of Tafahi (Turner et al. 1997).

3. One of the first accounts to marry the marine and terrestrial geology of Vanuatu was that by Greene and Wong (1988). As more data have been gathered, so our understanding of the effects of collision between the D'Entrecasteaux Ridge and the Vanuatu Island Arc has improved (Calmant et al. 2003).

4. All this information is from Hill and Glasby (1996).

5. Coulbourn et al. (1989).

6. Yonekura (1983).

7. Corals have been present on earth since the Cambrian era, about 570 million years ago (Veron 2000).

8. The Mauna Loa Volcano on Hawai'i Island, sixty-six kilometers from Lō'ihi, is still active but near the end of its active life.

9. The information about Meiji Seamount comes from the synthesis by Duncan and Clague (1985), with the revised date of eighty-five million years ago for pillow basalts from its cap given in Regelous et al. (2003).

10. Keigwin et al. (1992).

11. Although such a reconstruction is a marvelous aid to scientific understanding, this may have added conviction to the popular belief that a huge continent had once existed in the middle of the Pacific and had then sunk. And in such ways can science be seen as often being its own enemy. For, par-

ticularly in the empirical sciences, the imperative of recording observations without the embellishment of unwarranted interpretation sometimes leaves those observations vulnerable to the interpretations of others who have far fewer scruples about objectivity and accuracy.

12. The Mid-Pacific Mountains remain a remarkable archive of Cretaceous-era landscape. On Allison Guyot, for example, the volcanic basement shows signs of having been eroded by rivers and waves. This was overlain by a reef capping, which shows signs of having been affected by large flank landslides (Winterer 1995), just like many modern limestone islands.

13. These evocative terms were used first by Neumann and Macintyre (1985). The term "give-up reefs" refers to those that did not succeed in either keeping up or catching up with rising postglacial sea level. A discussion of these types of reefs is given in Nunn (1994).

14. Fairbridge and Stewart (1960:100).

15. Ibid., 108.

16. The volume edited by Brocher (1985) is the most detailed to target this area, renamed the Northern Melanesian Borderland. The interpretation of Alexa Bank and the other underwater atolls having been drowned largely by rapid sea-level rise was tentatively advanced by Nunn (1994).

17. This explanation is essentially one of those (b on page 112) given by Fairbridge and Stewart (1960).

18. Brousse et al. (1978).

19. Data on CREs come from Blanchon and Shaw (1995). It must be stressed that "rapid" does not necessarily equate with "catastrophic." In most parts of the Pacific, the most rapid rate of sea-level rise within a CRE may have been 50 centimeters in one week. There is no suggestion, except in very restricted areas, that large wave fronts in the vanguard of higher sea level advanced across the ocean at such times.

20. Lincoln and Schlanger (1987).

21. After Darwin proposed his Subsidence Theory of Atoll Formation, it was Sir John Murray, who had traveled the Pacific during the voyage of HMS *Challenger* in 1872–1876, who championed the alternative theory outlined here: that atolls grew upward and outward from submerged platforms (Murray 1880). No examples of such atolls are known to me.

22. Keating (1998).

23. Keating (1987).

24. The Johnston Atoll Chemical Agent Destruction System (JACADS) was terminated at the end of 2003, and the island is currently under the jurisdiction of the United States Fish and Wildlife Service.

25. Interestingly it has been suggested that Johnston is an island in the process of becoming submerged and is therefore instructive in terms of guyot formation. Even when islands like Johnston become fully submerged, they will still be liable to large-scale flank collapse. Analogs are found elsewhere in the Pacific. Chapman Seamount (Line Islands, eastern Kiribati) and Horizon Guyot (Marshall Islands) were both affected by a major flank collapse after they had become submergent (Keating et al. 1991, Bergersen 1995).

26. For example, when novelist Robert Louis Stevenson first saw the Marquesas in 1888, he wrote that "the customary thrill of landfall [was] heightened by the strangeness of the shores that we were then approaching. Slowly they took shape in the attenuating darkness. Ua Huka, piling up to a truncated summit, appeared the first upon our starboard bow; almost abeam arose our destination, Nuku Hiva, whelmed in cloud; and between and to the southward, the first rays of the sun displayed the needles of Ua Pou. These pricked about the line of the horizon; like the pinnacles of some ornate and monstrous church, they stood there, in the sparkling brightness of the morning, the fit signboard of a world of wonders" (1900:3, with modern island names substituted).

27. Wolfe et al. (1994).

28. Holcomb and Searle (1991).

29. Menard (1983).

30. Mark and Moore (1987). Slopes as steep as 15°–20° are common between sea level and the ocean floor 5 kilometers below.

31. Ward (2001).

32. Radiometric dates are imprecise but suggest that the ʻĀlika Slide occurred more than 13,000 years ago but less than the time of formation of the Nīnole Basalt, a few hundred thousand years old (Lipman et al. 1988).

33. Moore et al. (1994b).

34. As with Oʻahu Island, the form of an island is generally a good guide as to whether or not it has been affected by a giant flank collapse, but it is not foolproof. The north coast of the island of Moʻorea in French Polynesia has a steep-sided arcuate scarp that resembles the headwall of a giant landslide and was once thought to be exactly that. More recent work has determined that this is not the case; the scarp is a result of caldera formation. The same conclusion applies to the island of Rapa, the southernmost in French Polynesia, whose form might otherwise be interpreted wrongly (Clouard et al. 2000).

35. Tracey et al. (1964).

36. Holcomb and Searle (1991:26).

37. Clouard et al (2001).

38. Clouard and Bonneville (2004).

39. Moore et al. (1989).

40. An account of Niue's geology and geological history was given by Nunn and Britton (2004). Niue is one of the world's smallest independent countries, with a permanent population of around 1,000 people. One day, I was waiting at the airport for my outbound flight to New Zealand when a man came and sat next to me. "Hello," he said, "I am the prime minister of Niue." I confess to being a bit dubious at first and acknowledged him somewhat circumspectly, but he was in fact the prime minister, and, having evidently heard of me, he proceeded then to tell me about a huge arcuate crack that appeared to be opening along the back of the capital, Alofi. He wondered whether the crack presaged a major flank collapse of that part of the island. I was unable to reassure him, although I pointed out that there was only a small chance that such a collapse would occur anytime soon.

41. A study of the shapes of ninety-nine atolls found that ninety-four had concave scallops that were most likely a result of flank landslides (Stoddart 1965).

42. Keating (1998).

43. Although reported by earlier writers, the case for the deposits on Lānaʻi being the product of a megatsunami was made by Moore and Moore (1984) and on Molokaʻi by Moore et al. (1994a).

44. Questions about the tsunamigenic nature of these deposits were raised by Felton et al. (2000). A study by Keating and Helsley (2002) of the Lānaʻi deposits found that they were in situ only below 190 meters, not above. Those authors suggested that the deposits were a result of island uplift, not a giant wave. This conclusion, seemingly settling the issue, has recently been questioned by Webster et al. (2007), whose research on submerged reef terraces off the island show that it has not been affected by significant uplift within the past million years or so, and that therefore the "giant-wave" hypothesis seems the most plausible.

45. Other examples include the imbricate deposits of large boulders on the Great Barrier Reef, off the coast of East Australia (Nott 1997).

46. The collapse of Fatu Huku was described by Filmer et al. (1994), and the tsunami deposits at Rangiroa by Talandier and Bourrouilh-Le-Jan (1988). None of these authors made the connection between

the two events, which was suggested by Keating and McGuire (2000) and on balance appears improbable, not least because island barriers exist in the Marquesas between Fatu Huku and Rangiroa.

47. Okal et al. (2002).

48. Ward and Day (2001:3397).

49. The Kuwae eruption was documented by Monzier et al. (1994), the Krakatau eruption by numerous authors including Winchester (2003). Note that Krakatau is actually just outside the Pacific Basin, as defined for this book.

50. The Kikai Caldera lies mostly submerged close to Yakushima and Tanegashima (islands) in the northernmost Ryukyus.

51. Machida and Arai (1983).

52. Ponomareva et al. (2004).

53. Coats (1952).

54. The Vatukoula gold deposit is described by Setterfield et al. (1991), that on Lihir by Corbett et al. (2001).

55. The formation of Myojin Knoll Caldera occurred long before many people were in the area and probably mostly below sea level. Key details are from Iizasa et al. (1999). This caldera has since spawned a shallow-water volcano that occasionally forms islands.

56. The case for the southeastern Polynesia biogeographic province was argued by Kingston et al. (2003).

57. Kingston et al. (2003). Notwithstanding that, Thor Heyerdahl's idea (1989) that Easter Islanders and their island's flora originated in South America, making Easter Island exceptional in the Pacific, appears untenable. Most of the native plants of Easter Island originated in Southeast Asia and reached the island via the islands of Polynesia. The *totora* (*Scirpus riparius*) reed living on Easter Island that Heyerdahl thought was a key South American species that must have been carried to the island since human settlement around AD 690 has in fact been growing there for around 30,000 years (Bahn and Flenley 1992).

58. The Hawaiian Drosophilinae were discussed by Beverley and Wilson (1985), the honeycreepers by Sibley and Ahlquist (1982).

59. Fossils of these snails are found in the Miocene rocks, now well below sea level, of Bikini, Enewetak, and Midway (Solem, 1976, Part I:118).

60. Solem (1976, Part II).

61. Beverley and Wilson (1985:4756).

62. A variety of possible scenarios for Pacific Island snail dispersal were discussed by Vagvolgyi (1975) including wind/storm dispersal and the attachment of snails to birds and large insects.

63. Thorne (1963).

64. Rehder (1980).

65. Pilger and Handschumacher (1981).

66. Newman and Foster (1983).

67. Rotondo et al. (1981).

68. Collette (1974) and Nelson (1978), respectively.

69. Zimmerman (1948).

70. This should not be taken to mean that no Hawaiian species have affinities with the continental rim. Most of the birds and bats are closely related to North American species, and the Hawaiian marine biota includes many species that apparently reached the islands from Japan (Rotondo et al. 1981).

71. The time of initial opening of the Tasman Sea is uncertain, but by eighty-two million years ago it was a significant ocean barrier to dispersal of most terrestrial biota (Waight et al. 1998).

72. Probably because the entire New Zealand platform was drowned by higher sea level during the early Tertiary, about fifty million years ago (Pole 2001). This interpretation has, however, been questioned by some biogeographers who regard the use of molecular clocks as problematic (Heads 2005).

73. The West Wind Drift (ocean current) that flows from East Australia to New Zealand has long been regarded as the major agent of plant dispersal between the two. For anyone doubting that the 1,800-kilometer-wide Tasman Sea is a barrier to the airborne distribution of birds and insects, there is ample documentation of them being blown across it in recent times (Michaux 1991).

74. This information comes from the synthesis by Sanmartín and Ronquist (2004). It should be noted that the differences in both plant and animal trans-Tasman dispersals were not significant in permutation tests.

75. Herzer et al. (1997).

76. These conclusions come from animal and plant cladogram analysis illustrated by Sanmartín and Ronquist (2004).

77. Cracraft (2001).

78. Mueller-Dombois and Fosberg (1998). The numbers involved are not trivial. The Hawaiian Islands have 956 native angiosperm species (Wagner et al. 1990).

79. The biogeography of *Metrosideros* was discussed by Wright et al. (2000). In Hawai'i the shrub *Metrosideros polymorpha* is called *ōhi'a lehua* and is used to make garlands of flowers. In New Zealand the tree *Metrosideros excelsa* is known as *pohutukawa,* and it is also prized as an ornamental, particularly in California where it has been introduced. Interestingly it has been suggested that the scarlet blossoms of the *pohutukawa* misled early visitors to New Zealand into thinking that the land was rich in red feathers, a high-status exchange item in Pacific prehistory, thereby adding to the land's reputation as a desirable place to settle (Finney 1992:48–55).

80. Wright et al. (2001).

81. The best synthesis is that of Whittaker and Fernández-Palacios (2007). Long-distance cross-ocean dispersal from continents to islands is not accepted by all biogeographers. A prominent critic of the use of molecular clocks in biogeography, Michael Heads, favors the alternative view that many Pacific Island species of plants like *Metrosideros* and animals like *Brachylophus* (the Fiji iguanas) are insular variants of allies along the continental rim, not necessarily deriving from there (Heads 2005).

82. Gillespie (2002).

83. Van Duzer (2004).

84. Cogger (1974).

85. Squires et al. (1991).

86. Martill et al. (1991).

87. For those who missed it, a couple who fall in love decide to part to see whether destiny brings them back together (it does).

88. For this information, I am grateful to Richard Boyle's analysis, accessed on September 15, 2005, at http://livingheritage.org/three_princes.htm.

89. Books like that of Kirch (2000) are compilations of the evidence for Pacific Islander exploration of the ocean, and those like that of Irwin (1992) explain the seafaring and navigational skills that were used in this extraordinary feat. The acceptance of Pacific Islander crossing of the entire ocean from west to east, proposed by geographers Gerard Ward and Muriel Brookfield (1992) on the basis of the presence of coconuts in Central America at the time Europeans arrived there in 1513, was regarded by eminent archaeologist Atholl Anderson as "probable" although he was referring to a crossing (some 3,800 kilometers) from perhaps Easter Island to southern South America (2003:76). The latter contact

event has recently been confirmed by DNA analysis of chicken bones found in coastal Chile (Storey et al. 2007).

90. The existence of emergent islands in this region is explicit in models for the origin of the Easter Island biota, which shows signs of having been in existence for some thirty-five million years, much longer than the two and a half million years that Easter Island itself has been emergent (Newman and Foster 1983).

91. Gibbons and Clunie (1986).

Chapter 4: Ancient Continents Hidden by Time

1. Such a turbidite is the 2,700-m-thick Zorritas Formation (Early Devonian to early Carboniferous in age) of northern Chile (Isaacson et al. 1985).

2. Examples are provided by the El Toco, Sierra del Tigre, and Las Tórtolas formations that were folded and truncated during the early part of the late Carboniferous (Bahlburg and Breitkreuz 1991).

3. The description comes from Burckhardt (1902).

4. An idea popularized by Garnier (1870).

5. Dalla Salda et al. (1992).

6. Key references are Bahlburg (1993) and Dalziel et al. (1994).

7. My friend Cliff Ollier is a major critic of this explanation of fold mountains, having written a book in 2000 with Colin Pain entitled *The Origin of Mountains,* which proposes that such mountain ranges are actually the products of block uplift.

8. Geologists commonly use the term "allochthonous" rather than exotic. The converse term, meaning local, is "autochthonous."

9. A detailed account of terranes around the Pacific Rim is the subject of the 1985 book edited by Howell.

10. Thiede et al. (1981). As an aside, the dating of the Pacifica breakup has been questioned by pseudoscience writers who deliberately misrepresent geological arguments in their efforts to demonstrate that it could have been rapidly submerged in the same way as Atlantis reputedly was (see Childress, 1988:241, for example).

11. The original association between the Manihiki Plateau and a triple plate junction was made by Winterer et al. (1974); its classification as a Large Igneous Province (LIP) was by Sandwell and Smith (1997).

12. Larson et al. (2002).

13. Gladczenko et al. (1997).

14. Phinney et al. (1999).

15. Hallam (1986).

16. For example, Melville (1966) showed a map of an elongate central Pacific continent (named Pacifica) that he believed existed at the end of the Jurassic.

17. These range from Darwin (1859) and Wallace (1892) to Carlquist (1974) and Whittaker and Fernández-Palacios (2007). A popular account is that of Quammen (1997).

18. Gregory (1930). This idea may have been bolstered by the supposed geologic evidence for such a continent presented by Garnier (1870).

19. For most of the history of biogeography it has been assumed that Gondwanan biotic groups such as marsupials and ratites were not able to cross ocean barriers, and that their current distribution is explainable largely by their locations within the Gondwana landmass after its breakup (vicariance

explanations). This works well for certain groups, including the early placental mammals (from which humans descended) (Eizirik et al. 2001). Molecular studies of other groups show that dispersal played a bigger role than previously envisaged. Examples include the ratites, whose fossil distribution is best explained by both vicariance and dispersal (Cooper et al. 2001), and various post-Gondwana–breakup genera like baobabs (Baum et al. 1998) and the New Zealand cicadas (Buckley et al. 2002), whose modern distribution is explained solely by long-distance cross-ocean dispersal.

20. Kristan-Tollmann (1986).

21. Nelson (1985).

22. The original inference was made by Leon Croizat (1958); a more recent account is that of Grehan (2001).

23. Csejtey et al. (1982).

24. de Laubenfels (1985).

25. The Vitiaz Arc became inactive about twenty-five to twenty million years ago (Yan and Kroenke 1993).

26. Cunningham and Anscombe (1985).

27. Yan and Kroenke (1993).

28. Kumar (2005).

29. Mueller-Dombois and Fosberg (1998).

30. Woolnough (1903).

31. Gill (1987).

32. Mueller-Dombois and Fosberg (1998).

Chapter 5: The Coming of Humans to the Pacific

1. Our species is *Homo sapiens,* of which the only extant subspecies is *Homo sapiens sapiens,* to which we all belong. Immediately before *Homo sapiens* came *Homo erectus;* both species appeared first in Africa. *Homo erectus* was the first hominid to spread significantly beyond Africa; remains have been found in Southeast Asia (as Java Man) and China (as Peking Man).

2. The earliest certain appearance of *Homo sapiens* in East Asia is represented by the 120,000-year-old Maba (Mapa) cranium, discovered near Guangdong (Wu and Poirier 1995).

3. The catastrophic Mount Toba eruption 71,000 years ago in Indonesia probably had a major effect on human populations in East Asia. The ensuing volcanic winter may have caused most humans on earth to die, leaving a total of perhaps just 15,000 surviving through this evolutionary bottleneck (Ambrose 1998).

4. The oldest parts of the earliest settlement sites near the Pacific coast in both East Asia and Southeast Asia contain little evidence of marine food consumption.

5. In contrast to earlier views that regarded the earliest *Homo sapiens* in Southeast Asia as predating those who settled along the coasts of East Asia, it now appears that these areas were actually occupied first and that Southeast Asia was, to judge from the available radiometric dates, colonized later from southern China (Bowdler 1997).

6. In what were once thought to be signs of cannibalism, the teeth marks on human bones of *Homo erectus* at the famous site of Zhoukoudien (China) are now known to have been made by this hyena, *Pachycrocuta brevirostris* (Boaz and Ciochon 2004).

7. Groube et al. (1986).

8. High levels of charcoal in sediments dated to as much as 128,000 years ago flooring Lake George near Canberra have been suggested as evidence of an early human presence in the area (Singh et al.

1981). Charcoal and pollen evidence from a sediment core through the ocean floor off the Queensland coast has been interpreted as signaling a human presence near the coast as much as 140,000 years ago (Kershaw et al. 1993). Although intriguing, neither suggestion has gained wide support, and few professional archaeologists today support dates earlier than 45,000 years ago for the first crossings from Sunda to Sahul (O'Connell and Allen 2004).

9. Birdsell (1977). This work, which has not been surpassed, identified two land/sea routes from Sunda to Sahul, both involving eight to seventeen separate ocean crossings, including at least one leg of more than 70 kilometers and at least three legs of 30 or more kilometers. The prevailing view is that, because of their similarities in age, these crossings were purposeful and comparatively rapid (O'Connell and Allen 2004).

10. This view is supported by the evidence that these people crossed not just one water gap but many, culminating in the colonization of New Ireland in Papua New Guinea and Buka in Solomon Islands about 30,000 years ago, a journey that involved a number of ocean crossings of as much as 65 kilometers (Bowdler 1997).

11. The difficulty with the "accidental crossing" scenario is that it would have been highly unlikely that enough people would have survived to sustain a colony in Sahul.

12. The earliest-known evidence for manipulation of Papua New Guinea rain forests to boost their food production was described from western New Britain Island (Pavlides and Gosden 1994).

13. Golson (1982).

14. Fredericksen (1997).

15. The Pleistocene began about 1.8 million years ago and ended around 12,000 years ago. The ensuing period of time is known as the Holocene, in which we are still living. The Pleistocene and Holocene are the two divisions of the Quaternary Period (see Figure 2.2).

16. The sites of the Diuktai Culture in Northeast Asia date from about 35,000 to 10,000 years ago —with a break in occupation at many during the coldest part of the last ice age (approximately 22,000–18,000 years ago)—and show that people in that area appear to have been subsisting largely on wild foods including large animals like mammoths and seals (Mochanov 1980). The idea that people spread along the western Pacific Rim and into the Americas on the basis of a maritime economy has been convincingly argued by Ackerman (1998).

17. One recent suggestion is that of Wyatt (2004). There has been, and remains still, much orthodox resistance to any suggestion that the first Americans did *not* reach the continents across the Bering Strait. Consider this comment of archaeologist E. James Dixon: "In the early 1980s I had published a popular article on the peopling of the Americas in which I had merely hinted that humans may have colonized the Americas via the Pacific. I was sharply and swiftly criticized by several of my colleagues. One senior associate suggested that I not pursue this further for fear of losing my credibility within the profession" (1993:129).

18. Diffusionist in this sense is used to mean an adherent of anthropological diffusionism, the idea that human cultures originated from a small number of culture centers (sometimes even just one) and then diffused across the world. Such diffusionism, although largely discarded by scientists today, is disingenuously peddled by many modern pseudoscientists as an explanation for many supposed "secrets" of ancient cultures. A good example of such charlatanry is Graham Hancock's *Fingerprints of the Gods* (1995).

19. Meggers et al. (1965).

20. McEwan and Dickson (1978).

21. The Austronesian language group has ten subgroups, of which nine are found only in Taiwan. The other is spoken from Madagascar in the western Indian Ocean to Easter Island in the eastern

Pacific and from Hawai'i in the North Pacific to New Zealand in the South Pacific. Such evidence is used to support an origin for Austronesian language speakers in Taiwan, although it is possible that, because some languages have died out, we do not have a sufficiently complete picture today to make that judgment (Blust 2004).

22. This view has been championed by Oppenheimer (1998) and has received equivocal support from mitochondrial DNA studies such as that of Underhill et al. (2001). The contribution of European genes to Polynesians, specifically associated with the men of the Spanish caravel *San Lesmes* that was wrecked in the Pacific on May 26, 1526, was advocated by Robert Langdon (1988).

23. Nelson (1990).

24. A general account was given by Nunn (1999). Area-specific studies suggest that much of this area (including Japan and Taiwan, which was attached to the Asian mainland because of the low sea levels prevailing at the time) was drier and consequently there were fewer forests at the time (Igarashi 1996).

25. Precise ages for the Younger Dryas event are difficult to obtain because the fluctuations in atmospheric radiocarbon at the time give large errors. The date range quoted is uncontroversial, given our current knowledge. The dates are given in years "before present," where present is the year AD 1950. In calendar years these dates for the Younger Dryas event are 9250–8250 BC.

26. See the general account by Nunn (1999), together with specific studies for East Asia (Jarvis 1993, Igarashi 1996).

27. Mannion (1999), for example.

28. The earliest date for rice cultivation in this region is about 8400 years BP (Higham and Lu 1998).

29. Pigs were probably among the earliest feral animals to be domesticated in lowland-coastal East Asia (Larson et al. 2005). Among the plant domesticates were rice, foxtail millet, beans, peas, bottle gourds, olive, and cucumber.

30. Blanchon and Shaw (1995).

31. Oppenheimer (1998).

32. This conclusion was reached by both Gibbons and Clunie (1986) and Thiel (1987). A more detailed scenario, similar to that just described, was developed by Meacham (1996) but was unknown to me at the time of my 2003 keynote address to the 20th Pacific Science Congress in Bangkok, entitled "Environmental Influences on the Dispersion of Humans throughout the Pacific Basin."

33. Sea nomads, or sea gypsies, feature in both the ethnographic record, particularly in Southeast Asia (Sopher 1977, Ivanoff 2005), and in prehistory in the Taiwan area (Chen Chung-Yu 2002). Their lifestyles involve living on boats most of the time, making landfall only occasionally (perhaps seasonally) when needed. The case for the first Pacific Islanders (Lapita people) being descended in some places from such sea nomads was put forward by Nunn (2007a).

34. Stephen Oppenheimer's (1998) magisterial book *Eden in the East* argued much the same thing as I have here, except with Southeast Asia and the drowned Sunda Shelf as the place from which people were driven out by rising postglacial sea level. Oppenheimer found ample evidence for the ancestors of not only the Lapita people originating in Southeast Asia (rather than Taiwan–southern China) but also those who went on to found the civilizations of India, Mesopotamia, Egypt, and the Mediterranean (Oppenheimer 2003).

35. Bellwood (1979, 1997).

36. Chazine (2005).

37. Best (1923:9).

38. The dates for incipient Lapita culture in the Bismarck Archipelago are 3600–3300 calendar years BP (1650–1350 BC) (Specht and Gosden 1997). An excellent account of the Lapita culture is that entitled *The Lapita Peoples* by Patrick Kirch (1997). Another book by Kirch, *On the Road of the Winds* (2000),

is more accessible to the nonspecialist and includes Lapita within an account of the prehistory of all Pacific Island peoples. The book by Geoff Irwin, *The Prehistoric Exploration and Colonisation of the Pacific* (1992), is also recommended for an account of why Lapita (and later) seafaring was so successful. For the specialist, the 2003 volume edited by Christophe Sand following a conference to celebrate the fiftieth anniversary of the first discovery of Lapita pottery contains some excellent statements on the state of our knowledge about various aspects of Lapita culture and the archipelagoes the Lapita people inhabited.

39. It was originally thought that the Lapita people sailed against the wind, actually at 45 degrees across it (Irwin 1992), but it has been suggested recently that eastward colonization of the Pacific Islands was accomplished during El Niño events, when the southeast trade winds were weak (Anderson et al. 2006).

40. I use the word "colonization" here to mean first sustained contact between people and an island.

41. Sharp (1963), for instance.

42. The region of the Pacific east of Tonga and Samoa is aceramic, in the sense that its earliest inhabitants did not use or manufacture pottery, in contrast to their counterparts on western tropical Pacific islands. The one exception comes from the Marquesas Islands in northeastern French Polynesia, where shards of pottery made using a quartzose sand as temper, not found in the region, are thought to have been made in Fiji (Dickinson and Shutler 1974).

43. The basic blitzkrieg model of avian extinction associated with island-to-island movements by early settlers was described by Steadman and Martin (2003), with a more detailed discussion of extinctions of birds and iguanas in Polynesia given by Steadman et al. (2002).

44. This date has since been revised to AD 1200 by Hunt and Lipo (2006).

45. Straight-line distance from Easter Island to South America is around 3,800 kilometers. From Rarotonga to the closest point of North Island, New Zealand, is around 2,800 kilometers. Like some other long-distance voyages of the first Pacific Islanders, it can be demonstrated that return voyaging took place between eastern Polynesia and New Zealand (Anderson and McFadgen 1990).

46. Although far from universal, of course, this position is exemplified by those who found evidence for a massive sunken continent in the Pacific in the otherwise inexplicable similarities of culture and language between the inhabitants of far-flung Pacific Islands (see chapter 7).

47. Interested readers might consult my recent book *Climate, Environment, and Society in the Pacific during the Last Millennium* (Nunn 2007b).

48. Such explanations have been advocated for the Pacific Islands by Kirch (1984) and are also implicit in the influential work *Easter Island, Earth Island* (Bahn and Flenley 1992), which argues that (rather than climate change) deforestation of Easter Island by humans was responsible for the profound societal crisis that affected it around AD 1300–1400. The manufacture of weapons of war on Easter Island began only around AD 1300 when the island's obsidian began to be mined for fashioning spearheads named *mata'a*.

49. More details are given in my recent book (Nunn 2007b) and an article in the journal *Human Ecology* (Nunn et al. 2007a).

50. Key references are Nunn (2000, 2007a) and Kumar et al. (2006). An estimate of the cooling during the AD 1300 Event from New Zealand is about 1.5°C (Wilson et al. 1979). The latest estimates of the sea-level fall are that it occurred in two stages: a fall of 75 centimeters occurred between AD 1270 and 1325, and a fall of 40 centimeters between AD 1455 and 1475. An opposing view, which regards last-millennium climate change and its effects on Pacific Island societies quite differently, is espoused by Allen (2006).

51. Good accounts of the effects of European contact with the Pacific Islands are those of Howe (1984) and Campbell (2003).

52. The winning entry was by Denicagilaba (1894), believed to be a pseudonym for Ilai Motonicocoka of Bau, who was colonial administrator Basil Thomson's secretary. The fictitious origins of the Kaunitoni myth were revealed by France (1966) and Geraghty (1977); at the other extreme, the myth was embellished to the point where it suggested that the ancestors of the Fijians who allegedly left Africa were one of the lost tribes of Israel (Zambucka 1978). It is a tragedy of modern Fiji that many Fijians believe such stories uncritically and inculcate their children accordingly.

53. This is not the only time that the status of "Polynesians" has been raised above that of other Pacific Islanders because of their skin color. For years the proselytizing activities of the Mormon Church (Church of Jesus Christ of Latter-Day Saints) in the Pacific never went farther west than Polynesian Tonga owing to the church's belief that the darker-skinned people of Melanesia to the west were sons of Ham and markedly inferior to the fairer-skinned sons of Shem. The situation has changed somewhat today.

54. Some such studies focus on material culture (Terrell 2004) and DNA (Matisoo-Smith and Robins 2004).

55. Williams (2001:82). In a similar vein, the excitable Childress (1988:143) approvingly quoted the eighteenth-century Dutch navigator Jacob Roggeveen as commenting that "human understanding was powerless to comprehend by what means they [Polynesians] could have been transported to the Pacific."

56. The eminent John Macmillan Brown, one-time chancellor of the University of New Zealand, was a firm believer in the white (Aryan) roots of Polynesians, supporting his ridiculous arguments by noting parallels in Maori myths with Viking voyages and traveling the Pacific Islands tracing Aryan explorations through megaliths (Brown 1924).

Chapter 6: Mythical Islands in Pacific Islander Traditions

1. This point is made explicitly by Geraghty, who noted that such "collectors and translators of oral traditions . . . did an admirable job of salvage ethnology under difficult circumstances" (1993:371).

2. Walter and Anderson (1995:478). See also Nunn (2004c) for an extended discussion of Niue myths and their origins.

3. There have been two compilations of myths about Māui (Westervelt 1910, Luomala 1949). I have analyzed Māui myths from throughout the Pacific Islands to identify probable geologic influences (Nunn 2003).

4. Griffis (1895:63–64), accessed through Project Gutenberg (www.gutenberg.org).

5. Tregear (1891:589) quoted the Marquesan aphorism *Oai tuto e tomi ia te Papanui Tinaku ma he tai toko e hetu, è?* (Who would have thought to bury the great earth in a roaring flood?).

6. Accounts of flank collapses in the Marquesas were given by Filmer et al. (1994). A discussion of the lack of coral reefs around the Marquesas is given by Brousse et al. (1978). Recent work by Hébert et al. (2001) has focused on the uncommon vulnerability of the Marquesas to long-range tsunamis.

7. Handy (1930:19–20).

8. Handy (1923:19).

9. Toko-eva may be a mistake for Toko'eva, a dialectal variant of *togareva.

10. The story by Titi-Ouoho is given in Christian (1895:188–189). The association of Toko-eva with Clarke's Reef is given in Christian (1910: 204).

11. Handy (1930:115).

12. Christian (1895).

13. Information from Christian (1910) and Luomala (1949).

14. Tonaeva and Toko-eva are dialectal variants of *togareva, so the chances are either that they refer to the same island or that traditions regarding two vanished islands have become interwoven.

15. See, for example, Hale (1846:128) and Williamson (1933, 1:309).

16. Quoted in English from Von den Steinen's collection by Williamson (1933, 2:36). A slightly different (and later) version of this myth states that dead souls leaped from the top of the cliff at Kiukiu onto a rock called Hi'ia, which then turned with them, propelling them into the sea (Handy 1930:250).

17. McNutt and Fisher (1987).

18. Beckwith (1940:75).

19. This interpretation is given by Buck (1938a:114). For another, also from Buck, involving Nuku-tere as Rarotonga (Cook Islands), see chapter 9.

20. Stimson (1937:39–41).

21. The name Tongareva is cognate with Toko'eva (Toko-eva) and Tonaeva in the Marquesas. Tonga can also mean "island in the south."

22. The reasoning behind this is that Penrhyn (Tongareva) Atoll is uncommonly isolated and unlikely to have been permanently settled before the Tuamotu island chain, where the name Tongareva occurs frequently in myths. In fact, it is possible that Tuamotuans colonized Penrhyn, naming it Tongareva because they believed it to be the vanished island of their myths.

23. The story was written in Tuamotuan by Paea a Avehe from Vahitahi and published in translation by Stimson (1937).

24. Stimson (1937).

25. Beckwith (1940:75).

26. Stimson (1937:34).

27. Montiton (1874:343).

28. Henry (1928:468).

29. From a Mangarevan text given by Ioane Mamatui, translated and abridged by Buck (1938b:311).

30. Nunn (2003).

31. The myth was reported by Montiton (1874:379), interpreted by Williamson (1933, 2:215). That Mangarevans imported the Pele tradition from Hawai'i was suggested by a tradition collected and interpreted by Young (1898).

32. The so-called Davis' Land was long believed to be an outpost of *Terra Australis* (see chapter 7). If the narrative of Lionel Wafer, who traveled with Davis, is to be believed, then it is likely that Davis may also have seen Easter Island, just some "12 leagues" (about 67 kilometers) from the sand cay called Davis' Land (see the start of chapter 8 for an extract from Wafer's narrative).

33. Brown (1924:41–42).

34. Childress (1988:290).

35. Maziére (1965).

36. Barthel (1978).

37. For example, Thomson stated that Hotu Matu'a had left his home on Marae-renga because of a battle with Oroi, who had been cuckolded by the brother of Hotu Matu'a (Thomson 1891:528–530). Routledge recorded that a man named Haumaka, also living on Marae-renga, had a dream of Easter Island and, upon waking, took six men and sailed off to find it (Routledge 1919:277–278). Métraux's informant recounted a story with elements of both these (1940:57–60).

38. Barthel (1978:9–10).

39. Scarr (2001:40).

40. Métraux (1940:55).

41. See, for example, Beckwith (1940:67) and Andersen (1969:48).

42. Fornander (1878, 2:9).

43. Ra'iatea is an island in the Society Islands Group of French Polynesia. The text and its interpretation comes from Teuira Henry's (1928) *Ancient Tahiti* and is based on material recorded by the Reverend J. M. Orsmond early in the nineteenth century.

44. This translation is by Henry (1928).

45. Porter (1823:93–94). Stories of *he fenua imi* (land seeking) from the Marquesas were related by Handy (1923:19–20).

46. Beckwith (1940:67).

47. Beckwith (1940:73) quoted Nauahi of Hilo as complaining that "the gods had hidden it" when he tried to relocate Pali-uli.

48. W. D. Alexander, quoted in Lyons (1893:162).

49. Beckwith (1940:11).

50. Rice (1923:46).

51. The accounts in this paragraph come from Beckwith (1940:68).

52. This is analogous to the situation claimed on submerged islands like Tolamp in Vanuatu and Vuniivilevu in Fiji (see chapter 9).

53. The full chant is given by Fornander (1878); the extract given here comes from Andersen's (1969) more recent translation.

54. The word *"kahiki"* etymologically means the east.

55. Andersen (1969).

56. Scarr (2001:39).

57. Newell (1895:233, 235).

58. The myth and the comment are both from Grimble (1972:87).

59. Matang is defined as such in the *Gilbertese-English Dictionary* (Sabatier 1971). Details about Matang are in Sir Arthur Grimble's (1989) *Tungaru Traditions,* his most detailed ethnographic work.

60. Grimble (1989:32).

61. If the etymology of *i nano* seems implausible, then consider that its second meaning given here is literally "where the sun goes into the depths."

62. Interisland contacts were maintained through traditional exchanges such as the *sawei* (Hunter-Anderson and Zan 1996).

63. Hunter-Anderson and Zan (1996).

64. The early human history of these islands is not as well known as that of the islands farther south; a recent synthesis is that of Rainbird (2004).

65. A point made forcefully by Dixon (1964:247), who castigated early European visitors to the region for their lack of curiosity about its inhabitants' traditions.

66. The existence of the Caroline Plate was first mooted by Weissel and Anderson (1978), but its boundaries are still "not clearly defined" (Lee 2002:64).

67. The myth about Fais comes from Flood et al. (2002:79–81). The geology and tectonics of the Fais region were outlined by Nedachi et al. (2001).

68. Ashby (1983:18–21).

69. The most recent information about the Sorol Trough comes from the work of Altis (1999).

70. Ashby (1983:10–12).

71. Ibid., 22–23.

72. Buck (1938a:187).

73. A Maori tradition quoted in Taylor (1870:109).

74. An early discussion was that of Smith (1904); a more recent one is by Kirch and Green (2001).

75. Andersen (1969:89–92).

76. Sorrenson (1979).

77. Tregear (1891:56).

78. Williamson (1933, 1:292–311).

79. A Maori aphorism is *Ehara i te mea poka hou mai: no Hawaiki mai ano* translated as *It is not a new thing done without proper cause: it has come to us all the way from Hawaiki* (Royal 2005).

80. Orbell (1985).

81. Gill (1876).

82. For example, Smith (1904).

83. Brown (1920:37).

84. Best (1923:12).

85. Clouard and Bonneville (2004).

86. Santos (2005).

87. Childress (1988), for example.

88. An idea discussed but not decided upon by Williamson (1933, 2:311).

89. Hale (1846:119–120).

90. The volcanic islands of southeastern Hawai'i are similar in size and form to Savai'i Island in Samoa. This explanation for the naming of Hawai'i was proposed by Paul Geraghty (personal communication, 2006).

91. Grey (1855), for example. Note that Ra'iatea is also called Rangiatea.

92. Best (1923:7) quoted an account that names several islands grouped as Hawaiki.

93. Smith (1904).

94. Santos (2005).

95. Heyerdahl (1952, 1968).

96. Williamson (1933), for example.

97. Anderson (1967:949–950).

98. Quoted in Langdon (1988:130). The visit was after the famous mutiny; Morrison was later acquitted of involvement. That Paroodtoo is Pulotu is beyond doubt: linguist Paul Geraghty explained that the first vowel being /a/ rather than /u/ is a result of "the English speaker's tendency to interpret all unstressed vowels as /a/" (1993:369).

99. In support of his idea that one of the Tongan myths mentioning Pulotu came from Papua New Guinea, Oppenheimer (1998:471) regards the two as synonymous. To a geographer, this seems less than plausible, given the numbers of landmasses in the region between the two, including many that existed in the past but have disappeared since people arrived in the region (see Figure 2.7, for example).

100. Geraghty (1993).

101. Ibid., 370.

102. An excellent account is given in Henry Stommel's (1984) book *Lost Islands*.

103. All the information about DeGreaves comes from Stommel (1984).

104. These two explanations of vigias were noted by Scarr (2001:11).

105. Van Duzer (2004).

Chapter 7: Mythical Continents of the Pacific

1. In the early 1950s a group of journalists was asked to rank the ten most important news stories they could imagine. The reemergence of Atlantis came fourth, five places ahead of the Second Coming of Christ (de Camp 1970:2–3).

2. A good example is the channeled description of Lemuria by Lazaris, who explained that "The continent itself is surrounded by the various beaches, and in the very north and very south also by jagged rock" (quoted by Williams [2001:10]). Such remarks make sense when applied to a smaller island but are nonsensical descriptors of something the size of North America or Australia, for example.

3. One view is that of the Baron de Montesquieu, who wrote in 1748 that "the inhabitants of oceanic islands have a higher relish for liberty than those of the continent. Islands are commonly of small extent; one part of the people cannot be so easily employed to oppress the other; the sea separates them from great empires" (1949, 18:v).

4. The idea that ancient Egyptians plied the waters of the Pacific Ocean thousands of years ago and built megalithic "cities" like Nan Madol is found in various books by David Childress and Barry Fell and others. The idea is manifest nonsense, no less.

5. Something proposed by Brown (1924): a ludicrous and misleading suggestion, as explained later in this chapter.

6. Among these things were root crops like taro and yam, coconuts, and a variety of commensal animals including chicken, dog, and rat. It has been suggested recently that rats were carried deliberately as a food item rather than being stowaways on Lapita watercraft (Matisoo-Smith and Robins 2004). The question of whether the Lapita people introduced pigs to Pacific Islands, or whether they arrived with later settlers, is unresolved. Chunks of obsidian were carried for their value in producing sharp-edged flakes, but it is so rare in Fiji that the piece shown in Figure 5.3A was probably kept for totemic rather than practical purposes.

7. Rather against the grain of this discussion, in an article that my wife thinks I shall come to regret writing, I have speculated that the earliest Lapita people were in fact sea nomads who lived on the ocean, making landfall only when necessary (Nunn 2007a). As ocean wanderers, they may have discovered many Pacific Islands much earlier than the scientific evidence suggests but had no particular inclination to settle them. It may have been the transformation of the coastal environments of Pacific Islands during the past 3,000 years or so, as sea level fell, that made them more attractive to such people and led to their colonization (Dickinson 2003).

8. In the company of his dog. An account of Balboa's expedition is given by Romoli (1953).

9. The evidence for culture contact between Polynesia and Chile was assembled by Ramírez (1990/1991).

10. The research was reported by Storey et al. (2007).

11. Klar and Jones (2005).

12. There have been claims to the contrary. One story says "that Kahoupo-o-Kane was the original continent that once connected all the Pacific Island groups, then came a great flood—Kaiaka-hina-alii [the sea that made all the chiefs go down]—that flooded the lowlands and left the mountain peaks" (quoted by Williams 2001:102). If this is a genuine expression of an indigenous oral tradition, and I am unconvinced of this because I have not been able to find the original source (neither name is mentioned in Beckwith's [1940] authoritative compilation, for example), then I consider it an indigenous version of a story introduced to the Pacific Islands by Europeans.

13. They almost certainly did not know the world was round. In the words of Ramsay, "they were aware of the reasons for supposing it to be so—including the nonscientific supposition that the sphere was the most perfect geometrical form" (1972:22).

14. A comparable pseudoscience observation, interesting only for its astonishing naiveté, is that there is an apparent imbalance between the Eastern and Western (Pacific) hemispheres (Williams 2001:70). There is not. What Williams labeled "a problem for geologists" is no such thing, simply rhetoric to stir the interest of his readers.

15. This and many other of my insights regarding issues of early European contact with the Pacific come from the work of Robert Langdon (1988:53).

16. Including Coatu, Qüen, and Acabana (Langdon 1988:53).

17. Quoted in Langdon (1988:58).

18. One of the forebears of current pseudoscience interest in lost continents in the Pacific, John Macmillan Brown, visited Easter Island and concluded in 1924 that it was not the land seen by Davis in 1687. Brown therefore reasoned that, between then and Roggeveen's traverse of the area in 1722, "considerable tracts of land in the south-east Pacific have gone down" (1924:44). This is not true but, especially because it was written by a renowned scholar, has bolstered recent speculations about lost continents in the Pacific (see, for example, Childress [1988:286]).

19. All references to Cook's voyages come from Cook (1968).

20. Dalrymple (1996). Dalrymple was actually earmarked to lead the expedition to search for *Terra Australis,* but, because the British Admiralty would not agree to his demand that he also captain the *Endeavour,* Dalrymple withdrew and was replaced by Cook.

21. Cited in Beaglehole (1967:279).

22. For anyone who assumes that such racially motivated ideas are long dead, some of the new-age and pseudoscience writing about lost continents in the Pacific with their invariably fair-skinned inhabitants will come as a shock. In his book *In Search of Lemuria,* Williams (2001:103) told his unsuspecting readers that pre-European migration from Melanesia to Hawai'i, a well-established scientific fact, is unlikely to have happened because "Melanesians were never good enough seafarers for such a challenging voyage," a repugnant echo of Spence's (1933) view.

23. This has not stopped modified versions of it appearing, one being that of Sharp (1963), who, although admitting the fact of long-distance cross-ocean voyaging by the first Pacific Islanders, averred that those voyages were accidental rather than intentional. This view is now recognized as inconsistent with the body of evidence suggesting purposeful voyaging (Kirch 2000).

24. Accounts of allegedly sunken continents in Pacific Island myths such as Kalu'a in Hawai'i given by Berlitz (1984) appear to be invented. There is no mention of such a place in Hawaiian dictionaries or in syntheses of Hawaiian mythology such as that of Beckwith (1940). A similar conclusion applies to the continent called Hiva that Childress (1988) asserted to be a fundamental part of Easter Island traditions, which oddly does not merit an entry in the authoritative work *Pacific Mythology* (Knappert 1992) or in the classic memoir about the island's traditions by Métraux (1940). Hiva (or Maori) is mentioned in the memoir of Barthel (1978) as a homeland of unspecified size; the traditions collected by Barthel are not regarded as authentic (pre-European contact) traditions (see chapter 6). It is likely that Berlitz and Childress claimed that Pacific Islanders had legends about former Pacific continents to bolster the credibility of their own outlandish theories.

25. Williams (2001:71).

26. Robertson (1948:135).

27. Howe (1999:317).

28. Dumont d'Urville (1832).

29. Moerenhout (1837).

30. The analogy with Europe and South America was originally drawn by Quatrefages (1864) in his criticism of sunken-continent explanations for the origins of Pacific Islanders.

31. Garnier (1870).

32. Dana quoted in Andersen (1969:24).

33. Jacolliot (1879:308).

34. Ibid.

35. For a thoughtful review of this question, see Douglas (1999). It still seems self-evident to proselytizing groups like the Mormons, who insist, contrary to every shred of pertinent scientific evidence, that "Polynesia" was colonized by fairer-skinned people from the eastern Pacific Rim rather than darker-skinned people from the west.

36. The quotes in this paragraph all come from Jacolliot (1879:308).

37. A fine account of Blavatsky and her legacy is given by Washington (1993).

38. Passages like this, from Blavatsky's (1888, 2:224) *The Secret Doctrine* are representative of her combative antiscience style that many modern pseudoscience writers emulate.

39. Garnier was a mining engineer whose legacy is largely the exploitation of the nickel and chromium resources of New Caledonia. His 1870 book, containing abundant geological detail, most likely convinced nongeologists like Brown that the idea of a sunken Pacific continent was accepted by most geologists at the time. Ironically, Garnier's work may also have led geologists like Burckhardt (1902) to their correct belief in the Paleozoic continent in the Southeast Pacific, discussed in chapter 4.

40. Darwin (1842) talked about atolls as being a result of subsiding islands and showed maps of the Pacific indicating regions of subsidence. But it is unclear whether Darwin conceived these as regions that had subsided or as regions in which there were many island edifices showing signs of subsidence.

41. Brown (1924:52).

42. Ibid.

43. Churchward (1931:15).

44. At the outset, I need to declare that, like most empirical scientists, I cannot believe in such received wisdom until such a time as its veracity is independently and conclusively demonstrated. But then, when I am asked "is everything known," my reply is of course negative, and I have to admit that (as in the past) things that today appear undeniably false may one day be shown to be true.

45. A compilation of Cayce's statements about Atlantis is found in Cayce (1968).

46. Williams (2001:13).

47. Blavatsky (1888) and Steiner (1911).

48. Newbrough (1932); this is a later edition of a book first published in the 1880s.

49. Spence (1926).

50. Ibid., 2.

51. In many ways symbols are like oral traditions. Even if the symbol has a meaning within it, then how can anyone not already aware of that meaning know which parts of the symbol have meaning and which are artistic embellishments?

52. Williams (2001:12).

53. Witness the splendid debunking of Blavatsky and the theosophists by Washington (1993).

54. Supposedly the Third Root Race to inhabit Lemuria (Blavatsky 1888).

55. Blavatsky (1888).

56. Winchester (2003).

57. Jacolliot (1879).

58. Spence (1933).

59. The 1931 book *Lemuria, The Lost Continent of the Pacific* was written by Wishar S. Cervé, a partial anagram of the author's real name, H. Spencer Lewis.

60. Cervé (1931).

61. Irwin (1981).

62. One of the most successful contemporary channelers of information about Lemuria is J. Z. Knight, who claims to act as the channel for Ramtha, a former occupant of the continent. Located in the Pacific, Lemuria had apparently been home to a great human civilization since the time of the

dinosaurs. For any objective educated person, this detail should be enough to dismiss such accounts: the dinosaurs became extinct about sixty-six million years ago, and the earliest remains of human ancestors (hominids) date from around four to five million years ago.

63. In an imaginative interpretation of two Mayan symbols in the Troano Codex as the letters M and U, Brasseur (1869) decided that these spelled the name of the land submerged in the volcanic catastrophe that he believed was described in the Troano; we now know that the Troano was about astrology. The name Mu was first used in 1886 by Le Plongeon as a synonym for Atlantis, which he believed to have been located in the Atlantic. Following his efforts excavating Mayan ruins in Yucatán, Le Plongeon created a story around Queen Móo of Mu (or Atlantis), who fled to Egypt as the continent began sinking.

64. Churchward (1931:15).

65. Kuhn (1996).

66. Washington (1993).

67. Blavatsky (1888) had uses for Lemuria (Indian Ocean), Mu (Pacific), and Atlantis (Atlantic Ocean), sometimes citing just one of them, usually Lemuria, as the home of a particular "root race," sometimes all.

68. Boas was one of the founders of diffusionism, the idea that human cultural traits found across the entire planet had diffused from a single place rather than developed independently in more than one place (see Boas [1929], for example).

69. Brown thought that the megaliths on Malden had been parts of once more-elaborate structures (great temple pyramids) visited by pilgrims from nearby islands that subsequently sank. Archaeologist Kenneth Emory's investigations of Malden's megaliths (as part of the 1924 Whippoorwill Expedition) showed that they are the remains of *marae* (temple platforms), houses, and graves — nothing like Brown imagined — and point to the former occupation of the island by a few hundred people for probably several generations before European contact (Emory 1934).

70. The quotation is from the 1962 book *The Ultimate Frontier* by Eklal Kueshana, referred to by Childress (1988:195–196), who echoed its idiotic conclusion that the roads "disappear into the sea" rather than having been built from the sea inland. Naturally, such inverted logic has been taken as supporting the idea that Malden is part of a huge, largely submerged continent, rather than an island from which people built "roads" from their landing places inland across the rough peripheral terrain to their habitation, gardening, and temple sites.

71. Di Piazza and Pearthree (2004:58).

72. A good discussion based on original field research is in Di Piazza and Pearthree (2004).

73. This discussion is largely from the pioneering research of Anne Di Piazza and Erik Pearthree on the mystery islands of eastern Kiribati and the surrounding area (2001a, 2001b). They argue that Malden would have been an important stopover on voyages between large Kiritimati (Christmas Island, eastern Kiribati) and Tahiti (French Polynesia), for example, during the pre-European history of the Pacific.

74. See Berlitz (1984), for example.

75. Heyerdahl (1952, 1968).

76. The earliest date for artificial-island construction obtained from Nan Madol is about AD 500. The high point of the site's history, when it was most fully developed, was about AD 1150. It fell into disrepair and was abandoned in the early part of the seventeenth century (Ayres 1990).

77. The ideas quoted come from Childress (1998). The idea that Nan Madol was built by Greek sailors more than 2,000 years ago was first proposed by Ballinger (1978).

78. As an extinct volcanic island, Pohnpei has probably been subsiding for thousands of years (Nunn

1994); the figure of 5.5 meters of subsidence in the past 6,000 years quoted by Spengler et al. (1992) would certainly put Pohnpei's subsidence rate at the faster end of the spectrum. The sea level has been rising overall throughout the Pacific since about AD 1800 (Nunn 2001a).

79. The only scientific report of the Yonaguni structures that I have been able to find is Kimura (2004). His age calculations are somewhat obscure, but the date of 10,000 years seems to come from the approximate time that postglacial sea-level rise would have begun to cover the Yonaguni Underwater Pyramid. Yet Kimura admits that the Yonaguni Underwater Pyramid possibly "slipped seawards with the help of gravity" (2004:948), in which case its alleged human modification might date from much more recent times. In support of his interpretation of the Yonaguni Underwater Pyramid as extensively human-modified, Kimura asserted that a nearby underwater "statue" looks like an Easter Island *moai* and is "definitely a man-made construction" (2004:950). This kind of rash judgment plays into the hands of the pseudoscience writers, who laud Kimura as a hero. Sadly, Kimura is mistaken.

80. Robert Schoch, personal communication, December 24, 2005.

81. Slobodin (1978).

82. Childress (1998).

83. Zambucka (1978).

84. Not included in this number is the idea of Perry (1923) that North America was colonized by the Mayas who were themselves the product of diffusion from Egypt.

85. Donnelly (1882).

86. A good example of such diffusionism was given by Braghine (1938), who found that Mexican hieroglyphics are "strikingly similar" to those of ancient Egypt and to some on ancient Chinese monuments, from which he inferred that the Mexicans, ancient Egyptians, and Chinese all came from the same root race — on Atlantis. Similar arguments are used by modern writers. The obvious retort to Braghine (and company) is that why, given the limited number of combinations of dots and lines available, would he not expect ancient symbol-making people in different parts of the world at different times to come up with symbols that appear superficially the same but that had different functions/meanings?

87. The largest part of Antarctica (East Antarctica) appears to have been completely and continuously covered with ice since at least twelve million years ago, long before the appearance of the first humans on earth; the smaller part (West Antarctica) may have temporarily lost its ice cover during the Last Interglacial period, approximately 130,000–120,000 years ago (Nunn 1999).

88. Hancock (1995:501).

89. For example, see Nunn (1999), and references therein. One of the most comprehensive summaries of the geologic history of Antarctica is that by Tingey (1991).

90. See Hearty (2002) for possible evidence from Hawai'i of Antarctic ice surges.

Chapter 8: Vanishing Islands: Processes of Island Disappearance Witnessed by Humans

1. The idea of islands as a universal archetype is one that is central to two marvelous expositions of the theme, John Fowles' *Islands* (1978) and Bill Holm's *Eccentric Islands* (2000).

2. This account is in the words of Lionel Wafer, who was with Davis on his ship *The Batchelor's Delight* when the island was discovered (Wafer 1934). The text quoted comes from Thomson (1891:447).

3. Ramsay (1972:103). Some writers claim Davis' Land was the last part of Lemuria to sink beneath the ocean surface.

4. Part of my 2007 book *Climate, Environment, and Society in the Pacific during the Last Millennium*

focuses on one such event, the AD 1300 Event (Nunn 2007b). The argument is that rapid sea-level fall of perhaps 80 centimeters decimated coastal food resources in the Pacific Islands (and elsewhere) to such an extent that conflict followed, and people abandoned exposed coastal sites in favor of upland, inland fortified sites.

5. Oppenheimer (1998:277).

6. Appreciate that the Australian islands were typically settled long before those in the Pacific Ocean, at times when sea level was much lower than today.

7. See Lampert (1981) and Draper (1988).

8. See Jones (1977) and Sim (1994).

9. Obtained by Gudgeon (1905:109).

10. Huggett (1989).

11. Oppenheimer (1998:279–286).

12. Oppenheimer (1998).

13. Nunn (1999, 2007b).

14. The distinctions between cays and *motu* are discussed in Nunn (1994).

15. Examples are noted in unpublished reports by the South Pacific Applied Geoscience Commission (SOPAC).

16. Dudley and Lee (1988).

17. Bird (1905:122–124).

18. The effects of the 1975 earthquake were described by Lipman et al. (1985); the likelihood of future collapse was predicted in Kerr (1994).

19. The original research into the November 2000 slide was reported by Cervelli et al. (2002), who avoided deriving any prediction of future collapse from this. Less restrained is the view of Ward (2002), who argued that, although such monitoring is the key to predicting future collapses of oceanic-island volcanoes like Kīlauea, this event should make hazard managers in the Pacific sit up and take account of the potential impact of a collapse of the entire flank, which Ward calculated could produce waves 30 meters high along the western seaboard of North America.

20. Morgan et al. (2003:20).

21. Details in this section about Oshima-Oshima come from Satake and Kato (2001) and those about Ritter Island from Ward and Day (2003).

22. Siebert (1992).

23. Filmer et al. (1994).

24. Keating and McGuire (2000).

25. Gragg (2000).

26. Carracedo et al. (1999).

27. Ward and Day (2001).

28. For example, submarine cables were cut in 1953 off the Samoa Islands as a result of submarine landsliding (Soloviev and Go 1984).

29. The devastating 2004 Indian Ocean tsunami originated here.

30. This story comes from Oppenheimer (1998:276), who quoted unpublished work by Aone van Engelenhoven.

31. St. Johnston (1921:67–68).

32. Ellis (1832, 1:332).

33. Clouard et al. (2001).

34. Crealock (1955:163).

35. Englert (1972:45).

36. Paton (1890:157).

37. Scarr (2001:39).

38. Todorovska et al. (2002).

39. Hayashi and Self (1992).

40. Walker and McCreery (1988).

41. Oppenheimer (1998:280–281). Oppenheimer regards most flood myths from the Pacific Islands farther east as derived from those of the western Pacific Rim rather than being autochthonous (283–285). Most myths from Taiwan (Formosa) reflect the island's coseismic uplift rather than the less-common subsidence; one account, for example, talks of sea-dwelling dragons that became playful and "heaved up a regular series of hills and mountains" (MacLeod 1923:1).

42. Beaglehole and Beaglehole (1938:386). Those authors refer to the wave as "seismic" even though its arrival was associated with a stormy night in the account they quote.

43. The quotations come from Beaglehole and Beaglehole (1938:386). Pukapuka has a land area of about 7 square kilometers and a current population of about 780.

44. Graham (1939).

45. Gudgeon quoted by Best (1923:11).

46. For the effects of large waves on coastal settlements in New Zealand, and their possible effects on settlement pattern, see Goff and McFadgen (2001). For an account of the Huahine village buried by sand driven onshore by a tsunami around AD 850, see Sinoto (1979).

47. Goff et al. (2003).

48. A review of these types of jack-in-the-box volcanoes was given by Nunn (1994).

49. Nunn (2003).

50. Oppenheimer (1998).

51. It is also memorable for the loss of the Japanese Hydrographic Department's vessel *No. 5 Kaiyo-Maru* and its crew sent to investigate Myojin-Sho that had the misfortune to be caught up in an unanticipated underwater eruption.

52. Fiske et al. (1991).

53. Oshima (2002).

54. Kato and Nishizawa (2002).

55. Monzier et al. (1994).

56. Nunn (1998b). Fonuafo'ou means "new land" in Tongan.

57. Information from the Smithsonian Institution's Global Volcanism Program accessed on November 20, 2005, at http://www.volcano.si.edu/.

58. The Reverend Baker arrived in Tonga as a Wesleyan missionary but became a close adviser to King Tupou I and eventually the country's premier.

59. Baker (1886:41–43).

60. Stair (1896:35).

61. Luomala (1949:222).

62. Ibid., 46–47.

63. Fison (1984:144–145).

64. Martin (1817:164–165).

65. Taylor (1995).

66. For an account of the relevant traditions of Ni-Vanuatu (the people of Vanuatu), see Galipaud (2002).

67. Winchester (2003).

68. The formation of the Santorini (Thera) Caldera was explained by Druitt and Francaviglia (1992).

Chapter 9: Recently Vanished Islands in the Pacific

1. Most information about Yomba comes from Mennis (1981).

2. Interview with Madmai of Mitibog on Kranket Island, quoted in Mennis (1981:96).

3. Interview with Daimon of Yabob on Kranket Island, quoted in Mennis (1981:98).

4. Interview with Majar of Bilibil on Kranket Island, quoted in Mennis (1981:96).

5. The suggestion that the Olgaboli Tephra was the product of the eruption that destroyed Yomba was made by Blong (1982:199); the dates are from Russell Blong (personal communication, 2001).

6. In all this, it could be noted that genealogies are often unreliable in Melanesia compared with Polynesia, where the need to track hereditary titles ensured more accurate record keeping.

7. For example, A. B. Lewis, whose work was summarized by Welsch (1998).

8. The extracts from the narrative of Neuhauss are quoted in Churchill (1916:12–13).

9. Phinney et al. (2004).

10. In oral traditions from Ulawa Island, the island is consistently named Teonimenu. In other traditions it is referred to as Teo.

11. Fox (1925:169–171).

12. Mead (1973:219).

13. Ibid.

14. The sighting was recorded by Surville, quoted by Ivens (1927:152).

15. Ivens (1927:163).

16. Kellington Simeon and Lysa Wini collected these traditions in 2004 as part of my research project; they have since been published (Nunn et al. 2006b).

17. Informants claimed that the islands disappeared many years before 1925, perhaps even before Alvaro de Mendaña came to Solomon Islands, the first European to do so, in 1568.

18. I am especially grateful to Chief Francis Sao and the elders of the Aiga Tatari clan of Santa Catalina for this story.

19. Tony Heorake collected this information from Ulawa in 2004.

20. In an earlier collection of traditions from Ulawa, the submerged land (Hanua Asi) is said to have comprised "Teonimenu, Ato Usuna, [and] Ere ni keula" (Ivens 1927:478).

21. Cook never reached Solomon Islands but passed through northern Vanuatu, to the southeast of this area, on his second voyage (1772–1775).

22. Ivens (1927:333).

23. Usuramo (2004:41). I am grateful to Dr. Paul Geraghty for alerting me to Usuramo's work.

24. These traditions were collected in 2004 by Esther Tegu for my research project; they have since been published (Nunn et al. 2006b).

25. The forest of tree trunks was described in December 2004 by Philip Hou'uwa and Andrew Suraru to Esther Tegu, who observed the submerged reef platforms for herself.

26. Grover (1960:30).

27. Usuramo (2004:41).

28. Phinney et al. (2004:239).

29. According to informant Phillip Masura'a, there were two islands called Taroapuala and Ta'arumahui. According to three other interviewees, there were three islands, named Ta'alumanule, Ta'aluto'o,

and Ahiramarama. The 2004 research by Bronwyn Oloni was carried out as part of my research project and has since been published (Nunn et al. 2006b).

30. Assuming a motor at the lower end of the range, this distance might be 40–50 kilometers, although obviously such conclusions are fraught with possible error.

31. Ivens (1927).

32. Information collected by Kellington Simeon and Lysa Wini in 2004 for my research project.

33. Galipaud (2002).

34. Gao et al. (2006).

35. Michelsen (1893:13–16).

36. Paraphrased from Clark's (1996) rendering of the versions of this myth collected by Jean Guiart in the 1950s (Guiart 1973).

37. Hébert (1963–1965: 90–91).

38. Clark (1996).

39. The connections between the Kuwae eruption and the siege of Constantinople were made originally by Pang (1993).

40. Briffa et al. (1998).

41. Gao et al. (2006).

42. A technical account of the origin of the Vanuatu islands is given by Greene et al. (1988). The uplift rates quoted come from the study by Jouannic et al. (1980).

43. Morris Harrison collected these traditions for me; results have been published recently (Nunn et al. 2006a).

44. Such details are likely to be apocryphal — a way of remembering the way the islands once were.

45. Benoit and Dubois (1971).

46. The details in this paragraph are from Dalsie Bani and her family, translated by John Lynch in 1998. The story was published in Nunn (2001b).

47. Quoted by Stommel (1984:60).

48. Mary Baniala collected these traditions for me; results have been published recently (Nunn et al. 2006a).

49. This event occurred about 100 years ago according to Warden (1970).

50. St. Johnston (1918:30).

51. See discussion in Nunn and Omura (1999).

52. This work has now been published (Nunn et al. 2005).

53. Geraghty (1993:347), although see his note 5.

54. A description, published in 1880, is representative. It portrays Bulotu as "well-stocked with all kinds of useful and ornamental plants, and the whole atmosphere was redolent with the scent of flowers. Birds of gorgeous plumage carolled ceaselessly . . . and of every variety of food there was an inexhaustible supply; for as soon as a hog was killed, another one immediately took his place" (Cooper 1880:154).

55. Geraghty (1993).

56. Ibid.

57. The only account of the geology of Matuku is that by Coulson (1976). Research on recent uplift and subsidence in the area was reported by Nunn (1995b).

58. Geraghty (1993:362).

59. Examples of accounts of Burotu that are almost certainly autochthonous include those in Hale (1846) and Waterhouse (1866).

60. Geraghty (1993:362).

61. Nunn et al. (2005).

62. The figure for Hawai'i comes from drowned coral reefs along the sides of the Hawaiian Ridge (Jones 1995), and that for Johnston Atoll comes from Keating (1987).

63. Sources of data are given in Nunn et al. (2005).

64. Hocart (1929:195).

65. Informants told of red flowers from Burotu being washed up on Matuku's beaches, together with red reeds (*gasau* in Fijian, the reed *Miscanthus floridulus*), red *kasari ni niu* (coconut exocarp), and a red-tailed bird named *lawedua* (*Phaethon rubricauda*) (Nunn et al. 2005).

66. Fison (1907:16).

67. Geraghty (1993:348).

68. Benson (1983:40–41). This tale, intended for schoolchildren, may be based on the white pig story.

69. Geraghty (1993:350).

70. Ibid., 360.

71. Mokunitulevu Na Rai (undated: 12).

72. Nunn et al. (2005).

73. The description of the 1953 tsunami appearing in the Suva reef passage as "a large brown bubble" was reported by Houtz (1962).

74. Morvan (1980:104–105).

75. The automatic reaction of a western-trained scientist such as I is to discount such traditions, but I have learned to suspend judgment. One fine afternoon in 2001, two boats belonging to the University of the South Pacific were transporting people from Moturiki Island across the site of Vuniivilevu back to Suva, a three-hour trip. I was in the first boat, in which at the appropriate place everyone was silent and removed their hats. Among the passengers in the second boat was a rowdy who insisted on singing loudly as the boat crossed the site of Vuniivilevu. The engine of that well-maintained boat stopped, and the engineer on board could not start it again. The boat drifted for six hours before coming ashore on a nearby island where they radioed for another engine and eventually limped back to Suva. Coincidence, you say. Of course it was.

76. This project was directed by me. The results summarized here come from Nunn et al. (2005).

77. In Fiji the *ivi* tree is the Tahitian chestnut, *Inocarpus fagiferus*.

78. The whale's tooth (*tabua*) is the highest form of offering in Fijian ceremonies. This particular anecdote comes from an unpublished 1996 manuscript by J. Vulava entitled *Ko Vuniivilevu se ko Davetalevu* in the possession of Dr. Paul Geraghty.

79. One prominent diarist who failed to mention Vuniivilevu or any other vanished island in the area that he repeatedly traversed in 1875–1877 was the ebullient Baron von Hugel (Roth and Hooper 1990).

80. Information about Bikeman and Tebua obtained from personal sources and from Howorth (2000) and Moore (2002).

81. Observations about Tebua come from Moore (2002).

82. The average rate of twentieth-century sea-level rise in the Pacific has been 1.0–1.5 millimeters/year (Nunn 2001a).

83. This research was carried out as part of my research project in this region.

84. Another name given for this island, Bikenimakin, was suggested by informants to imply that it was used as a beacon, an ocean landmark, by sailors in these waters. Such sea marks (*betia n borau*) have been in common use in Kiribati for generations. More realistically, it is clear that the word *"bike"* means simply sandbank in Kiribati, so that *Te Bike* is translated as "the sandbank" and *Bikenimakin* as "the sandbank of Makin [Island]" (Paul Geraghty, personal communication, 2006).

85. Grimble (1972).

86. Interview at Antekana on Butaritari, December 12, 2004, by Teena Kum Kee and Vicky Claude.

87. Sea-level rise generally leads to erosion along sandy shorelines (Mimura and Nobuoka 1995).

88. Some of the best accounts of this in English are Grimble's books (1956, 1972).

89. Interview at Matairawa on Makin, December 18, 2004, by Teena Kum Kee and Vicky Claude.

90. Strippel (1972).

91. Grimble (1972). This tradition may owe something to that of Mone Island, which was described in chapter 6.

92. All quotations are from Ward (1985:9).

93. Luomala (1949).

94. The details of the story quoted here come from Gill (1876:15–16). Its early date makes it likely to be an authentic myth.

95. Guppy (1888:2).

96. Buck (1938a:114). Nuku-tere is also placed in the Tuamotus (see chapter 6).

97. William Gill (1856:72–73).

98. Ibid., 72.

99. Ibid.

100. Wyatt Gill (1911:145).

101. I do not want to foment mystery here (there are dependable maps of the ocean floor in this region), but there are also still a number of seamounts that are unmapped and perhaps incorrectly located. Haymet Rocks is sometimes described as "existence doubtful."

102. Stommel (1984:60).

103. The research by Kori Raumea was carried out as part of my research project on vanished islands in the Cook Islands and has not been published before.

104. In Cook Islands Maori, *papa* is a coral reef exposed at low tide. The informant was Tuaiva Mautairi, once mayor of Mangaia Island, who said he was told that the *papa* of Tuanaki could be seen from Tua o Koropa in southwestern Mangaia. In dictionaries of Cook Island Maori, *papa* is translated as bedrock, flat slab, or shelf of rock (Paul Geraghty, personal communication, 2006).

105. The informant was Mataora Harry, *kavana* (headman) for the village of Keia on Mangaia and a qualified teacher. He speculated that the role of Tuanaki as a refuge is one reason why so much about the island was kept secret.

106. Stommel (1984). In another critical review of lost-island tales, Ramsay (1972:215) concluded that "there is no reasonable doubt that Tuanaki did exist."

107. Most of the details about Marsters' visit to Victoria come from Percival (1964).

108. Stommel (1984).

109. Helm and Percival (1973:67).

110. This detail is from Frisbie's excellent biography (Frisbie 1961:28).

111. The interviews carried out by Kori Raumea in 2003–2004 were part of my research project on vanished islands in the Cook Islands. A direct descendant of William Marsters, William Powell was not able to name the island that vanished, but he stated that it was visited by his relatives from Palmerston in the 1900s to collect "food, mainly coconuts."

112. Stommel (1984).

113. Smith (1899:72–73).

114. Best (1923:11) quoting Colonel Gudgeon, Resident at Rarotonga.

115. The cartographic evidence is presented in Filmer et al. (1994).

116. St. Johnston (1921:67–68).

117. Christian (1895:199).

118. Bryan (1940).

119. Kerr (2004).

120. The discussion of the sightings by Villalobos, Marshall, and the unnamed whaler are from Stommel (1984).

121. Levesque (1997). A sure sign of long isolation from humans is that Minami-tori was inhabited, like many other Pacific Islands that never before had a resident human population, by a naive bird population, in this case, large, heavy and so little inclined to fly that Arriola's men named them "fool birds."

122. I have been unable to trace the record of this discovery and depend on Bryan's (1940) account.

123. Bryan (1940:523).

124. Bryan (1940).

125. Hayami et al. (2001).

126. A handful were. These include most in Papua New Guinea and a few of the larger ones in the western Solomon Islands.

127. See chapters in the book edited by Arnold (2001).

128. Porcasi et al. (1999).

129. Keller et al. (2001).

130. It perhaps needs stating clearly that, for an island to be submerged (or drowned) by rising sea level, it does not need to sink. So much of the current media rhetoric that focuses today on Pacific Islands threatened by future sea-level rise talks, inaccurately, about these islands sinking.

Chapter 10: Vanished Islands of the Future

1. Most of the projections are given in the Intergovernmental Panel on Climate Change (IPCC) Third Assessment Report (IPCC WGI, 2001). Data for temperature and sea-level rise during the 1900-2000 period in the Pacific are from various sources, summarized in Nunn (2001a). Recent research suggests that even higher temperatures are likely for 2100 than the IPCC estimated in 2001 (Stainforth et al. 2005).

2. The variation (the short-term intra-annual and interannual ups and downs of sea level) is referred to as the noise around the signal, where the signal is the long-term trend of sea-level change. To be able to separate the signal from the noise, at least thirty years of sea-level data are needed. Notwithstanding this, some scientists have shamelessly drawn conclusions that are politically expedient yet scientifically flawed from much shorter time series of data (see Nunn [2004b] for an example from Tuvalu).

3. Examples from the Pacific Islands are in Sherwood and Howorth (1996).

4. Barnett and Adger (2003).

5. Dickinson (1999).

6. An excellent case study of the Tuvalu Islands is that by Aalbersberg and Hay (1992). A more popular account is the 2004 work by Mark Lynas entitled *High Tide* that includes a chapter about Tuvalu.

7. The danger in characterizing sea-level rise as *the* island problem is that other serious problems facing islands may be overlooked or underrated by the international community and even by cash-strapped island governments themselves (Nunn 2004b).

8. The relationship between rising temperatures and increasing tropical-cyclone frequency has been understood for nearly twenty years (Emanuel 1987) and validated with empirical data (Emanuel 2005).

9. Hoegh-Guldberg (1999).

10. Pelling and Uitto (2001).

11. The scenarios were described by Nakicenovic et al. (2000).

12. Nicholls (2004).

13. The land-tenure system of Tonga is governed by the monarch and the nobility, who, when population was lower in the past, were able to allocate a plot of land to every eligible adult male. Now, because of population growth, there is insufficient land available for allocation. See Campbell (2001) for further details.

14. Mimura and Pelesikoti (1997).

15. Wyrtki (1990).

16. A detailed account of Lau tectonics was given by Nunn (1996), case studies of south-central Lau by Nunn (1995a), and of the Vanua Balavu Group in northern Lau by Nunn et al. (2002).

17. Nunn et al. (2006a).

18. Masson et al. (2002).

19. Ward and Day (2001).

20. Nunn and Pastorizo (2006).

21. Ward (2001:11201).

22. Ward (2002).

23. Nisbet and Piper (1998).

24. The Clathrate Gun Hypothesis was developed by Kennett et al. (2003). It appears to be correct for at least the past forty-five million years (Maslin et al. 2004).

25. McMurtry et al. (2004).

26. McGuire et al. (1997).

27. Johnson (1987).

28. Ambae is considered the most dangerous volcano in Vanuatu because deposits derived from its large explosive eruptions are unusually widespread (Robin and Monzier 1994).

29. Chen et al. (1995:577).

30. Wolfe et al. (1994).

31. Ibid.

32. The 1975 Kalapana earthquake was described by Lipman et al. (1985), the 2000 aseismic event by Cervelli et al. (2002).

33. Morgan et al. (2003).

34. Ibid.

35. Ward (2002). Unfortunately for the purposes of scientific debate, this interpretation has been eagerly taken up by alarmists and millennialists whose Web sites are littered with incorrect interpretations and improbable scenarios.

36. The phrases come from Keating and McGuire (2000). A good account of Cumbre Vieja and the threat posed by its future collapse is by Ward and Day (2001).

37. Ward and Day (2003).

38. Ward and Day (2001:3397).

39. Such details may not be explainable solely by the rapidity and unexpectedness of island disappearance but also by the distance to nearby land, the means available to reach it, and the reception likely to be accorded to refugees there. Some of the oral traditions about people fleeing from the sinking of Teonimanu report that when some of them reached Makira Island, they hid in the hills for fear that the coastal peoples would attack and kill them. In the event, the coastal peoples succored them and helped them move along the coast to the offshore islands of Santa Catalina and Santa Ana. But their fears were evidently not groundless. In the early twentieth century, "an old chief" of Paama Island in

central Vanuatu recalled that when he was young, refugees from an eruption on nearby Lopevi Island had sought refuge on Paama but "we ate them" (Frater 1922:27).

Chapter 11: Vanished Islands and Hidden Continents in the World's Oceans: Last Thoughts

1. Crichton (1988:x).

2. de Camp (1970:287).

3. Ibid., 77.

4. "No fervent believer in [the lost continent of] Mu, it seems, will give up his belief merely for the sake of a few facts" (de Camp 1970:52).

5. de Camp (1970:77).

6. The Lemurian Fellowship, founded in 1936 in Chicago and now in California, believes that one day Mu will rise again from the depths as a new continent called Pacifica, and it is apparently preparing groups of suitable colonists.

7. The term "boundedness" appears in the discussion by Edmond and Smith (2003:2) from which the quotation also comes.

8. The best books about Easter Island are by Bahn and Flenley (1992) and McCall (1994). The ethnological study by Métraux (1940) is regarded as sound but others less so. Most of the views of Thor Heyerdahl on the origins of Easter Island biota, island peoples, and cultural affinities are not accepted by scientists.

9. Just in case any reader should miss the moral of their story, Bahn and Flenley subtitled their book "a message from our past for the future of our planet." Reviewers generally praised the message. One noted that "the authors use the evidence for prehistoric environmental degradation leading to cultural collapse as an example of humanity's propensity to regard natural resources as limitless" (Saunders 1992:42), and another, reading between the lines, stated that "human selfishness lay behind the disaster" (Macrae 1992:41), a conclusion used to bolster the global arguments in *Collapse* by Jared Diamond (2005).

10. That said, I and others have argued, contrary to most scientific opinion, that the interpretation of Easter Island's history on which the message was based was speculative and probably wrong (McCall 1993, Nunn 1994, Hunter-Anderson 1998, Nunn et al. 2007a). That of course does not invalidate the message.

Appendix 1: The Story of Teonimenu, Central Solomon Islands, and How It Vanished

1. This oral tradition was rendered in the Ulawan language, in which the name Teonimenu is used instead of Teonimanu.

2. The *ngoli* was a kind of charm made by inserting the leaves of the cabbage tree (*Cordyline terminalis*) into the tip of a rod for fishing for bonito (skipjack tuna [*Katsuwonus pelamis*] in Solomon Islands), a high-status fish to catch, and then sealing it with lime to make it magically powerful (Ivens 1929).

Appendix 2: Extracts from Maretu's Narrative of Cook Islands History Pertaining to the Vanished Island Tuanaki

1. The honorific name for Aitutaki Island.

2. Probably a carpenter's plane.

3. It was common for people, particularly of rank, to have multiple names, usually given by various ancestral lines. The fact that he had a name from Rarotonga suggests that one of his ancestors came from there.

4. In the vernacular of the time, one night meant twenty-four hours (Crocombe 1974).

5. Probably a reference to stars used for navigation. In this case, Tau is probably Tau-ono (or Matariki-tau-ono), which was the Cook Islands name for the Pleiades.

6. This conversation was taking place on Mangaia where Maretu was highly regarded by the people and was consequently greatly valued by the European missionaries. The people of Mangaia wanted Maretu to stay with them, but Gill, the leading missionary, clearly valued his presence when they reached Tuanaki, hitherto unvisited by missionaries.

7. Visitors who were disliked or begrudged often had their boat overturned on the way to the ship that was to take them away, a custom still practiced (Crocombe 1974).

References

Aalbersberg, W., and J. Hay. 1992. Implications of climate change and sea-level rise for Tuvalu. *SPREP (South Pacific Regional Environment Programme) Reports and Studies* 54.

Ackerman, R. E. 1998. Early maritime traditions in the Bering, Chukchi and East Siberian seas. *Arctic Anthropology* 35:247–262.

Ali, J. R., R. Hall, and S. J. Baker. 2001. Palaeomagnetic data from a Mesozoic Philippine Sea Plate ophiolite on Obi Island, eastern Indonesia. *Journal of Asian Earth Sciences* 19:535–546.

Allen, M. S. 2006. New ideas about Late Holocene climate variability in the central Pacific. *Current Anthropology* 47:521–535.

Altis, S. 1999. Origin and tectonic evolution of the Caroline Ridge and the Sorol Trough, western tropical Pacific, from admittance and a tectonic modeling analysis. *Tectonophysics* 313:271–292.

Alvarez, L. W., W. Alvarez, F. Asaro, and H. V. Michel. 1980. Extraterrestrial cause for the Cretaceous-Tertiary extinction. *Science* 208:1095–1108.

Ambrose, S. H. 1998. Late Pleistocene human population bottlenecks. *Journal of Human Evolution* 34:623–651.

Andersen, J. C. 1969. *Myths and legends of the Polynesians.* Tokyo: Tuttle.

Anderson, A. 2003. Initial human dispersal in Remote Oceania: Pattern and explanation. In *Pacific archaeology: Assessments and prospects (Proceedings of the International Conference for the 50th Anniversary of the First Lapita Excavation, Koné-Nouméa 2002),* ed. C. Sand, 71–84. Nouméa: Services des Musées et du Patrimoine.

Anderson, A., J. Chappell, M. Gagan, and R. Grove. 2006. Prehistoric maritime migration in the Pacific islands: An hypothesis of ENSO forcing. *The Holocene* 16:1–6.

Anderson, A., and B. McFadgen. 1990. Prehistoric two-way voyaging between New Zealand and East Polynesia: Mayor Island obsidian on Raoul Island, and possible Raoul Island obsidian in New Zealand. *Archaeology in Oceania* 25:24–37.

Anderson, W. 1967. A journal of a voyage made in His Majesty's Sloop Resolution. In *The voyage of the Resolution and the Discovery 1776–1780,* ed. J. C. Beaglehole, 723–986. Hakluyt Society Extra Series 36. Cambridge: Cambridge University Press.

Arnold, J. E., ed. 2001. *The origins of a Pacific coast chiefdom: The Chumash of the Channel Islands.* Salt Lake City: University of Utah Press.

Ashby, G., ed. 1983. *Never and always: Micronesian stories of the origins of islands, landmarks and customs.* Eugene: Rainy Day Press.

Ayres, W. 1990. Mystery islets of Micronesia. *Archaeology* January–February: 59–63.

Bahlburg, H. 1993. Hypothetical Southeast Pacific continent revisited: New evidence from the middle Paleozoic basins of northern Chile. *Geology* 21:909–912.

Bahlburg, H., and C. Breitkreuz. 1991. The evolution of marginal basins in the southern central Andes of Argentina and Chile during the Paleozoic. *Journal of South American Earth Sciences* 4:171–188.

Bahn, P. G., and J. Flenley. 1992. *Easter Island, Earth island*. London: Thames and Hudson.

Baker, W. S. 1886. A description of the new volcano in the Friendly Islands. *Transactions of the New Zealand Institute* 18:41–46.

Ballinger, B. S. 1978. *Lost city of stone: The story of Nan Madol, the "Atlantis" of the Pacific*. New York: Simon and Schuster.

Barnett, J., and W. N. Adger. 2003. Climate dangers and atoll countries. *Climatic Change* 61:321–337.

Barnett, T. P. 1984. The estimation of "global" sea level change: A problem of uniqueness. *Journal of Geophysical Research* 89:7980–7988.

Barthel, T. S. 1978. *The eighth land: The Polynesian discovery and settlement of Easter Island*. Honolulu: University of Hawai'i Press.

Baum, D. A., R. L. Small, and J. F. Wendel. 1998. Biogeography and floral evolution of baobabs (Adansonia: Bombaceae) as inferred from multiple data sets. *Systematic Biology* 47:181–207.

Beaglehole, E., and P. Beaglehole. 1938. Ethnology of Pukapuka. *Bernice P. Bishop Museum Bulletin* 150.

Beaglehole, J. C. 1967. *The journals of Captain Cook on his voyages of discovery*. Vol. 3, *The voyage of the Resolution and Discovery, 1776–1780*. Cambridge: Hakluyt Society.

Beckwith, M. 1940. *Hawaiian mythology*. New Haven: Yale University Press.

Beer, G. 1990. The island and the aeroplane: The case of Virginia Woolf. In *Nation and Narration*, ed. H. Bhabha, 265–290. London: Routledge.

Bellwood, P. 1979. *Man's conquest of the Pacific: The prehistory of Southeast Asia and Oceania*. New York: Oxford University Press.

———. 1997. *Prehistory of the Indo-Malayan Archipelago*. Honolulu: University of Hawai'i Press.

Benoit, M., and J. Dubois. 1971. The earthquake swarm in the New Hebrides Archipelago, August 1965. *Bulletin of the Royal Society of New Zealand* 9:141–148.

Benson, C. 1983. *Pacific folk tales*. Suva: Oceania Printers.

Bergersen, D. D. 1995. Physiography and architecture of Marshall Islands guyots drilled during Leg 144: Geophysical constraints on platform development. *Proceedings of the Ocean Drilling Program, Scientific Results* 144:561–583.

Berlitz, C. 1984. *Atlantis: The eighth continent*. New York: Fawcett Crest.

Best, E. 1923. *Polynesian voyagers: The Maori as a deep-sea navigator, explorer and colonizer*. Wellington: Dominion Museum.

Beverley, S. M., and C. A. Wilson. 1985. Ancient origin for Hawaiian Drosophilinae inferred from protein comparisons. *Proceedings of the National Academy of Sciences of the United States of America* 82:4753–4757.

Bevis, M., F. W. Taylor, B. E. Schutz, J. Recy, B. L. Isacks, S. Helu, R. Singh, E. Kendrick, J. Stowell, B. Taylor, and S. Calmant. 1995. Geodetic observations of very rapid convergence and back-arc extension at the Tonga arc. *Nature* 374:249–251.

Bird, I. L. 1905. *The Hawaiian Archipelago: Six months among the palm groves, coral reefs and volcanoes of the Sandwich Islands*. London: John Murray.

Birdsell, J. B. 1977. The recalibration of a paradigm for the first peopling of Greater Australia. In *Sunda and Sahul*, ed. J. Allen, J. Golson, and R. Jones, 113–168. London: Academic Press.

Blanchon, P., and J. Shaw. 1995. Reef drowning during the last deglaciation: Evidence for catastrophic sea-level rise and ice-sheet collapse. *Geology* 23:4–8.

Blavatsky, H. P. 1888. *The secret doctrine: The synthesis of science, religion and philosophy*, 2nd ed. 2 vols. London: Theosophical Society.

Blong, R. 1982. *The time of darkness: Local legends and volcanic reality in Papua New Guinea*. Canberra: Australian National University Press.

Bloom, A. L., W. S. Broecker, J. M. A. Chappell, R. K. Matthews, and K. J. Mesolella. 1974. Quaternary sea level fluctuations on a tectonic coast: New ^{230}Th/^{234}U dates from the Huon Peninsula, New Guinea. *Quaternary Research* 4:185–205.

Blust, R. A. 2004. *Austronesian languages.* Cambridge: Cambridge University Press.

Boas, F. 1929. *Anthropology and modern life.* London: Allen and Unwin.

Boaz, N. T., and R. L. Ciochon. 2004. *Dragon Bone Hill.* Oxford: Oxford University Press.

Bonnemaison, J. 1996. *Gens de Pirogue et Gens de la Terre.* Paris: Editions de l'ORSTOM.

Bowdler, S. 1997. The Pleistocene Pacific. In *The Cambridge history of the Pacific Islanders*, ed. D. Denoon, 41–50. Cambridge: Cambridge University Press.

Braghine, A. 1938. *The shadow of Atlantis.* London: Rider.

Brasseur, C.-E. 1869. *Manuscrit Troano (Études sur la système et langue des Mayas).* 2 vols. Paris: Impériale.

Briffa, K. R., P. D. Jones, F. H. Schweingruber, and T. J. Osborn. 1998. Influence of volcanic eruptions on Northern Hemisphere summer temperature over the past 600 years. *Nature* 393:450–455.

Brocher, T. M., ed. 1985. *Geological investigations of the Northern Melanesian borderland.* Houston: Circum-Pacific Council for Energy and Mineral Resources.

Brousse, R., J.-P. Chevalier, M. Denizot, and B. Salvat. 1978. Etude géomorphologique des Iles Marquises. *Cahiers du Pacifique* 21:9–74.

Brown, J. M. 1920. Mythology of the Pacific. *Mid-Pacific Magazine (Honolulu)* 19:33–38, 147–152, 253–258, 345–348, 447–452.

———. 1924. *The riddle of the Pacific.* Boston: Small, Maynard and Co.

Bryan, G. S. 1940. Los Jardines (E.D.). *United States Naval Institute Proceedings* April: 520–523.

Buck, P. H. (Te Rangi Hiroa). 1938a. *Vikings of the sunrise.* Philadelphia: Lippincott.

———. 1938b. Ethnology of Mangareva. *Bernice P. Bishop Museum Bulletin* 157.

Buckley, T. R., P. Arensburger, C. Simon, and G. K. Chambers. 2002. Combined data, Bayesian phylogenetics, and the origins of the New Zealand cicada genera. *Systematic Biology* 51:4–18.

Burckhardt, C. 1902. Traces géologiques d'un ancient continent Pacifique. *Revista del Museo La Plata* 10:179–192.

Calmant, S., B. Pelletier, P. Lebellegard, M. Bevis, F. W. Taylor, and D. A. Phillips. 2003. New insights on the tectonics along the New Hebrides subduction zone based on GPS results. *Journal of Geophysical Research* 108:2319, doi:10.1029/2001JB000644.

Campbell, I. C. 2001. *Island kingdom: Tonga ancient and modern.* Christchurch: Canterbury University Press.

———. 2003. *Worlds apart: A history of the Pacific islands*, rev. ed. Christchurch: Canterbury University Press.

Campbell, J. F. 1984. Rapid subsidence of Kohala Volcano and its effect on coral reef growth. *Geo-Marine Letters* 4:31–36.

Carlquist, S. 1974. *Island biology.* New York: Columbia University Press.

Carracedo, J. C., S. J. Day, H. Guillou, and F. J. P. Torrado. 1999. Giant Quaternary landslides in the evolution of La Palma and El Hierro, Canary Islands. *Journal of Volcanology and Geothermal Research* 94:169–190.

Casamiquela, R. M. 1980. Considérations écologiques et zoogéographiques sur les vertébrés de la zone littorale de la mer du Maestrichien dans le Nord de la Patagonie. *Mémoires de la Société Géologique de France* 139:53–55.

Cayce, E. E. 1968. *Edgar Cayce on Atlantis.* New York: Paperback Library.

Cervé, W. S. 1931. *Lemuria, the lost continent of the Pacific.* San Jose: Rosicrucian Press.

Cervelli, P., P. Segall, K. Johnson, M. Lisowski, and A. Miklius. 2002. Sudden aseismic fault slip on the south flank of Kilauea Volcano. *Nature* 415:1014–1018.

Chazine, J.-M. 2005. Rock art, burials, and habitations: Caves in East Kalimantan. *Asian Perspectives* 44:219–230.

Chen, J. K., F. W. Taylor, R. L. Edwards, H. Cheng, and G. S. Burr. 1995. Recurrent emerged reef terraces of the Yenkahe resurgent block, Tanna, Vanuatu: Implication for volcanic, landslide and tsunami hazards. *Journal of Geology* 103:577–590.

Chen, Y. G., and T. K. Liu. 2000. Holocene uplift and subsidence along an active tectonic margin, southwestern Taiwan. *Quaternary Science Reviews* 19:923–930.

Chen Chung-Yu, J. 2002. Sea nomads in prehistory on the southeast coast of China. *Indo-Pacific Prehistory Association Bulletin* 22:51–54.

Childress, D. H. 1988. *Lost cities of ancient Lemuria and the Pacific.* Stelle: Adventures Unlimited.
———. 1998. *Ancient Micronesia and the lost city of Nan Madol.* Kempton: Adventures Unlimited.

Christian, F. W. 1895. Notes on the Marquesas. *Journal of the Polynesian Society* 4:187–202.
———. 1910. *Eastern Pacific lands: Tahiti and the Marquesas Islands.* London: Robert Scott.

Churchill, W. 1916. *Sissano: Movements of migration within and through Melanesia.* Washington: Carnegie Institution.

Churchward, J. 1926. *Lost continent of Mu: The motherland of man.* New York: Rudge.
———. 1931. *The children of Mu.* New York: Ives Washburn.

Clark, R. 1996. Linguistic consequences of the Kuwae eruption. In *Oceanic culture history: Essays in honour of Roger Green,* ed. J. M. Davidson, G. Irwin, B. F. Leach, A. Pawley, and D. Brown, 275–285. New Zealand Journal of Archaeology Special Publication..

Clouard, V., and A. Bonneville. 2004. Submarine landslides in French Polynesia. In *Oceanic hotspots: Intraplate submarine magmatism and tectonism,* ed. R. Hekinian, P. Stoffers, and J.-L. Cheminé, 209–238. Berlin: Springer-Verlag.

Clouard, V., A. Bonneville, and H. G. Barsczus. 2000. Size and depth of frozen magma chambers under atolls and islands of French Polynesia using detailed gravity studies. *Journal of Geophysical Research* 105:8173–8192.

Clouard, V., A. Bonneville, and P.-Y. Gillot. 2001. A giant landslide on the southern flank of Tahiti Island, French Polynesia. *Geophysical Research Letters* 28:2253–2256.

Coats, R. R. 1952. Magmatic differentiation in Tertiary and Quaternary volcanic rocks from Adak and Kanaga islands, Aleutian Islands, Alaska. *Geological Society of America, Bulletin* 63:485–514.

Cogger, H. G. 1974. Voyage of the banded iguana. *Australian Natural History* 18:144–149.

Collette, B. B. 1974. Geographic variation in the central Pacific halfbeak, *Hyporhamphus acutus* (Günther). *Pacific Science* 28:111–122.

Collins, L. S., A. G. Coates, W. A. Berggren, M.-P. Aubry, and J. Zhang. 1996. The late Miocene Panama isthmian strait. *Geology* 24:687–690.

Cook, J. C. 1968. *The journals of Captain James Cook on his voyages of discovery,* ed. J. C. Beaglehole. 2 vols. Cambridge: Cambridge University Press for the Hakluyt Society.

Cooper, A., C. Lalueza-Fox, S. Anderson, A. Rambaut, J. Austin, and R. Ward. 2001. Complete mitochondrial genome sequences of two extinct moas clarify ratite evolution. *Nature* 409:704–707.

Cooper, H. S. 1880. *Coral lands.* London: Bentley.

Corbett, G., S. Hunt, A. Cook, P. Tamaduk, and T. Leach. 2001. Geology of the Ladolam gold deposit, Lihir Island, from exposures in the Minifie open pit. In *Geology, Exploration and Mining Conference, July 2001, Port Moresby, Papua New Guinea, Proceedings,* ed. G. Hancock, 69–78. Parkville, The Australasian Institute of Mining and Metallurgy.

Coulbourn, W. T., P. J. Hill, and D. D. Bergersen. 1989. Machias Seamount, Western Samoa: Sediment remobilization, tectonic dismemberment and subduction of a guyot. *Geo-Marine Letters* 9:119–125.

Coulson, F. I. E. 1976. Geology of the Lomaiviti and Moala Island groups. *Fiji Mineral Resources Department, Bulletin* 2.

Cracraft, J. 2001. Avian evolution, Gondwana biogeography and the Cretaceous-Tertiary mass extinction event. *Proceedings of the Royal Society of London B Biological Sciences* 268:459–469.

Crealock, W. I. B. 1955. *Towards Tahiti*. London: Peter Davies.

Crichton, M. 1988. *Travels*. New York: Ballantine.

Crocombe, M. T. 1974. *Maretu's narrative of Cook Islands history*. Port Moresby: University of Papua New Guinea, Department of History.

———, ed. 1983. *Cannibals and converts: Radical change in the Cook Islands by Maretu*. Suva: Institute of Pacific Studies, The University of the South Pacific.

Croizat, L. 1958. *Panbiogeography: Or, an introductory synthesis of zoogeography, phytogeography, and geology*. Caracas: published by the author.

Crough, S. T. 1984. Seamounts as recorders of hot-spot epeirogeny. *Geological Society of America, Bulletin* 95:3–8.

Csejtey, B., D. P. Cox, R. C. Evarts, G. D. Stricker, and H. L. Foster. 1982. The Cenozoic Denali fault system and the Cretaceous accretionary development of southern Alaska. *Journal of Geophysical Research* 87:3741–3754.

Cunningham, J. K., and K. J. Anscombe. 1985. Geology of ʻEua and other islands, Kingdom of Tonga. In *Geology and offshore resources of Pacific island arcs — Tonga Region,* ed. D. W. Scholl and T. L. Vallier, 221–257. Houston: Circum-Pacific Council for Energy and Mineral Resources.

Dalla Salda, L. H., I. W. D. Dalziel, C. Cingolani, and R. Varela. 1992. Did the Taconic Appalachians continue into southern South America? *Geology* 20:1059–1062.

Dalrymple, A. 1996. *An account of the discoveries made in the South Pacifick Ocean*. Sydney: Hordern House for the Australian National Maritime Museum.

Dalziel, I. W. D. 1991. Pacific margins of Laurentia and East Antarctica–Australia as a conjugate rift pair: Evidence and implications for an Eocambrian supercontinent. *Geology* 19:598–601.

Dalziel, I. W. D., L. H. Dalla Salda, and L. M. Gahagan. 1994. Paleozoic Laurentia-Gondwana interaction and the origin of the Appalachian-Andean mountain system. *Geological Society of America, Bulletin* 106:243–252.

Darwin, C. R. 1842. *Structure and distribution of coral reefs*. London: Smith, Elder.

———. 1859. *On the origin of species by means of natural selection*. London: John Murray.

Davies, H. L., J. M. Davies, R. C. B. Perembo, and W. Y. Lus. 2003. The Aitape 1998 tsunami: Reconstructing the event from interviews and field mapping. *Pure and Applied Geophysics* 160:1895–1922.

Davis, W. M. 1913. Dana's confirmation of Darwin's theory of coral reefs. *American Journal of Science* 35:173–188.

de Camp, L. S. 1970. *Lost continents: The Atlantis theme*, 2nd rev. ed. New York: Ballantine.

de Laubenfels, D. J. 1985. A taxonomic revision of the genus Podocarpus. *Blumea* 30:51–278.

Denicagilaba. 1894. *Ai tukuni ni vanua eda vu maikina. Na Mata* (Suva, Fiji).

Diamond, J. M. 2005. *Collapse: How societies choose to fail or succeed*. New York: Viking.

Dickinson, W. R. 1999. Holocene sea-level record on Funafuti and potential impact of global warming on central Pacific atolls. *Quaternary Research* 51:124–132.

———. 2003. Impact of mid-Holocene hydro-isostatic highstand in regional sea level on habitability of islands in Pacific Oceania. *Journal of Coastal Research* 19:489–502.

Dickinson, W. R., and R. Shutler Jr. 1974. Probable Fijian origin of quartzose temper sands in prehistoric pottery from Tonga and the Marquesas. *Science* 185:454–457.

Dietz, R. S. 1961. Continent and ocean basin evolution by spreading of the sea floor. *Nature* 190:854–857.

Di Piazza, A., and E. Pearthree. 2001a. An island for gardens, an island for birds and voyaging: A settlement pattern for Kiritimati and Tabuaeran, two "mystery islands" in the northern Lines, Republic of Kiribati. *Journal of the Polynesian Society* 110:149–170.

———. 2001b. Voyaging and basalt exchange in the Phoenix and Line archipelagoes: The view point from three mystery islands. *Archaeology in Oceania* 36:146–152.

———. 2004. Sailing routes of old Polynesia: The prehistoric discovery, settlement and abandonment of the Phoenix Islands. *Bishop Museum Bulletin in Anthropology* 2.

Dixon, E. J. 1993. *Quest for the origins of the first Americans.* Albuquerque: University of New Mexico Press.

Dixon, R. B. 1964. *Oceanic mythology.* New York: Cooper Square Publishing.

Donnelly, I. 1882. *Atlantis: The antediluvian world.* New York: Harper.

Douglas, B. 1999. Science and the art of representing "Savages": Reading "race" in text and image in South Seas voyage literature. *History and Anthropology* 11:157–201.

Draper, N. 1988. Stone tools and cultural landscapes: Investigating the archaeology of Kangaroo Island. *South Australian Geographic Journal* 88:15–36.

Druitt, T. H., and V. Francaviglia. 1992. Caldera formation on Santorini and the physiography of the islands in the late Bronze Age. *Bulletin Volcanologique* 54:484–493.

Dudley, W. C., and M. Lee. 1988. *Tsunami!.* Honolulu: University of Hawai'i Press.

Dumont d'Urville, J.-S.-C. 1832. Sur les îles du Grand Océan. *Bulletin de la Société de Géographie* 17:3–21.

Duncan, R. A., and D. A. Clague. 1985. Pacific plate motions recorded by linear volcanic chains. In *The ocean basins and margins.* Vol. 7A, *The Pacific,* ed. A. E. M. Nairn, F. G. Stehli, and S. Uyeda, 89–121. New York: Plenum Press.

Duncan, R. A., and I. McDougall. 1976. Linear volcanism in French Polynesia. *Journal of Volcanology and Geothermal Research* 1:197–227.

Dupont, J., and R. H. Herzer. 1985. Effect of subduction of the Louisville Ridge on the structure and morphology of the Tonga arc. In *Geology and offshore resources of Pacific island arcs — Tonga Region,* ed. D. W. Scholl and T. L. Vallier, 323–332. Houston: Circum-Pacific Council for Energy and Mineral Resources.

Edmond, R., and V. Smith. 2003. Editors' introduction. In *Islands in history and representation,* ed. R. Edmond and V. Smith, 1–18. London: Routledge.

Eissen, J. P., M. Monzier, and C. Robin. 1994. Kuwae, l'éruption volcanique oubliée. *La Recherche* 270:1200–1202.

Eizirik, E., W. J. Murphy, and S. J. O'Brien. 2001. Molecular dating and biogeography of the early placental mammal radiation. *Journal of Heredity* 92:212–219.

Ellis, W. 1832. *Polynesian researches: During a residence of nearly eight years in the Society and Sandwich islands,* 2nd ed. 4 vols. London: Fisher, Son and Jackson.

Emanuel, K. A. 1987. The dependence of hurricane intensity on climate. *Nature* 326:483–485.

———. 2005. Increasing destructiveness of tropical cyclones over the past 30 years. *Nature* 436:686–688.

Emory, K. P. 1934. Archaeology of the Pacific equatorial islands. *Bernice P. Bishop Museum Bulletin* 123.

Englert, S. 1972. *Island at the centre of the world: New light on Easter Island.* London: Robert Hale.

Fairbridge, R. W., and H. B. Stewart. 1960. Alexa Bank, a drowned atoll on the Melanesian Border Plateau. *Deep-Sea Research* 7:100–116.

Felton, E. A., K. A. W. Crook, and B. H. Keating. 2000. The Hulupoe Gravel, Lanai, Hawaii: New sedimentological data and their bearing on the "giant wave" (mega-tsunami) emplacement hypothesis. *Pure and Applied Geophysics* 157:1257–1284.

Filmer, P. E., M. K. McNutt, H. F. Webb, and D. J. Dixon. 1994. Volcanism and archipelagic aprons in the Marquesas and Hawaiian islands. *Marine Geophysical Researches* 16:385–406.

Finney, B. 1992. *From sea to space.* Palmerston North: Massey University Press.

Fiske, R. S., S. Oshima, and K. V. Cashman. 1991. Shallow submarine eruptions of Myojin-sho: Growth of a pyroclastic cone on an active volcanic arc. *EOS (Transactions of the American Geophysical Union)* 72:248.

Fison, L. 1907. *Tales from old Fiji.* London: Moring.

———. 1984. *Tales from old Fiji,* rev. ed. Papakura: McMillan.

Flood, B., B. E. Strong, and W. Flood. 2002. *Micronesian legends.* Honolulu: Bess Press.

Fornander, A. 1878. *An account of the Polynesian race.* Vol. 1. London: Trubner.

Fowles, J. 1978. *Islands.* Boston: Little, Brown and Co.

Fox, C. E. 1925. *The threshold of the Pacific.* New York: Knopf.

France, P. 1966. The Kaunitoni migration: Notes on the genesis of a Fijian tradition. *Journal of Pacific History* 1:107–113.

Frater, M. 1922. *Midst volcanic fires: An account of missionary tours among the volcanic islands of the New Hebrides.* London: Clarke.

Fredericksen, C. 1997. The maritime distribution of Bismarck Archipelago obsidian and island Melanesian prehistory. *Journal of the Polynesian Society* 106:375–393.

Frisbie, J. 1961. *The Frisbies of the South Seas.* Birkenhead: Willmer Brothers and Haram.

Galipaud, J.-C. 2002. Under the volcano: Ni-Vanuatu and their environment. In *Natural disasters and cultural change,* ed. R. Torrence and J. Grattan, 162–171. London: Routledge.

Gao, C., A. Robock, S. Self, J. Witter, J. P. Steffenson, H. B. Clausen, M.-L. Siggaard-Andersen, S. Johnsen, P. A. Mayewski, and C. Ammann. 2006. The 1452 or 1453 A.D. Kuwae eruption signal derived from multiple ice core records: Greatest volcanic sulfate event of the past 700 years. *Journal of Geophysical Research* doi:10.1029/2005JD006710.

Garnier, J. 1870. *Les migrations humaines en Océanie d'après les faits naturels.* Paris: Martinet.

Geraghty, P. 1977. How a myth is born — the story of the Kaunitoni story. *Mana* 2:25–29.

———. 1993. Pulotu, Polynesian homeland. *Journal of the Polynesian Society* 102:343–384.

Gibbons, J. R. H., and F. G. A. U. Clunie. 1986. Sea level changes and Pacific prehistory. *Journal of Pacific History* 21:58–82.

Gill, J. B. 1987. Early geochemical evolution of an oceanic island arc and backarc: Fiji and the South Fiji Basin. *Journal of Geology* 95:589–615.

Gill, W[illiam]. 1856. *Gems from the coral islands.* London: Ward.

Gill, W[illiam]. W[yatt]. 1876. *Myths and songs from the South Pacific.* London: King.

———. 1911. Extracts from Dr. Wyatt Gill's papers. *Journal of the Polynesian Society* 20:116–151.

Gillespie, R. G. 2002. Biogeography of spiders on remote oceanic islands of the Pacific: Archipelagoes as stepping stones? *Journal of Biogeography* 29:655–662.

Gladczenko, T. P., M. F. Coffin, and O. Eldholm. 1997. Crustal structure of the Ontong Java Plateau: Modelling of new gravity and existing seismic data. *Journal of Geophysical Research* 102:22711–22729.

Goff, J., K. Hulme, and B. McFadgen. 2003. "Mystic fires of Tamaatea": Attempts to creatively rewrite New Zealand's cultural and tectonic past. *Journal of the Royal Society of New Zealand* 33:795–809.

Goff, J. R., and B. G. McFadgen. 2001. Catastrophic seismic-related events and their impact on prehistoric human occupation in coastal New Zealand. *Antiquity* 74:155–162.

Golson, J. 1982. Kuk and the history of agriculture in the New Guinea highlands. In *Melanesia: Beyond diversity,* ed. R. J. May and H. Nelson, 297–307. Canberra: Australian National University Press.

Gragg, L. 2000. The Port Royal earthquake. *History Today* 50:28–34.

Graham, G. 1939. The ancestral canoe of Ngati-whatua (Kaipara). *Journal of the Polynesian Society* 48:186–191.

Greene, H. G., A. Macfarlane, D. P. Johnson, and A. J. Crawford. 1988. Structure and tectonics of the central New Hebrides arc. In *Geology and offshore resources of Pacific island arcs — Vanuatu Region,* ed. H. G. Greene and F. L. Wong, 377–412. Houston: Circum-Pacific Council for Energy and Mineral Resources.

Greene, H. G., and F. L. Wong, eds. 1988. *Geology and offshore resources of Pacific island arcs — Vanuatu Region.* Houston: Circum-Pacific Council for Energy and Mineral Resources.

Gregory, J. W. 1930. The geological history of the Pacific Ocean. *Proceedings of the Geological Society* 86:72–136.

Grehan, J. R. 2001. Biogeography and evolution of the Galapagos: Integration of the biological and geological evidence. *Biological Journal of the Linnean Society* 74:267–287.

Grey, G. 1855. *Polynesian mythology and ancient traditional history of the New Zealand race.* London: Routledge.

Griffis, W. E. 1895. *The religions of Japan: From the dawn of history to the era of Méiji.* New York: Charles Scribner's Sons.

Grimble, A. F. 1956. *A pattern of islands.* London: Murray.

———. 1989. *Tungaru traditions: Writings on the atoll culture of the Gilbert Islands.* Pacific Islands Monograph Series, No. 7. Honolulu: University of Hawai'i Press.

Grimble, R. 1972. *Migrations, myth and magic from the Gilbert Islands: Early writings of Sir Arthur Grimble.* London: Routledge and Kegan Paul.

Grossman, E., C. Fletcher, and B. Richmond. 1998. The Holocene sea-level highstand in the equatorial Pacific: Analysis of the insular paleosea-level database. *Coral Reefs* 17:309–327.

Groube, L., J. Chappell, J. Muke, and D. Price. 1986. A 40,000 year-old human occupation site at Huon Peninsula, Papua New Guinea. *Nature* 324:453–455.

Grover, J. C. 1960. Reports on investigations into the geology and mineral resources of the Protectorate. *British Solomon Islands Geological Record* 1957–1958, Vol. 1.

Gudgeon, W. E. 1905. Maori religion. *Journal of the Polynesian Society* 55:107–130.

Guiart, J. 1973. Le dossier rassemblé. In *Système des titres, électifs ou héréditaires, dans les Nouvelles-Hébrides centrales d'Éfate aux Iles Shepherd,* ed. J.-J. Espirat, J. Guiart, M.-S. Lagrange, and M. Renaud, 149–370. Paris: Mémoires de l'Institut d'Ethnologie 10.

Guppy, H. B. 1888. *Coral islands and savage myths.* London: Victoria Institute and Philosophical Society of Great Britain.

Haeckel, E. 1874. *Anthropogenie oder Entwickelungsgeschichte des Menschen.* Leipzig: Engelmann.

Hale, H. 1846. *United States Exploring Expedition: Ethnography and philology.* Philadelphia: Lea and Blanchard.

Hallam, A. 1986. Evidence of displaced terranes from Permian to Jurassic faunas around the Pacific margins. *Journal of the Geological Society of London* 143:209–216.

Hamilton, E. L. 1956. *Sunken islands of the mid-Pacific mountains.* New York: Geological Society of America.

Hancock, G. 1995. *Fingerprints of the gods: A quest for the beginning and the end.* London: Heinemann.

Handy, E. S. C. 1923. The native culture in the Marquesas. *Bernice P. Bishop Museum Bulletin* 9.

———. 1930. Marquesan legends. *Bernice P. Bishop Museum Bulletin* 69.

Hau'ofa, E., ed. 1993. *Our sea of islands.* Suva: School of Social and Economic Development, The University of the South Pacific.

Hayami, K., K. Nagakura, H. Abe, S. Masumi, F. Kumasaka, M. Hayashida, M. Ushijima, S. Sameshima, S. Ikeda, S. Kanda, A. Uchida, M. Ogawa, M. Iizuka, M. Nakagawa, S. Yamazaki, W. Inoue, and K. Oikawa. 2001. Preliminary report of continental shelf survey of "western sea area off Minami-tori Shima," "Minami-tori Shima," "eastern sea area off Minami-tori Shima." *Technical Bulletin on Hydrography* 19:30–50 [in Japanese].

Hayashi, J. N., and S. Self. 1992. A comparison of pyroclastic flow and debris avalanche mobility. *Journal of Geophysical Research* 97:9063–9071.

Heads, M. 2005. Dating nodes on molecular phylogenies: A critique of molecular biogeography. *Cladistics* 21:62–78.

Hearty, P. 2002. The Ka'ena highstand of O'ahu, Hawai'i: Further evidence of Antarctic ice collapse during the Middle Pleistocene. *Pacific Science* 56:65–81.

Hébert, B. 1963–1965. Contribution à l'étude archéologique de l'île Éfaté et des îles avoisinantes. *Études Mélanésiennes* 18–20:71–98.

Hébert, H., P. Heinrich, F. Schindelé, and A. Piatanesi. 2001. Far-field simulation of tsunami propagation in the Pacific Ocean: Impact on the Marquesas Islands (French Polynesia). *Journal of Geophysical Research* 106:9161–9178.

Helm, A. S., and W. H. Percival. 1973. *Sisters in the sun: The story of Suwarrow and Palmerston atolls.* London: Robert Hale.

Henderson, R. S. 1980. Observations on colonization of subsided shorelines at Halape and Keauhou, Hawai'i. In *Proceedings of the Third Conference in Natural Sciences, Hawaii Volcanoes National Park,* ed. C. W. Smith, 155–163. Honolulu: University of Hawai'i Press.

Henry, T. 1928. Ancient Tahiti. *Bernice P. Bishop Museum Bulletin* 48.

Herzer, R. H., G. C. H. Chaproniere, A. R. Edwards, C. J. Hollis, B. Pelletier, J. I. Raine, G. H. Scott, V. Stagpoole, C. P. Strong, P. Symonds, G. J. Wilson, and H. Zhu. 1997. Seismic stratigraphy and structural history of the Reinga Basin and its margins, southern Norfolk Ridge system. *New Zealand Journal of Geology and Geophysics* 40:425–451.

Heyerdahl, T. 1952. *American Indians in the Pacific.* London: Allen and Unwin.

———. 1968. *Sea routes to Polynesia.* London: Allen and Unwin.

———. 1989. *Easter Island: The mystery solved.* London: Souvenir Press.

Higham, C. F. W., and T. L.-D. Lu. 1998. The origins and dispersal of rice cultivation. *Antiquity* 72:867–877.

Hildebrand, A. R., G. T. Penfield, D. A. Kring, M. Pilkington, Z. A. Carmargo, S. B. Jacobsen, and W. V. Boynton. 1991. Chicxulub Crater: A possible Cretaceous/Tertiary boundary impact crater on the Yucatán Peninsula, Mexico. *Geology* 19:867–871.

Hill, P. J., and G. P. Glasby. 1996. Capricorn Seamount — Geology and geophysics of a subducting guyot. In *Manihiki Plateau, Machias and Capricorn seamounts, Niue, and Tonga Trough: Results of Tui cruises,* ed. M. A. Meylan and G. P. Glasby, 17–29. SOPAC Technical Bulletin 10. Suva: SOPAC.

Hocart, A. M. 1929. Lau Islands, Fiji. *Bernice P. Bishop Museum Bulletin* 62.

Hoegh-Guldberg, O. 1999. Coral bleaching, climate change and the future of the world's coral reefs. *Review of Marine and Freshwater Research* 50:839–866.

Holcomb, R. T., and R. C. Searle. 1991. Large landslides from oceanic volcanoes. *Marine Geotechnology* 10:19–32.

Holm, B. 2000. *Eccentric islands*. Minneapolis: Milkweed.

Houtz, R. E. 1962. The 1953 Suva earthquake and tsunami. *Seismological Society of America, Bulletin* 52:1–12.

Howe, K. R. 1984. *Where the waves fall: A new South Seas history from first settlement to colonial rule*. Sydney: Allen and Unwin.

———. 1999. Maori/Polynesian origins and the "new learning." *Journal of the Polynesian Society* 108:305–325.

Howell, D. G., ed. 1985. *Tectonostratigraphic terranes of the circum-Pacific region*. Earth Science Series, No 1. Houston: Circum-Pacific Council for Energy and Mineral Resources.

Howorth, R. 2000. Waves and vanishing islands. *Islands Business* June: 48–49.

Huang, C. Y., W. Y. Wu, C. P. Chang, S. Tsao, P. B. Yuan, C. W. Lin, and X. Kuan-Yuan. 1997. Tectonic evolution of accretionary prism in the arc-continent collision terrane of Taiwan. *Tectonophysics* 281:31–51.

Huggett, R. 1989. *Cataclysms and earth history*. Oxford: Clarendon.

Hunt, T. L., and C. P. Lipo. 2006. Late colonization of Easter Island. *Science* 311:1603–1606.

Hunter-Anderson, R. L. 1998. Human vs. climatic impacts at Easter Island: Did the people really cut down all those trees? In *Easter Island in Pacific context, South Seas Symposium, Proceedings of the Fourth International Conference on Easter Island and East Polynesia, University of New Mexico*, ed. C. M. Stevenson, G. Lee, and F. J. Morin, 85–99. Los Osos: The Easter Island Foundation.

Hunter-Anderson, R., and Y. Zan. 1996. Demystifying the sawei, a traditional inter-island exchange system. *ISLA: A Journal of Micronesian Studies* 4:1–45.

Igarashi, Y. 1996. A late glacial climatic reversion in Hokkaido, Northeast Asia, inferred from the *Larix* pollen record. *Quaternary Science Reviews* 15:989–995.

Iizasa, K., R. S. Fiske, O. Ishizuka, M. Yuasa, J. Hashimoto, J. Ishibashi, J. Naka, Y. Horii, Y. Fujiwara, A. Imai, and S. Koyama. 1999. A Kuroko-type polymetallic sulfide deposit in a submarine silicic caldera. *Science* 283:975–977.

IPCC WGI. 2001. *Climate change 2001, scientific basis*. Cambridge: Cambridge University Press.

Irwin, G. 1992. *The prehistoric exploration and colonisation of the Pacific*. Cambridge: Cambridge University Press.

Irwin, W. P. 1981. Tectonic accretion of the Klamath Mountains. In *The geotectonic development of California*. Vol. 1, ed. W. G. Ernst, 29–49. Englewood Cliffs: Prentice-Hall.

Isaacson, P. E., L. Fisher, and J. Davidson. 1985. Devonian and Carboniferous stratigraphy of Sierra del Almeida, northern Chile, preliminary results. *Revista Geológica de Chile* 25–26:113–121.

Ivanoff, J. 2005. Sea gypsies of Myanmar. *National Geographic* April: 36–55.

Ivens, W. G. 1927. *Melanesians of the South-east Solomon Islands*. New York: Benjamin Blom.

———. 1929. *Dictionary and grammar of the language of Sa'a and Ulawa, Solomon Islands*. Washington: Carnegie Institution of Washington.

Jacolliot, L. 1879. *Histoire des Vierges*. Paris: Lacroix.

Jarvis, D. I. 1993. Pollen evidence of changing Holocene monsoon climate in Sichuan Province, China. *Quaternary Research* 39:325–337.

Jeffers, R. 2004. Robinson Jeffers — Poems, poemhunter.com. (accessed March 2008).

Johnson, R. W. 1987. Large-scale volcanic cone collapse: The 1888 slope failure of Ritter Volcano, and other examples from Papua New Guinea. *Bulletin Volcanologique* 49:666–679.

Jones, A. T. 1995. Geochronology of drowned Hawaiian coral reefs. *Sedimentary Geology* 99:233–242.

Jones, R. 1977. Man as an element of a continental fauna: The case of the sundering of the Bassian bridge. In *Sunda and Sahul*, ed. J. Allen, J. Golson, and R. Jones, 317–386. London: Academic Press.

Jouannic, C., F. W. Taylor, A. L. Bloom, and M. Bernat. 1980. Late Quaternary uplift history from emerged reef terraces on Santo and Malekula islands, central New Hebrides Island Arc. *United Nations ESCAP, CCOP/SOPAC Technical Bulletin* 3:91–108.

Joyner, T. 1992. *Magellan.* Camden: International Marine.

Kato, Y., and A. Nishizawa. 2002. The 1986 submarine eruption and topography of the Fukutokuoka-no-ba at the southern end of the Izu-Ogasawara Arc, Japan. Unpublished abstract, 38. Chapman Conference on Explosive Subaqueous Volcanism, Dunedin, New Zealand, 21–25 January 2002, www.agu.org/meetings/cc02aabstracts/Kato.pdf (accessed February 2008).

Keating, B. H. 1987. Structural failure and drowning of Johnston Atoll, central Pacific Basin. In *Seamounts, islands, and atolls,* ed. B. H. Keating, P. Fryer, R. Batiza, and G. W. Boehlert, 49–59. Geophysical Monograph Series 43. Washington: American Geophysical Union.

———. 1998. Nuclear testing in the Pacific from a geologic perspective. In *Climate and environmental change in the Pacific Basin,* ed. J. Terry, 113–144. Suva: School of Social and Economic Development, The University of the South Pacific.

Keating, B. H., and C. E. Helsley. 2002. The ancient shorelines of Lanai, Hawaii, revisited. *Sedimentary Geology* 150:3–15.

Keating, B. H., I. Karogodina, and C. E. Helsley. 1991. West Chapman Seamount, central Line Islands, Kiribati. *Pacific seafloor atlas, sheet 23.* Honolulu: University of Hawai'i.

Keating, B. H., and W. J. McGuire. 2000. Island edifice failures and associated tsunami hazards. *Pure and Applied Geophysics* 157:899–955.

Keigwin, L. D. 1980. Palaeoceanographic change in the Pacific at the Eocene-Oligocene boundary. *Nature* 287:722–725.

Keigwin, L. D., G. A. Jones, and P. N. Froelich. 1992. A 15,000 year paleoenvironmental record from Meiji Seamount, far northwestern Pacific. *Earth and Planetary Science Letters* 111:425–440.

Keller, E. A., M. Kamerling, P. Eichhubl, T. Hill, and J. Kennett. 2001. Isla Calafia, the fifth Santa Barbara Channel Island? Unpublished abstract, Paper 29-0, Geological Society of America Annual Meeting, 5–8 November 2001, gsa.confex.com/gsa/2001AM/finalprogram/abstract_22386.htm (accessed February 2008).

Keller, G., S. D'Hondt, C. J. Orth, J. S. Gilmore, P. Q. Oliver, E. M. Shoemaker, and E. Molina. 1987. Late Eocene impact microspherules: Stratigraphy, age and geochemistry. *Meteoritics* 22:25–60.

Kennett, J. P., K. G. Cannariato, I. L. Hendy, and R. J. Behl. 2003. *Methane hydrates in Quaternary climate change: The Clathrate Gun Hypothesis.* Washington: American Geophysical Union.

Kerr, R. 2004. *A general history and collection of voyages and travels.* Vol. 2, *arranged in systematic order: Forming a complete history of the origin and progress of navigation, discovery, and commerce, by sea and land, from the earliest ages to the present time.* Project Gutenberg EBook #10803.

Kerr, R. A. 1994. Volcanoes with bad hearts are tumbling down all over. *Science* 264:660.

Kershaw, A. P., G. M. McKenzie, and A. McMinn. 1993. A Quaternary vegetation history of northeastern Queensland from pollen analysis of ODP site 820. *Proceedings of the Ocean Drilling Program, Scientific Results* 133:107–114.

Kimura, M. 2004. Ancient megalithic construction beneath the sea off Ryukyu Islands in Japan, submerged by postglacial sea-level change. In *Proceedings of Oceans '04 MTS/IEEEE Techno-Ocean '04, Kobe, Japan, 9–12 November 2004,* 947–953.

King, L. C. 1983. *Wandering continents and spreading sea floors on an expanding Earth.* New York: Wiley.

Kingdon, J. 1993. *Self-made man: Human evolution from Eden to extinction?* New York: Wiley.

Kingston, N., S. Waldren, and U. Bradley. 2003. The phytogeographical affinities of the Pitcairn Islands — A model for south-eastern Polynesia? *Journal of Biogeography* 30:1311–1328.

Kirch, P. V. 1984. *The evolution of the Polynesian chiefdoms.* Cambridge: Cambridge University Press.

———. 1997. *The Lapita peoples: Ancestors of the oceanic world.* Oxford: Blackwell.

———. 2000. *On the road of the winds: An archaeological history of the Pacific Islands before European contact.* Berkeley: University of California Press.

———. 2001. Three Lapita villages: Excavations at Talepakemalai (ECA), Etakosarai (ECB), and Etapakengaroasa (EHB), Eloaua and Emananus islands. In *Lapita and its transformations in Near Oceania: Archaeological investigations in the Mussau Islands, Papua New Guinea, 1985–88,* ed. P. V. Kirch, 68–145. Contribution 59. Archaeological Research Facility, University of California at Berkeley.

Kirch, P. V., and R. C. Green. 2001. *Hawaiki, ancestral Polynesia: An essay in historical anthropology.* Cambridge: Cambridge University Press.

Klar, K .A., and T. L. Jones. 2005. Linguistic evidence for a prehistoric Polynesia–Southern California contact event. *Anthropological Linguistics* 47:369–400.

Knappert, J. 1992. *Pacific mythology.* London: Aquarian Press.

Knoll, A. 1991. End of the Proterozoic era. *Scientific American* 265:64–73.

Knoll, A., and M. R. Walter. 1992. Latest Proterozoic stratigraphy and earth history. *Nature* 356:673–678.

Kristan-Tollmann, E. 1986. Triassic of the Tethys and its relations with the Triassic of the Pacific realm. In *Shallow Tethys 2,* ed. K. G. McKenzie, 169–186. Rotterdam: Balkema.

Kuhn, T. S. 1996. *The structure of scientific revolutions,* 3rd ed. Chicago: The University of Chicago Press.

Kumar, R. 2005. Geology, climate, and landscape of the PABITRA Wet-Zone Transect, Viti Levu Island, Fiji. *Pacific Science* 59:141–157.

Kumar, R., P. D. Nunn, J. E. Field, and A. de Biran. 2006. Human responses to climate change around AD 1300: A case study of the Sigatoka Valley, Viti Levu Island, Fiji. *Quaternary International* 151:133–143.

Kuno, H. 1962. *Catalogue of the active volcanoes of the world including solfatara fields. Part XI, Japan, Taiwan and Marianas.* Rome: International Association of Volcanology.

Lampert, R. J. 1981. *The great Kartan mystery.* Terra Australis 5. Canberra: Research School of Pacific and Asian Studies, Australian National University.

Langdon, R. 1988. *The lost caravel re-explored.* Canberra: Brolga Press.

Langridge, M., and J. Terrell. 1988. *Von den Steinen's Marquesan myths.* Canberra: Journal of Pacific History.

Larson, G., K. Dobney, U. Albarella, M. Fang, E. Matisoo-Smith, J. Robins, S. Lowden, H. Finlayson, T. Brand, E. Willerslev, P. Rowley-Conwy, L. Andersson, and A. Cooper. 2005. Worldwide phylogeography of wild boar reveals multiple centers of pig domestication. *Science* 307:1618–1621.

Larson, R. L., R. A. Pockalny, R. F. Viso, E. Erba, L. J. Abrams, B. P. Luyendyk, J. M. Stock, and R. W. Clayton. 2002. Mid-Cretaceous tectonic evolution of the Tongareva triple junction in the southwestern Pacific basin. *Geology* 30:67–70.

Lecolle, J., J. E. Bokilo, and M. Bernat. 1990. Soulèvement et tectonique de l'ile d'Efaté (Vanuatu) arc insulaire des Nouvelles-Hébrides, au cours du Quaternaire récent. Datations de terrasses soulevées par la methode U/Th. *Marine Geology* 94:251–270.

Lee, S.-M. 2002. Rotation of the Caroline Plate and change in its boundary type: Constraints from recent observations. *EOS (Transactions of the American Geophysical Union)* 83, Western Pacific Geophysics Meeting Supplement, Abstract SE22C-11.

Le Plongeon, A. 1886. *Sacred mysteries among the Mayas and Quiches 11,500 years ago.* New York: Macoy.

Levesque, R. 1997. The odyssey of Captain Arriola and his discovery of Marcus Island in 1694. *Journal of Pacific History* 32:229–233.

Lewis, D. 1994. *We, the navigators: The ancient art of landfinding in the Pacific,* 2nd ed. Honolulu: University of Hawai'i Press.

Lincoln, J. M., and S. O. Schlanger. 1987. Miocene sea-level falls related to the geologic history of Midway Atoll. *Geology* 15:454–457.

Lipman, P. W., J. P. Lockwood, R. T. Okamura, D. A. Swanson, and K. M. Yamashita. 1985. Ground deformation associated with the 1975 magnitude-7.2 earthquake and resulting changes in activity of Kilauea Volcano, Hawaii. *U.S. Geological Survey Professional Paper* 1276.

Lipman, P. W., W. R. Normark, J. B. Wilson, and C. E. Gutmacher. 1988. The giant submarine Alika debris slide, Mauna Loa, Hawaii. *Journal of Geophysical Research* 93:4279–4299.

Longfellow, H. W. 1881. *Ultima Thule.* Boston: Houghton, Mifflin and Co.

Luomala, K. 1949. Maui-of-a-thousand tricks: His Oceanic and European biographers. *Bernice P. Bishop Museum Bulletin* 198.

Lynas, M. 2004. *High tide: News from a warming world.* London: Harper Collins.

Lyons, C. J. 1893. The song of Kualii, of Hawaii. *Journal of the Polynesian Society* 2:160–178.

Machida, H., and F. Arai. 1983. Extensive ash falls in and around the Sea of Japan from large late Quaternary eruptions. *Journal of Volcanology and Geothermal Research* 18:151–164.

MacLeod, D. 1923. *The island beautiful: The story of fifty years in North Formosa.* Toronto: Board of Foreign Missions of the Presbyterian Church in Canada.

Macrae, E. 1992. Book briefing: Easter Island, Earth Island. *Geographical Magazine* 64:41.

Mannion, A. M. 1999. Domestication and the origins of agriculture. *Progress in Physical Geography* 23:37–56.

Mark, R. K., and J. G. Moore. 1987. Slopes of the Hawaiian Ridge. *U.S. Geological Survey Professional Paper* 1350:101–108.

Marshall, J. F., and G. Jacobson. 1985. Holocene growth of a mid-Pacific atoll: Tarawa, Kiribati. *Coral Reefs* 4:11–17.

Martill, D. M., A. R. I. Cruickshank, and M. A. Taylor. 1991. Dispersal via whale bones. *Nature* 351:193.

Martin, J. 1817. *Tonga Islands: William Mariner's account.* London: Printed for the author. Repr., 1981, Neiafu, Tonga: Vava'u Press.

Maslin, M., M. Owen, S. J. Day, and D. Long. 2004. Linking continental-slope failures and climate change: Testing the Clathrate Gun Hypothesis. *Geology* 32:53–56.

Masson, D. G., A. B. Watts, M. J. R. Gee, R. Urgeles, N. C. Mitchell, T. P. Le Bas, and M. Canals. 2002. Slope failures on the flanks of the western Canary Islands. *Earth-Science Reviews* 57:1–35.

Matisoo-Smith, E,. and J. H. Robins. 2004. Origins and dispersals of Pacific peoples: Evidence from mtDNA phylogenies of the Pacific rat. *Proceedings of the National Academy of Sciences of the United States of America* 101:9167–9172.

Matsumoto, T., D. Tappin, and SOS Onboard Scientific Party. 2003. Possible coseismic large-scale landslide off the northern coast of Papua New Guinea in July 1998: Geophysical and geological results from SOS cruises. *Pure and Applied Geophysics* 160:1923–1943.

Maude, H. C., and H. E. Maude, eds. 1984. *An anthology of Gilbertese oral tradition.* Suva: Institute of Pacific Studies, The University of the South Pacific.

Maziére, F. 1965. *Fantastique Île de Pâques.* Paris: Laffont.

McCall, G. 1993. Little Ice Age: Some speculations for Rapanui. *Rapa Nui Journal* 7:65–70.

———. 1994. *Rapanui: Tradition and survival on Easter Island.* St. Leonards: Allen and Unwin.

McEwan, G. F., and B. Dickson. 1978. Valdivia, Jomon fishermen, and the nature of the North Pacific: Some nautical problems with Meggers, Evans, and Estrada's (1965) transoceanic contact thesis. *American Antiquity* 43:362–371.

McGuire, W. J., R. J. Howarth, C. R. Firth, A. R. Solow, A. D. Pullen, S. J. Saunders, I. S. Stewart, and C. Vita-Finzi. 1997. Correlation between rate of sea-level change and the frequency of explosive volcanism in the Mediterranean. *Nature* 389:473–476.

McMurtry, G. M., E. Herrero-Bervera, M. D. Cremer, J. R. Smith, J. Resig, C. Sherman, and M. E. Torresan. 1999. Stratigraphic constraints on the timing and emplacement of the Alika 2 giant Hawaiian submarine landslide. *Journal of Volcanology and Geothermal Research* 94:35–58.

McMurtry, G. M., P. Watts, G. J. Fryer, J. R. Smith, and F. Imamura. 2004. Giant landslides, mega-tsunamis, and paleo-sealevel in the Hawaiian Islands. *Marine Geology* 203:219–233.

McNutt, M. K., and K. M. Fisher. 1987. The South Pacific Superswell. In *Seamounts, islands, and atolls,* ed. B. H. Keating, P. Fryer, R. Batiza, and G. W. Boehlert, 25–34. Geophysical Monograph Series 43. Washington: American Geophysical Union.

McSaveney, M. J., J. R. Goff, D. J. Darby, P. Goldsmith, A. Barnett, S. Elliott, and M. Nongkas. 2000. The 17 July 1998 tsunami, Papua New Guinea: Evidence and initial interpretation. *Marine Geology* 170:81–92.

Meacham, W. 1996. Defining the hundred yue. *Indo-Pacific Prehistory Association, Bulletin* 15:93–99.

Mead, S. M. 1973. Folklore and place names in Santa Ana, Solomon Islands. *Oceania* 43:215–237.

Meggers, B. J., C. Evans, and E. Estrada. 1965. Early formative period of coastal Ecuador: The Valdivia and Machalilla phases. *Smithsonian Contributions to Anthropology* 1.

Melville, R. 1966. Continental drift, Mesozoic continents and migrations of the angiosperms. *Nature* 211:116–120.

Menard, H. W. 1983. Insular erosion, isostasy, and subsidence. *Science* 220:913–918.

———. 1984. Origin of guyots: The Beagle to Seabeam. *Journal of Geophysical Research* 89:11117–11123.

———. 1986. *Islands.* New York: Scientific American Books.

Mennis, M. R. 1978. The existence of Yomba Island near Madang: Fact or fiction. *Oral History* 6:2–81.

———. 1981. Yomba Island: A real or mythical volcano. In *Cooke-Ravian volume of volcanological papers,* ed. R. W. Johnson, 95–99. Memoir 10. Port Moresby: Geological Survey of Papua New Guinea.

Métraux, A. 1940. Ethnology of Easter Island. *Bernice P. Bishop Museum Bulletin* 160.

Michaux, B. 1991. Distributional patterns and tectonic development in Indonesia: Wallace reinterpreted. *Australian Systematic Botany* 4:25–36.

Michelsen, O. 1893. *Cannibals won for Christ: A story of missionary perils and triumphs in Tongoa, New Hebrides.* London: Morgan and Scott.

Mimura, N., and H. Nobuoka. 1995. Verification of the Bruun Rule for the estimation of shoreline retreat caused by sea-level rise. In *Coastal dynamics 95,* ed. W. R. Dally and R. B. Zeidler, 607–616. New York: American Society of Civil Engineers.

Mimura, N., and N. Pelesikoti. 1997. Vulnerability of Tonga to future sea-level rise. *Journal of Coastal Research, Special Issue* 24:117–132.

Mochanov, Y. A. 1980. Early migrations to America in the light of study of the Dyuktai Paleolithic Culture in Northeast Asia. In *Early Native Americans: Prehistoric demography, economy, and technology,* ed. D. L. Browman, 174–177. The Hague: Mouton.

Moerenhout, J. A. 1837. *Voyages aux Îles du Grand Océan.* 2 vols. Paris: Bertrand.

Mokunitulevu Na Rai. undated [1928?]. *Ai Tukutuku kei Viti*. Suva: Methodist Church in Fiji. (Unpublished manuscript, National Archives, Suva.)

Montesquieu, C. L. J. de S. 1949. *L'esprit des lois (The spirit of laws)*, trans. T. Nugent. New York: Hafner.

Montiton, A. 1874. *Les Paumotous*. Bullétin 6. Lyon: Les Mission Catholiques.

Monzier, M., C. Robin, and J. P. Eissen. 1994. Kuwae (c. 1425): The forgotten caldera. *Journal of Volcanology and Geothermal Research* 59:207–218.

Moore, C. A. 2002. Awash in a rising sea — How global warming is overwhelming the islands of the tropical Pacific. *International Wildlife* January–February: 1–5.

Moore, J. G., and G. W. Moore. 1984. Deposit from a giant wave on the island of Lanai, Hawaii. *Science* 226:1312–1315.

Moore, J. G., W. B. Bryan, and K. R. Kudwig. 1994a. Chaotic deposition by a giant wave, Molokai, Hawaii. *Geological Society of America, Bulletin* 106:962–967.

Moore, J. G., D. A. Clague, R. T. Holcomb, P. W. Lipman, W. R. Normark, and M. E. Torresan. 1989. Prodigious submarine landslides on the Hawaiian Ridge. *Journal of Geophysical Research* 94:17,465–17,484.

Moore, J. G., W. R. Normark, and R. T. Holcomb. 1994b. Giant Hawaiian landslides. *Annual Review of Earth and Planetary Sciences* 22:119–144.

Morgan, J. K., G. F. Moore, and D. A. Clague. 2003. Slope failure and volcanic spreading along the submarine south flank of Kilauea Volcano, Hawaii. *Journal of Geophysical Research* 108:2415, doi:10.1029/2003JB002411.

Morvan, F. 1980. *Legends of the sea*. Geneva: Minerva.

Mueller-Dombois, D., and F. R. Fosberg. 1998. *Vegetation of the tropical Pacific Islands*. New York: Springer.

Murray, J. 1880. On the structure and origin of coral reefs and islands. *Proceedings of the Royal Society of Edinburgh* 10:505–518.

Nakicenovic, N., J. Alcamo, G. Davis, B. de Vries, J. Fenhann, S. Gaffin, K. Gregory, A. Grubler, T. Y. Jung, T. Kram, E. L. La Rovere, L. Michaelis, S. Mori, T. Morita, W. Pepper, H. Pitcher, L. Proce, K. Riahi, A. Roehrl, H.-H. Rogner, A. Sankovski, M. Schlesinger, P. Shukla, S. Smith, R. Swart, S. van Rooijen, N. Victor, and Z. Dadi. 2000. *Special report on emissions scenarios: A special report of Working Group III of the Intergovernmental Panel on Climate Change*. Cambridge: Cambridge University Press.

Nance, R. D., T. R. Worsley, and J. B. Moody. 1988. The supercontinent cycle. *Scientific American* 259:72–79.

Nedachi, M., T. Maeda, R. Shigeyoshi, A. Inoue, K. Shimada, M. Higashi, A. Habano, T. Azuma, and A. Hatta. 2001. Geological environments of Yap Islands, Micronesia. *Kagoshima University Research Center for the Pacific Islands, Occasional Paper* 34:69–76.

Nelson, G. 1985. A decade of challenge: The future of biogeography. *Earth Sciences History* 4:187–196.

Nelson, J. S. 1978. *Limnichthys polyactis,* a new species of blennioid fish from New Zealand, with notes on the taxonomy and distribution of other Creediidae (including Limnichthyidae). *New Zealand Journal of Zoology* 5:351–364.

Nelson, S. M. 1990. Neolithic sites in northeastern China and Korea. *Antiquity* 64:234–248.

Neumann, A. C., and I. Macintyre. 1985. Reef response to sea-level rise: Keep-up, catch-up or give-up. *Proceedings of the 5th International Coral Reef Congress* 3:105–110.

Newbrough, J. B. 1932. *Oahspe: A Kosmon revelation in the words of Jehovih and his angel ambassadors*. Los Angeles: Kosmon.

Newell, J. E. 1895. The legend of the coming of Nareau from Samoa to Tarawa, and his return to Samoa. *Journal of the Polynesian Society* 4:231–235.

Newman, W. A., and B. A. Foster. 1983. The Rapanuian faunal district (Easter and Sala y Gómez): In search of ancient archipelagos. *Bulletin of Marine Science* 33:633–644.

Nicholls, R. J. 2004. Coastal flooding and wetland loss in the 21st century: Changes under the SRES climate and socio-economic scenarios. *Global Environmental Change* 14:69–86.

Nisbet, E. G., and D. J. W. Piper. 1998. Giant submarine landslides. *Nature* 392:329–330.

Nott, J. 1997. Extremely high-energy wave deposits inside the Great Barrier Reef, Australia: Determining the cause — Tsunami or tropical cyclone. *Marine Geology* 141:193–207.

Nunn, P. D. 1994. *Oceanic islands.* Oxford: Blackwell.

———. 1995a. Holocene tectonic histories for five islands in the south-central Lau Group, South Pacific. *The Holocene* 5:160–171.

———. 1995b. Lithospheric flexure in Southeast Fiji consistent with the tectonic history of islands in the Yasayasa Moala. *Australian Journal of Earth Sciences* 42:377–389.

———. 1996. *Emerged shorelines of the Lau Islands.* Memoir 4. Suva: Fiji Mineral Resources Department.

———. 1998a. Late Quaternary tectonic change on the islands of the northern Lau-Colville Ridge, Southwest Pacific. In *Coastal tectonics,* ed. I. S. Stewart and C. Vita-Finzi, 269–278. Special Publication 146. London: Geological Society of London.

———. 1998b. *Pacific Island landscapes.* Suva: Institute of Pacific Studies, The University of the South Pacific.

———. 1999. *Environmental change in the Pacific Basin: Chronologies, causes, consequences.* London: Wiley.

———. 2000. Environmental catastrophe in the Pacific Islands about AD 1300. *Geoarchaeology* 15:715–740.

———. 2001a. Sea-level change in the Pacific. In *Sea-level changes and their effects,* ed. J. Noye and M. Grzechnik, 1–23. Singapore: World Scientific Publishing.

———. 2001b. On the convergence of myth and reality: Examples from the Pacific Islands. *The Geographical Journal* 167:125–138.

———. 2003. Fished-up or thrown-down: The geography of Pacific Island origin myths. *Annals of the Association of American Geographers* 93:350–364.

———. 2004a. Through a mist on the ocean: Human understanding of island environments. *Tijdschrift voor Economische en Sociale Geografie* 95:311–325.

———. 2004b. Understanding and adapting to sea-level change. In *Global environmental issues,* ed. F. Harris, 45–64. Chichester: Wiley.

———. 2004c. Myths and the formation of Niue Island, central South Pacific. *The Journal of Pacific History* 39:99–108.

———. 2007a. Space and place in an ocean of islands: Musings on the attitudes of the Lapita people towards islands and colonization. *South Pacific Studies* 27:24–35.

———. 2007b. *Climate, environment, and society in the Pacific during the last millennium.* Amsterdam: Elsevier.

Nunn, P. D., M. Baniala, M. Harrison, and P. Geraghty. 2006a. Vanished islands in Vanuatu: New research and a preliminary geohazard assessment. *Journal of the Royal Society of New Zealand* 36:37–50.

Nunn, P. D., and J. M. R. Britton. 2004. The long-term evolution of Niue Island. In *Niue Island: Geographical perspectives on the rock of Polynesia,* ed. J. Terry and W. Murray, 31–74. Paris: INSULA (International Scientific Council for Island Development).

Nunn, P. D., P. Geraghty, E. Nakoro, A. Nasila, and S. Tukidia. 2005. Location and palaeogeography of allegedly vanished islands in Fiji. *People and Culture in Oceania* 21:83–109.

Nunn, P. D., T. Heorake, E. Tegu, B. Oloni, K. Simeon, L. Wini, S. Usuramo, and P. Geraghty. 2006b. Geohazards revealed by myths in the Pacific: A study of islands that have disappeared in Solomon Islands. *South Pacific Studies* 27:37–49.

Nunn, P. D., R. Hunter-Anderson, M. T. Carson, F. Thomas, S. Ulm, and M. Rowland. 2007a. Times of plenty, times of less: Chronologies of last-millennium societal disruption in the Pacific Basin. *Human Ecology: An Interdisciplinary Journal* 35:385–401.

Nunn, P. D., T. Ishimura, W. R. Dickinson, K. Katayama, F. Thomas, R. Kumar, S. Matararaba, J. Davidson, and T. Worthy. 2007b. The Lapita occupation at Naitabale, Moturiki Island, central Fiji. *Asian Perspectives* 46:96–132.

Nunn, P. D., C. D. Ollier, G. S. Hope, P. Rodda, A. Omura, and W. R. Peltier. 2002. Late Quaternary sea-level and tectonic changes in Northeast Fiji. *Marine Geology* 187:299–311.

Nunn, P. D., and A. Omura. 1999. Penultimate Interglacial emerged reef around Kadavu Island, Southwest Pacific: Implications for late Quaternary island-arc tectonics and sea-level history. *New Zealand Journal of Geology and Geophysics* 42:219–227.

Nunn, P. D., and M. R. Pastorizo. 2006. Geological histories and geohazard potential of Pacific Islands illuminated by myths. In *Myth and geology,* ed. L. Piccardi and W. B. Masse, 143–163. Special Publication 273. London: Geological Society of London.

Nunn, P. D., and W. R. Peltier. 2001. Far-field test of the ICE-4G (VM2) model of global isostatic response to deglaciation: Empirical and theoretical Holocene sea-level reconstructions for the Fiji Islands, Southwest Pacific. *Quaternary Research* 55:203–214.

Nunn, P. D., and E. Waddell. 1992. *Implications of climate change and sea-level rise for the Kingdom of Tonga.* Reports and Studies 58. Apia: South Pacific Regional Environment Programme.

Nur, A., and Z. Ben-Avraham. 1977. Lost Pacifica continent. *Nature* 270:41–43.

O'Connell, J. F., and J. Allen. 2004. Dating the colonization of Sahul (Pleistocene Australia–New Guinea): A review of recent research. *Journal of Archaeological Science* 31:835–853.

Okal, E. A., C. E. Synolakis, G. J. Fryer, P. Heinrich, J. C. Borrero, C. Ruscher, D. Arcas, G. Guille, and D. Rousseau. 2002. A field survey of the 1946 Aleutian tsunami in the far field. *Seismological Research Letters* 73:490–503.

Ollier, C. D., and C. Pain. 2000. *The origin of mountains.* London: Routledge.

Oppenheimer, S. 1998. *Eden in the east: The drowned continent of Southeast Asia.* London: Weidenfeld and Nicolson.

———. 2003. *Out of Eden: The peopling of the world.* London: Constable.

Orbell, M. R. 1985. *Hawaiki: A new approach to Maori tradition.* Christchurch: University of Canterbury.

Oshima, O. 2002. An overview of the recent submarine and nearshore eruptions in Japan. Unpublished abstract, 18. Chapman Conference on Explosive Subaqueous Volcanism, Dunedin, New Zealand, 21–25 January 2002, www.otago.ac.nz/Geology/staff/jdlw/program_w_abstracts.pdf (page 16, accessed February 2008).

Ota, Y., J. Chappell, R. Kelley, N. Yonekura, T. Nishimura, E. Matsumoto, and J. Head. 1993. Holocene coral reef terraces and coseismic uplift of Huon Peninsula, Papua New Guinea. *Quaternary Research* 40:177–188.

Ota, Y., T. Miyauchi, R. Paskoff, and M. Koba. 1995. Plio-Quaternary marine terraces and their deformation along the Altos de Talinay, North-central Chile. *Revista Geológica de Chile* 22:89–102.

Pang, K. D. 1993. Climatic impact of the mid-fifteenth century Kuwae caldera formation as reconstructed from historical and proxy data. *EOS (Transactions of the American Geophysical Union)* 74:106.

Paton, J. G. 1890. *John G. Paton, missionary to the New Hebrides: An autobiography*, 6th ed. London: Hodder and Stoughton.

Pavlides, C., and C. Gosden. 1994. 35,000-year-old sites in the rainforests of West New Britain, Papua New Guinea. *Antiquity* 68:604–610.

Pelling, M., and J. I. Uitto. 2001. Small island developing states: Natural disaster vulnerability and global change. *Environmental Hazards* 3:49–62.

Percival, W. H. 1964. People lived there last century — Now they've disappeared. *Pacific Islands Monthly* 77:79–80.

Perry, W. J. 1923. *The children of the sun (a study in the early history of civilization)*. New York: Dutton.

Philip, G. 1932. *Handy-volume atlas of the world*, 18th rev. ed. London: George Philip and Son.

Phinney, E. J., P. Mann, M. F. Coffin, and T. H. Shipley. 1999. Sequence stratigraphy, structure, and tectonic history of the southwestern Ontong Java Plateau adjacent to the North Solomon Trench and Solomon Islands arcs. *Journal of Geophysical Research* 104:20449–20466.

———. 2004. Sequence stratigraphy, structural style, and age of deformation of the Malaita accretionary prism (Solomon Arc–Ontong Java Plateau convergent zone). *Tectonophysics* 389:221–246.

Pickering, W. H. 1924. The separation of the continents by fission. *Geological Magazine* 62:31–34.

Pilger, R. H., and D. W. Handschumacher. 1981. The fixed hot-spot hypothesis and origin of the Easter–Sala y Gomez–Nazca trace. *Geological Society of America, Bulletin* 92:437–446.

Plafker, G. 1972. Alaskan earthquake of 1964 and Chilean earthquake of 1960: Implications for arc tectonics. *Journal of Geophysical Research* 77:901–925.

Pole, M. 2001. Can long-distance dispersal be inferred from the New Zealand fossil plant record? *Australian Journal of Botany* 49:357–366.

Ponomareva, V. V., I. V. Melekestsev, P. R. Kyle, P. G. Rinkleff, O. V. Dirksen, L. D. Sulerzhitsky, N. E. Zaretskaia, and R. Rourke. 2004. The 7600 (^{14}C) year BP Kurile Lake caldera-forming eruption, Kamchatka, Russia: Stratigraphy and field relationships. *Journal of Volcanology and Geothermal Research* 136:199–222.

Porcasi, P., J. F. Porcasi, and C. O'Neill. 1999. Early Holocene coastlines of the California Bight: The Channel Islands as first visited by humans. *Pacific Coast Archaeological Society Quarterly* 35:1–24.

Porter, D. 1823. *A voyage in the South Seas, in the years 1812, 1813, and 1814*. London: Phillips.

Quammen, D. 1997. *The song of the dodo: Island biogeography in an age of extinctions*. New York: Simon and Schuster.

Quatrefages, A. de. 1864. *Les Polynésiens et leurs migrations*. Paris: A. Bertrand.

Rainbird, P. 2004. *The archaeology of Micronesia*. Cambridge: Cambridge University Press.

Ramage, E. S. 1978. Perspectives ancient and modern. In *Atlantis: Fact or fiction?*, ed. E. S. Ramage, 3–45. Bloomington: Indiana University Press.

Ramírez, J. M. 1990/1991. Transpacific contacts: The Mapuche connection. *Rapa Nui Journal* 4:53–55.

Ramsay, R. H. 1972. *No longer on the map: Discovering places that never were*. New York: Viking.

Rawling, T. J., and G. S. Lister. 1999. Oscillating modes of orogeny in the Southwest Pacific and the tectonic evolution of New Caledonia. In *Exhumation processes: Normal faulting, ductile flow and erosion*, ed. U. Ring, M. T. Brandon, G. S. Lister, and S. D. Willett, 109–127. Special Publication 154. London: Geological Society of London.

Regelous, M., A. W. Hofmann, W. Abouchami, and S. J. G. Galer. 2003. Geochemistry of lavas from the Emperor Seamounts, and the geochemical evolution of Hawaiian magmatism from 85 to 42 Ma. *Journal of Petrology* 44:113–140.

Rehder, H. A. 1980. The marine mollusks of Easter Island and Sala y Gómez. *Smithsonian Contributions to Zoology* 167.

Repenning, C. A., and C. E. Ray. 1977. The origin of the Hawaiian monk seal. *Proceedings of the Biological Society of Washington* 89:667–688.

Rice, W. H. 1923. Hawaiian legends. *Bernice P. Bishop Museum Bulletin* 3.

Robertson, G. 1948. *The discovery of Tahiti: A journal of the second voyage of H.M.S. Dolphin round the world.* London: Hakluyt Society.

Robin, C., and M. Monzier. 1994. Volcanic hazards in Vanuatu. Government of Vanuatu, Department of Geology, Mines and Water Resources, report. 15 pp.

Romoli, K. 1953. *Balboa of Darién, discoverer of the Pacific.* New York: Doubleday.

Roth, J., and S. Hooper. 1990. *Fiji journals of Baron von Hugel, 1875–1877.* Suva: Fiji Museum.

Rotondo, G. M., V. G. Springer, G. A. J. Scott, and S. O. Schlanger. 1981. Plate movement and island integration—A possible mechanism in the formation of endemic biotas, with special reference to the Hawaiian Islands. *Systematic Zoology* 30:12–21.

Routledge, K. P. 1919. *The mystery of Easter Island.* London: Hazell, Watson, and Viney.

Royal, T. A. C. 2005. Hawaiki. In *The encyclopedia of New Zealand,* www.teara.govt.nz (accessed October 6, 2005).

Sabatier, E. 1971. *Gilbertese-English dictionary.* Nouméa: South Pacific Commission Publications Bureau.

Sand, C., ed. 2003. *Pacific archaeology: Assessments and prospects (Proceedings of the International Conference for the 50th Anniversary of the First Lapita Excavation, Koné-Nouméa 2002).* Nouméa: Services des Musées et du Patrimoine.

Sandwell, D. T., and W. F. Smith. 1997. Marine gravity anomaly from Geosat and ERS-1 satellite altimetry. *Journal of Geophysical Research* 102:10039–10054.

Sanmartín, I., and F. Ronquist. 2004. Southern Hemisphere biogeography inferred by event-based models: Plant versus animal patterns. *Systematic Biology* 53:216–243.

Sano, Y., K. Terada, H. Hidaka, K. Yokoyama, and A. P. Nutman. 1999. Palaeoproterozoic thermal events recorded in the ~4.0 Ga Acasta gneiss, Canada: Evidence from SHRIMP U-Pb dating of apatite and zircon. *Geochimica et Cosmochimica Acta* 63:899–905.

Santos, A. 2005. *Atlantis, the lost continent finally found.* Miami: Atlantis Publications.

Satake, K., and Y. Kato. 2001. The 1741 Oshima-Oshima eruption: Extent and volume of submarine debris avalanche. *Geophysical Research Letters* 28:427–430.

Saunders, N. 1992. Prehistoric myths demolished. *New Scientist* 135:42.

Savin, S. M,. and R. G. Douglas. 1985. Sea level, climate, and the central American land bridge. In *The great American biotic interchange,* ed. F. G. Stehli and S. D. Webb, 303–324. New York: Plenum.

Scarr, D. 2001. *A history of the Pacific Islands.* Richmond: Curzon.

Setterfield, T. N., P. C. Eaton, W. J. Rose, and R. S. J. Sparks. 1991. The Tavua Caldera, Fiji: A complex shoshonitic caldera formed by concurrent faulting and downsagging. *Journal of the Geological Society of London* 148:115–127.

Sharp, A. 1963. *Ancient voyages in Polynesia.* Sydney: Angus and Robertson.

Sherwood, A., and R. Howorth, eds. 1996. Coasts of Pacific Islands. *SOPAC Miscellaneous Report* 222. 39 pp.

Sibley, C. G., and J. E. Ahlquist. 1982. The relationships of the Hawaiian honeycreepers (Drepaninini) as indicated by DNA-DNA hybridization. *Auk* 99:130–140.

Siebert, L. 1992. Threats from debris avalanches. *Nature* 356:658–659.

Sim, R. 1994. Prehistoric human occupation in the King and Furneaux island regions, Bass Strait. In *Archaeology in the north: Proceedings of the 1993 Australian Archaeological Association Conference,* ed. M. Sullivan, S. Brockwell, and A. Webb, 358–374. Darwin: North Australia Research Unit, Australian National University.

Singh, G., A. P. Kershaw, and R. Clark. 1981. Quaternary vegetation and fire history in Australia. In *Fire and the Australian biota,* ed. A. M. Gill, R. H. Groves, and I. R. Noble, 23–54. Canberra: Australian Academy of Science.

Sinoto, Y. H. 1979. Excavations on Huahine, French Polynesia. *Pacific Studies* 3:1–40.

Slobodin, R. 1978. *W. H. R. Rivers.* New York: Columbia University Press.

Smith, S. P. 1899. History and traditions of Rarotonga. *Journal of the Polynesian Society* 8:61–88.

———. 1904. *Hawaiki: The original home of the Maori.* Christchurch: Whitcombe and Tombs.

Snow, C. P. 1959. *The two cultures and the scientific revolution.* New York: Cambridge University Press.

Solem, A. 1976. *Endodontoid land snails from Pacific Islands (Mollusca: Pulmonata: Sigmurethra). Part I. Family Endodontidae. Part II. Families Punctidae and Charopidae, Zoogeography.* Chicago: Field Museum of Natural History.

Soloviev, S. L., and C. N. Go. 1984. *Catalog of tsunamis on the western shore of the Pacific Ocean.* Ottawa: National Research Council.

Sopher, D. E. 1977. *The sea nomads.* Singapore: National Museum of Singapore.

Sorrenson, M. P. K. 1979. *Maori origins and migrations.* Auckland: Auckland University Press.

Specht, J., and C. Gosden. 1997. Dating Lapita pottery in the Bismarck Archipelago. *Asian Perspectives* 36:175–199.

Spence, L. 1926. *The history of Atlantis.* Philadelphia: McKay.

———. 1933. *The problem of Lemuria (The sunken continent of the Pacific).* Philadelphia: McKay.

———. 1942. *Will Europe follow Atlantis?* London: Rider and Co.

Spencer, T., D. R. Stoddart, and C. D. Woodroffe. 1987. Island uplift and lithospheric flexure: Observations and cautions from the South Pacific. *Zeitschrift für Geomorphologie, Supplementband* 63:87–102.

Spengler, S. R., F. L. Peterson, and J. F. Mink. 1992. *Geology and hydrogeology of the island of Pohnpei, Federated States of Micronesia.* Water Resources Research Center, Technical Report 189. Honolulu: University of Hawai'i at Mānoa.

Squires, R. L., J. L. Goedert, and L. G. Barnes. 1991. Whale carcasses. *Nature* 349:574.

Stainforth, D. A., T. Aina, C. Christensen, M. Collins, N. Faull, D. J. Frame, J. A. Kettleborough, S. Knight, A. Martin, J. M. Murphy, C. Piani, D. Sexron, L. A. Smith, R. A. Spicer, A. J. Thorpe, and M. R. Allen. 2005. Uncertainty in predictions of the climate response to rising levels of greenhouse gases. *Nature* 433:403–406.

Stair, J. B. 1896. Jottings on the mythology and spirit-lore of old Samoa. *Journal of the Polynesian Society* 5:33–57.

Steadman, D. W., and P. S. Martin. 2003. The late Quaternary extinction and future resurrection of birds on Pacific Islands. *Earth-Science Reviews* 61:133–147.

Steadman, D. W., G. K. Pregill, and D. V. Burley. 2002. Rapid prehistoric extinction of birds and iguanas in Polynesia. *Proceedings of the National Academy of Sciences of the United States of America* 99:3673–3677.

Stehli, F. G., and S. D. Webb, eds. 1985. *The great American biotic interchange.* New York: Plenum.

Steiner, R. 1911. *The submerged continents of Atlantis and Lemuria: Their history and civilization, being chapters from the Akashic records.* London: Theosophical Publishing Company.

Stevenson, R. L. 1900. *In the South Seas.* London: Chatto and Windus.

Stimson, J. F. 1937. Tuamotuan legends (island of Anaa): Part I: The demigods. *Bernice P. Bishop Museum Bulletin* 148.

St. Johnston, T. R. 1918. *The Lau Islands (Fiji) and their fairy tales and folk-lore.* London: Times Book Company.

———. 1921. *The islanders of the Pacific, or the children of the sun.* London: Fisher Unwin.

Stoddart, D. R. 1965. The shape of atolls. *Marine Geology* 3:369–383.

———. 1971. Environment and history in Indian Ocean reef morphology. *Symposium of the Zoological Society of London* 28:3–38.

Stommel, H. 1984. *Lost islands: The story of islands that have vanished from nautical charts.* Vancouver: University of British Columbia Press.

Storey, A. A., J. M. Ramírez, D. Quiroz, D. V. Burley, D. J. Addison, R. Walter, A. J. Anderson, T. L. Hunt, J. S. Athens, L. Huynen, and E. A. Matisoo-Smith. 2007. Radiocarbon and DNA evidence for a pre-Columbian introduction of Polynesian chickens to Chile. *Proceedings of the National Academy of Sciences of the United States of America* 104:10335–10339.

Strippel, D. 1972. *Amelia Earhart: The myth and the reality.* New York: Exposition Press.

Taira, A., P. Mann, and R. Rahardiawan. 2004. Incipient subduction of the Ontong Java Plateau along the North Solomon Trench. *Tectonophysics* 389:247–266.

Talandier, J., and F. Bourrouilh-Le-Jan. 1988. High energy sedimentation in French Polynesia: Cyclone or tsunami? In *Natural and man-made hazards,* ed. M. I. El-Sabh and T. S. Murty, 193–199. Rimiuski: Department of Oceanography, University of Quebec.

Taylor, F. W., B. L. Isacks, C. Jouannic, A. L. Bloom, and J. Dubois. 1980. Coseismic and Quaternary vertical tectonic movements, Santo and Malekula islands, New Hebrides Island Arc. *Journal of Geophysical Research* 85:5367–5381.

Taylor, P. W. 1995. Myths, legends and volcanic activity: An example from northern Tonga. *Journal of the Polynesian Society* 104:323–346.

Taylor, R. 1870. *Te Ika a Maui, or New Zealand and its inhabitants,* 2nd ed. Wanganui: Macintosh.

Te-ariki-tara-are. 1920. History and traditions of Rarotonga, trans. S. P. Smith, Part XIV. *Journal of the Polynesian Society* 29:165–188.

Terrell, J. E. 2004. 'Austronesia' and the great Austronesian migration. *World Archaeology* 36:586–590.

Thiede, J., W. E. Dean, D. K. Rea, T. L. Vallier, and C. G. Adelsack. 1981. The geologic history of the mid-Pacific mountains in the central North Pacific Ocean: A synthesis of deep sea drilling studies. In *Initial reports of the Deep-Sea Drilling Project.* Vol. 62, ed. J. Thiede and T. L. Vallier, 1073–1120. Washington: U.S. Government Printing Office.

Thiel, B. 1987. Early settlement of the Philippines, eastern Indonesia, and Australia–New Guinea: A new hypothesis. *Current Anthropology* 28:236–241.

Thomson, K. S. 1977. The pattern of diversification among fishes. In *Patterns of evolution as illustrated by the fossil record,* ed. A. Hallam, 377–404. Amsterdam: Elsevier.

Thomson, W. J. 1891. *Te Pito te Henua; or, Easter Island.* Washington: Smithsonian Institution.

Thorne, R. F. 1963. Biotic distribution patterns in the tropical Pacific. In *Pacific Basin biogeography,* ed. J. L. Gressitt, 311–350. Honolulu: Bishop Museum Press.

Tingey, R. J. 1991. *The geology of Antarctica.* Oxford: Clarendon Press.

Todorovska, M. I., A. Hayir, and M. D. Trifunac. 2002. A note on tsunami amplitudes above submarine slides and slumps. *Soil Dynamics and Earthquake Engineering* 22:129–141.

Tracey, J. I., Jr., S. O. Schlanger, J. T. Stark, D. B. Doan, and H. G. May. 1964. General geology of Guam. *U.S. Geological Survey Professional Paper* 403-A: 104 pp.

Tregear, E. 1891. *The Maori-Polynesian comparative dictionary.* Wellington: Lyon and Blair.

Turner, S., C. Hawkesworth, N. Rogers, J. Bartlett, T. Worthington, J. Hergt, J. Pearce, and I. Smith. 1997. ^{238}U-^{230}Th disequilibria, magma petrogenesis and flux rates beneath the depleted Tonga-Kermadec Island Arc. *Geochimica et Cosmochimica Acta* 61:4855–4884.

Underhill, P. A., G. Passarino, A. A. Lin, S. Marzuki, P. J. Oefner, L. L. Cavalli-Sforza, and G. K. Chambers. 2001. Maori origins, Y-chromosome haplotypes and implications for human history in the Pacific. *Human Mutation* 17:271–280.

Usuramo, S. 2004. A dictionary of place names in Mara Masike (Small Malaita), Solomon Islands. Graduate dissertation in linguistics, The University of the South Pacific, Suva, Fiji.

Vagvolgyi, J. 1975. Body size, aerial dispersal, and origin of the Pacific Island snail fauna. *Systematic Zoology* 24:465–488.

Van Duzer, C. 2004. *Floating islands: A global bibliography.* Los Altos Hills: Cantor Press.

Veron, J. E. N. 2000. *Corals of the world.* Townsville: Australian Institute of Marine Science.

Vulava, J. 1996. *Ko Vuniivilevu se ko Davetalevu.* Unpublished manuscript in possession of Paul Geraghty, The University of the South Pacific, Suva, Fiji.

Wafer, L. 1934. *A new voyage and description of the Isthmus of America.* 2nd Series, no. 73. Oxford: Hakluyt Society.

Wagner, W. L., D. R. Herbst, and S. H. Somer. 1990. *Manual of the flowering plants of Hawai'i.* Honolulu: University of Hawai'i Press, Bishop Museum Press.

Waight, T. E., S. D. Weaver, R. Maas, and G. N. Eby. 1998. French Creek granite and Hohonu dyke swarm, South Island, New Zealand: Late Cretaceous alkaline magmatism and the opening of the Tasman Sea. *Australian Journal of Earth Sciences* 45:823–835.

Walker, D. A., and C. S. McCreery. 1988. Deep ocean seismicity: Seismicity of the northwestern Pacific Basin interior. *EOS (Transactions of the American Geophysical Union)* 69:737, 742–743.

Wallace, A. R. 1892. *Island life: Or the phenomena and causes of insular faunas and floras.* London: Macmillan.

Walter, R., and A. Anderson. 1995. Archaeology of Niue Island: Initial results. *Journal of the Polynesian Society* 104:471–481.

Wang, Y. 1998. Sea-level changes, human impacts and coastal responses in China. *Journal of Coastal Research* 14:31–36.

Ward, E. V. 1985. Local lore and the Earhart case. *Pacific Islands Monthly* 56:9.

Ward, R. G. 1989. Earth's empty quarter? The Pacific Islands in the Pacific century. *The Geographical Journal* 155:235–246.

Ward, R. G., and M. Brookfield. 1992. The dispersal of the coconut: Did it float or was it carried to Panama? *Journal of Biogeography* 19:467–480.

Ward, S. N. 2001. Landslide tsunami. *Journal of Geophysical Research* 106:11201–11215.

———. 2002. Slip-sliding away. *Nature* 415:973–974.

Ward, S. N., and S. J. Day. 2001. Cumbre Vieja Volcano: Potential collapse and tsunami at La Palma, Canary Islands. *Geophysical Research Letters* 28:3397–3400.

———. 2003. Ritter Island Volcano—Lateral collapse and tsunami of 1888. *Geophysical Journal International* 154:891–902.

Warden, A. J. 1970. Evolution of Aoba Caldera Volcano. *Bulletin Volcanologique* 34:107–140.

Warrick, R. A., E. M. Barrow, and T. M. L. Wigley, eds. 1993. *Climate and sea level change: Observations, projections, and implications.* Cambridge: Cambridge University Press.

Washington, P. 1993. *Madame Blavatsky's baboon: Theosophy and the emergence of the western guru.* London: Secker and Warburg.

Waterhouse, J. 1866. *The king and people of Fiji.* London: Wesleyan Conference Office.

Webster, J. M., D. A. Clague, and J. C. Braga. 2007. Support for the giant wave hypothesis: Evidence from submerged terraces off Lanai, Hawaii. *International Journal of Earth Sciences* 96:517–524.

Wegener, A. 1915. *Die Entstehung der Kontinente und Ozeane.* Braunschweig: Vieweg.

Weissel, J. K., and R. N. Anderson. 1978. Is there a Caroline Plate? *Earth and Planetary Science Letters* 41:143–158.

Welsch, R. L. 1998. *An American anthropologist in Melanesia: A. B. Lewis and the Joseph N. Field South Pacific Expedition 1909–1913.* 2 vols. Honolulu: University of Hawai'i Press.

Westervelt, W. D. 1910. *Legends of Ma-ui — A demi-god of Polynesia and of his mother Hina.* Honolulu: Hawaiian Gazette Co.

Whittaker, R. J., and J. M. Fernández-Palacios. 2007. *Island biogeography: Ecology, evolution, and conservation,* 2nd ed. Oxford: Oxford University Press.

Williams, M. R. 2001. *In search of Lemuria: The lost Pacific continent in legend, myth and imagination.* San Mateo: Golden Era Books.

Williamson, R. W. 1933. *Religious and cosmic beliefs of central Polynesia.* 2 vols. Cambridge: Cambridge University Press.

Wilson, A. T., C. H. Hendy, and C. P. Reynolds. 1979. Short-term climate change and New Zealand temperatures during the last millennium. *Nature* 279:315–317.

Winchester, S. 2003. *Krakatoa: The day the world exploded, 27 August 1883.* London: Viking.

Winterer, E. L. 1995. Karst morphology and diagenesis of top of Albian limestone platforms, mid-Pacific mountains. *Proceedings of the Ocean Drilling Program, Scientific Results* 143 *(Western Pacific Guyots):* 433–468.

Winterer, E. L., P. F. Lonsdale, J. L. Matthews, and B. R. Rosendahl. 1974. Structure and acoustic stratigraphy of the Manihiki Plateau. *Deep-Sea Research* 21:793–814.

Wolfe, C., M. McNutt, and R. S. Detrick. 1994. The Marquesas archipelagic apron: Seismic stratigraphy and implications for volcano growth, mass wasting and crustal underplating. *Journal of Geophysical Research* 99:13591–13608.

Woolnough, W. G. 1903. The continental origin of Fiji. Part I — General geology. *Proceedings of the Linnean Society of New South Wales* 1903:457–496.

Wright, S. D., C. G. Yong, J. W. Dawson, D. J. Whittaker, and R. C. Gardner. 2000. Riding the ice age El Niño? Pacific biogeography and evolution of *Metrosideros* subgenus *Metrosideros* (Myrtaceae) inferred from nrDNA. *Proceedings of the National Academy of Sciences of the United States of America* 97:4118–4123.

———. 2001. Stepping stones to Hawaii: A trans-equatorial dispersal pathway for *Metrosideros* (Myrtaceae) inferred from nrDNA (ITA+ETS). *Journal of Biogeography* 28:769–774.

Wu, X., and F. E. Poirier. 1995. *Human evolution in China.* Oxford: Oxford University Press.

Wyatt, S. 2004. Ancient transpacific voyaging to the New World via Pleistocene South Pacific Islands. *Geoarchaeology* 19:511–529.

Wyrtki, K. 1990. Sea level rise: The facts and the future. *Pacific Science* 44:1–16.

Yan, C., and L. W. Kroenke. 1993. A plate tectonic reconstruction of the Southwest Pacific, 0–100 Ma. *Proceedings of the Ocean Drilling Project (ODP), Scientific Results* 130:697–709.

Yonekura, N. 1983. Late Quaternary vertical crustal movements in and around the Pacific as deduced from former shoreline data. In *Geodynamics of the western Pacific–Indonesian Region,* ed. T. W. C. Hilde and S. Uyeda, 41–50. Washington: American Geophysical Union.

Young, J. L. 1898. The origin of the name Tahiti: As related by Marerenui, a native of Faaiti Island, Paumotu Group. *Journal of the Polynesian Society* 7:109–110.

Zambucka, K. 1978. *Vuda: The source.* Honolulu: Mana Publishing.

Zimmerman, E. C. 1948. *Insects of Hawaii.* Vol. 1. Honolulu: University of Hawai'i Press.

Index

AD 1300 Event, 81–82, 221n.50, 231n.4

agriculture, 73, 75–77, 128, 160

Alaska, 9, 15, 18, 44, 66, 68, 139–140

Aleutian Islands, 15, 47, 50, 140, 190; Bogoslof Island, 140–141

Alexa Bank, 39–40, 213n.16

'Alika Slide, 44–45, 214n.32

American Group (of lost islands), 106

Antarctica, 5, 9, 17, 19–21, 31, 63–65, 129, 131, 160, 230nn.87–89

archetype, 130, 198, 230n.1

Atlantis, 1–3, 17, 118–119, 128–129, 196–198, 217n.10, 225n.1, 228n.45, 229nn.63, 67, 230n.86; Plato on, 1, 108–109, 145; Spence on, 121, 199, 228n.58

atoll: drowned, 30; formation, 27–29, 102, 117; islands, 92, 99, 133, 156, 170, 183, 185

Australia, 5, 9–10, 20–21, 46–47, 63–66, 70, 72, 110, 131–132, 231n.6; Flinders Island, 132; Kangaroo Island, 131; King Island, 132; Tasmania, 9, 132

Belau. See Palau

Bellona Platform, 32, 39, 181

Bering Strait, 20, 73, 219n.17

Beveridge Reef, 25, 139, 177

bikenikarakara, 89, 173–174

Burotu, 4, 89, 97, 105, 137, 163–168, 198, 234n.59, 235n.65

caldera, 48, 50–51, 142, 145–146, 160, 214n.34, 215n.55

California, 66, 110, 121, 131, 180, 216n.79

Canary Islands, 5, 188, 193

cannibalism, 82, 118, 169, 218n.6

catastrophic rise event, 76–77, 190

Central America, 11, 81, 216n.89

Channel Islands (California), 89, 180–181

channeling, 118–119, 197

China, 30, 64, 71–72, 75, 178, 187, 218nn.1, 5, 6, 220n.34

Christianity, 8, 27, 83, 132, 164, 205

Circum-Antarctic Current, 19, 129

Clathrate Gun Hypothesis, 189, 238n.24

climate change, 82, 182, 184, 221nn.48, 50, 237n.1

coconuts, 110, 171, 176, 178, 206, 216n.89, 226n.6

collapse (land): island-flank, 43–49, 133–138, 188–191

colonization: human, 74, 78–79, 81, 100, 103, 109, 219n.10, 221n.39, 221n.40; non-human, 52–53, 58

conflict, 81–82, 106, 165, 231n.4

continental drift, 11, 13–14, 33

continental island, 10, 22

Cook Islands, 81, 174–176, 205, 236nn.103, 104, 111, 240n.5; Mangaia Island, 23, 103, 174–175, 205–207, 236n.104,

240n.6; Manihiki Atoll, 174, 177; Palmerston Atoll, 175–177, 236n.111; Penrhyn (Tongareva) Atoll, 92, 176–177, 223n.22; Pukapuka Atoll, 95–96, 139, 232n.43; Rakahanga Atoll, 174, 177; Rarotonga Island, 23, 30, 174–175, 177, 206, 221n.45, 223n.19, 236n.114, 240n.3; Tuanaki Island (see Tuanaki); Victoria Island (see Victoria)

coral reefs: barrier reef, 3, 28–29, 176, 214n.45; catch-up reef, 38–39; drowned reef, 39, 155; fringing reef, 27–29, 175; give-up reef, 39, 213n.13; keep-up reef, 38–39

CRE. See catastrophic rise event

Creation (act of), 8, 132, 171

crossover point, 33

crust: earth's, 7, 13–17, 23, 64, 190, 210n.19; continental, 13, 16, 17, 56, 67, 210n.20; oceanic, 10, 13–14, 16, 36, 64, 66, 210n.20

Darwinism, 124

Davis' Land, 88, 94–95, 112, 130–131, 223n.32, 230n.3

debris apron/slide, 43, 45, 135, 191

delta, 30, 76–77

diffusion, 123, 128; diffusionism, 74, 125, 127–128, 219n.18, 229n.68, 230n.86

dinosaur, 18–19, 229n.62

disease, 83

dispersal (biotic), 42, 52–59, 68

DNA, 54, 58, 75, 217n.89, 220n.22, 222n.54

Drake Passage, 19–20

drosophilids (flies), 53–54, 215n.58

earthquakes, 15–16, 134–140, 162–163, 167–168, 231n.18; Hawke's Bay Earthquake, 23; Kalapana Earthquake, 134, 193, 238n.32; Prince William Sound Earthquake, 31, 50

Easter Island, 30, 81, 112, 123–127, 198, 221n.48, 239n.8; endemic biota of, 53–55, 215n.57, 216n.89, 217n.90; traditions of, 94–95, 135–138, 223n.37, 227n.24

El Niño, 170, 183, 221n.39

emergence (of land), 20–21, 23, 31, 92, 211n.41

endodontoid snails, 42, 54, 59

eruption. See volcano

Europeans in the Pacific, 61, 83, 86, 95, 99, 111, 216n.89, 226n.12

evolution (biotic), 17, 31, 58, 124, 128, 218n.3

Federated States of Micronesia, 88, 99–101, 138; Chuuk, 28, 88, 99, 102; Kosrae, 138; Pohnpei (Ponape), 99, 109, 118, 127, 229n.78; Yap, 88, 100–102

Fiji, 2, 26–27, 70, 105–106, 163–170, 187, 209n.2, 212n.62, 221n.42, 222n.52, 226n.6; Burotu Island, 4, 89, 97, 105, 137, 163–168, 198, 235n.65; Davetalevu (passage), 3, 168–169, 235n.78; Lau Islands, 185; Leleuvia Island, 2–3, 168; Ma-

tuku Island, 4, 106, 164–168, 198, 234n.57, 235n.65; Moturiki Island, 2–3, 79, 168–169, 209n.2, 235n.75; Viti Levu Island, 27, 50, 69–70, 80, 83, 168–169; Vuniivilevu Island, 3, 89, 168–170, 224n.52, 235n.75

fixism, 8, 13–14, 27, 63

floating island, 59, 92, 97, 107, 174, 203

fold mountains. See mountains, fold

French Polynesia, 43–45, 52, 175; Fangataufa Atoll, 45, 190; Gambier Islands, 52–53, 92–94; Huahine Island, 139, 232n.46; Marquesas Islands, 39, 58, 89, 137, 194, 224n.45; Moruroa Atoll, 45, 190; Ra'iatea Island, 224n.43, 225n.91; Rangiroa Atoll, 47, 136, 214n.46; Rapa Island, 67, 214n.34; Society Islands, 57–58, 67, 98, 224n.43; Tahiti Island, 44–45, 98, 115, 137–138, 224n.43, 229n.73; Tuamotu Islands, 23, 53, 88, 92–93, 95, 170, 211n.45, 223n.22, 236n.96

FSM. See Federated States of Micronesia

Galápagos Islands, 44, 54, 59, 68

geohazards. See hazards

glacial (period). See ice age

global warming. See climate change

gods, 87, 97, 103–106, 164, 224n.47

Gondwana, 13, 21, 64–70, 129, 217n.19, 218n.19; biota of, 56–57

gravity collapse. See collapse

Greenland, 10, 63

Guam, 22, 44

guyot, 28, 36–38, 55, 60, 211n.54, 213n.25

Hawai'i, 26, 44, 88, 116, 188–194, 230n.90; biotic (non-human) colonization of, 53–58, 215n.70, 216n.78; Hawai'i (Big) Island, 10, 30, 133–135, 185, 212n.8, 235n.62; human colonization of, 81, 220n.21, 227n.22; Koko Seamount, 54; Maui Island, 97; Meiji Seamount, 38, 212n.9; Midway Atoll, 41, 54, 215n.59; Moloka'i Island, 44–45, 214n.43; Necker Island, 56; traditions of, 83, 94–95, 97–98, 102–106, 223n.31, 225n.90, 227n.24; Wentworth Seamount, 56

Hawaiki, 4, 91, 102–106, 116, 164, 225n.92

Haymet Rocks, 175, 236n.101

hazards, 16, 138, 193–196, 231n.19; hazard management, 136

Hess Rise, 66–67

Hilina Slump, 133–134, 193

hot spot, 16, 38, 54, 56, 170

humans: arrival (see colonization); origins of, 11, 109, 120–121, 124, 128; skin color, 71, 222n.53; subsistence, 72, 75–76

ice age, 20, 31, 41, 52, 72, 183, 189–190; last, 44, 61, 74, 180, 219n.16

ice core, 160

iguana, 59, 216n.81, 221n.43

Indonesia, 77, 104, 109, 120, 218n.3

interglacial (period), 31, 52, 190, 230n.87

Intergovernmental Panel on Climate Change (IPCC), 184, 237n.1

inter-tropical convergence zone (ITCZ), 58
IPCC. *See* Intergovernmental Panel on Climate Change
irrigation. *See* agriculture
islands: appearance (traditions about), 87, 97–98, 140, 167–168; artificial, 118, 125, 156–157, 184, 229n.76; continental (*see* continental island); disappearance (traditions about), 35, 87, 95, 105, 137, 139, 153–158, 161–177; floating (*see* floating island); limestone (*see* limestone islands); lost (*see* lost island); mythical, 6, 86–99, 103–106, 131; oceanic (*see* oceanic island)

Jamaica, 136
Japan, 20, 50–51, 73–74, 87, 138, 187, 215n.70; Fukutokuoka-no-ba, 141–142; Hokkaido Island, 134; Kyushu Island, 51, 138; Minami-tori Shima, 178–179, 237n.121; Myojin-Sho, 141–142, 232n.51; Oshima-Oshima Island, 134–135, 163, 231n.21; Ryukyu Islands (*see* Ryukyu Islands); Yonaguni (*see* Yonaguni)
Johnston Atoll, 43–44, 166, 213n.25, 235n.62
Juan Fernandez Island, 54, 112

Kiribati, 98–99, 170–174, 229n.73, 235n.84; Abemama Atoll, 173; Banaba Island, 88, 98, 170; Bikeman Island, 89, 133, 170–171, 235n.80; Butaritari Atoll, 88, 99, 171, 173, 236n.86; Enderbury Atoll, 125, 127; Makin Atoll, 88–89, 99, 171–173, 235n.84, 236n.89; Malden Atoll, 109, 125, 229n.70; Niku-maroro Atoll, 173; Tarawa

Atoll, 89, 99, 133, 170–171, 211n.52; Tebua Island, 89, 133, 170–171, 235n.80
Kodiak Island, 31
Krakatau Island, 48, 120, 145, 215n.49
K-T Boundary Event, 19
Kuril Islands, 15

land bridge, 18, 52, 121
landslide: flank, 30, 213n.12, 214n.41; submarine, 43–45, 136, 149–151, 155, 163,
language group: Austronesian, 72–73, 75, 78, 109, 219n.21; Papuan, 72
Lapita, 77–81, 100, 109–110, 128, 220nn.33, 38, 221n.39, 226n.6
Large Igneous Province, 19, 66–67, 217n.11
Last Glacial Maximum. *See* ice age
Lemuria, 84–85, 104, 118–125, 212n.60, 226n.2, 228nn.54, 62, 229n.67, 230n.3
limestone islands, 22, 42, 44, 52, 68, 100, 160, 213n.12
Little Ice Age, 183
Lituya Bay, 139–140, 166
Lord Howe Rise, 57
Los Jardines, 89, 137, 178–180
lost island, 3, 95, 102, 175, 196, 205, 236n.106

magic, 87, 167, 202, 239n.2
Malaita Accretionary Prism, 151, 155–157
mammals, 11, 19, 218n.19
Mana (reconstructed head), 79, 85
Mangareva. *See* French Polynesia, Gambier Islands
Manihiki Plateau, 19, 65–66, 217n.11
Maori. *See* New Zealand
marae, 109, 125, 229n.69

Marcus Island. *See* Japan, Minami-tori Shima
Mariana Islands, 10, 15, 22, 38, 123, 179
Marshall Islands, 100, 173, 178, 213n.25; Bikini Atoll, 54; Enewetak Atoll, 37, 54, 178
Maui (demigod), 87, 91–93, 106, 137, 144–145, 174, 222n.3
megalith, 118, 123–127, 222n.56, 226n.4, 229n.69
megatsunami. *See under* tsunami
Melanesia, 61, 84–85, 128, 222n.53, 227n.22, 233n.6. *See also individual island groups*
methane, 189
Metrosideros, 58, 216n.81
Micronesia, 28, 84, 99, 109, 118, 127, 133, 138. *See also individual island groups*
mid-ocean ridge, 13–15, 30, 209n.9
mirage, 98, 130
missionaries, 4, 83, 132, 158, 166, 192, 205–207, 232n.58, 240n.6
moai, 125, 127, 198–199, 230n.79
moon, 16, 132
Mount Shasta, 121, 123
mountains, fold, 9, 21, 63–64, 217n.7
Mu, 84–85, 104, 118–123, 125, 127–128, 229nn.63, 67, 239n.4
mummification, 128
Myojin-Sho. *See* Japan, Myojin-Sho
myth-motif, 76, 87, 132, 141
myths, 4, 76, 86–87, 91–120, 132–177, 189, 198, 222n.2, 223n.22, 225n.99, 227n.24, 232n.41

Nan Madol, 109, 118, 125–127, 199, 226n.4, 229nn.76, 77
new age, 1, 16, 84, 118, 121–128, 196–197

New Caledonia, 22, 57, 109–110, 228n.39; Loyalty Islands, 22
New Guinea. *See* Papua New Guinea
New Zealand, 23, 56–58, 81, 102–104, 216n.79, 221n.50, 232n.46; North Island, 132, 139, 221n.45
Niue, 22, 25, 44, 214n.40; traditions of, 86, 139, 222n.2
nomadism, 73, 75–76; sea nomads, 77, 220n.33, 226n.7
Norfolk Ridge, 57, 70
North America, 18, 20, 63, 230n.84, 231n.19; biota of, 18, 215n.70
Nu'uanu Slide, 44, 46

obsidian, 73, 79, 102, 221n.48, 226n.6
ocean currents, 19, 39, 57, 68, 110, 204, 216n.73
ocean trench. *See* trench
ocean voyaging. *See* voyaging
oceanic island, 3, 43, 120, 137, 189, 193, 231n.19; biotic colonization of, 53, 56, 69–70; origins of, 10, 14, 180
Ontong Java Plateau, 17, 19, 21–22, 65–67, 151
ophiolite, 22, 25
oral traditions, 35, 117, 131, 147–149, 198, 222n.1, 228n.51, 238n.39; Pacific Islander, 82–105, 137–139, 144, 151–177, 188, 193, 233n.10
Oyashio (current), 20

Pacifica, 19, 64–68, 210n.30, 217nn.10, 16, 239n.6
Palau, 42, 54, 100
palolo, 107
Panama, 20, 74, 106, 110
Pangea, 9, 12–14, 18–19, 64–69
Panthalassa, 9, 12–13, 18, 65, 68
Papua New Guinea, 72–73, 147,

188–190, 211n.41, 219n.12; Bismarck Archipelago, 78–79, 109, 136, 148, 190, 194, 220n.38; human colonization of, 77–80; Lihir Island, 50, 215n.54; Long (Arop) Island, 135, 147–149; New Britain Island, 79, 190, 219n.12; Ritter Island, 48, 134–136, 149, 189–190, 194, 231n.21; Yomba Island, 89, 135, 145, 148–149, 189, 233n.1
Philippines, 15, 77–78, 98, 109, 112
Pitcairn Island (group), 52–54, 67, 93
plate. *See* Plate Tectonics
Plate Tectonics, 13–21, 33, 63, 121; convergence, 16–17, 36–37, 100–101, 123, 148–149, 164, 210n.15; divergence, 16, 65, 100, 209n.13; transform movement, 16, 210n.15
Plato, 1–2, 16, 108–109, 199
Polynesia, 84–85, 102–106, 215n.57, 221n.43, 226n.9, 228n.38, 233n.6. *See also individual island groups*
pottery, 2, 77–78, 81, 209n.2, 221n.42
pseudoscience, 2, 6, 114, 125–129, 195–199, 226n.14, 227nn.18, 22
Pulotu, 105–106, 163–164, 166, 225n.99. *See also* Burotu
pumice, 51, 59, 95, 97–98, 107, 131

Rapa Nui. *See* Easter Island
reef. *See* coral reefs
Ring of Fire, 14–15, 158, 209n.13
Rodinia, 14, 17–18, 63
root race, 84–85, 121, 124, 228n.54, 229n.67, 230n.86
Rosicrucians, 121
Russia, 9, 64, 73

Rutas, 116–117, 120–121
Ryukyu Islands (Japan), 15, 48, 215n.50; Yonaguni (*see* Yonaguni)

Sahul, 72, 219n.9
Saipan, 22
Sala-y-Gómez, 53–55
Samoa, 59, 79, 164, 231n.28; Savai'i Island, 37, 92, 99, 104, 164, 225n.90; Upolu Island, 92
San Andreas Fault, 16
sea-floor spreading, 13–14, 16, 65, 100–101, 113, 127
sea-level changes, 31, 131, 182–185, 190, 211n.41, 237n.2
seamount, 22–28, 180, 211n.54, 236n.101; Capricorn Seamount, 23, 25, 36–37; Fabert Seamount, 175; Machias Seamount, 37
sedentism, 73, 76
seismicity. *See* earthquakes
serendipity, 59, 61, 75
Shatsky Rise, 66–67
sinking. *See* subsidence
Solomon Islands, 66, 73, 78–79, 111, 151–157, 184, 189, 201, 204; Choiseul Island, 22; Guadalcanal Island, 69, 204; Kavachi Volcano, 142–143; Makira Island, 88, 151–153, 157, 204, 238n.39; Malaita Island, 88, 151–153, 156–157, 202, 204; Olu Malau Islands, 151–155; Teonimanu Island, 89, 137, 151–155, 189, 194, 238n.39, 239n.1; as Teonimenu, 201–204, 233n.20, 239n.1; Ulawa Island, 88, 151–155, 201, 203–204, 233nn.10, 19, 20, 239n.1
Sorol Trough, 101, 224n.69
South America, 17–20, 62–68, 73–74, 105, 125, 215n.57, 216n.89, 221n.45

South Pacific Superswell. *See* swell
South Pole, 18–19
stilt houses, 78, 80
stone tools, 72–73
submergence, 21, 26–28, 98, 100–104, 132, 154–156, 161–185, 203–204; submerged continents, 57, 113, 117–118, 121, 124, 127–128, 217, 229n.70; submerged islands, 38–41, 52–57, 93, 162, 173–174, 224n.52
subsidence, 19–21, 26–31, 37–39, 117, 156–157, 162, 185, 211n.58, 228n.40, 230n.78, 234n.57; coseismic, 21, 28, 30–31, 151, 157, 164, 187
Subsidence Theory of Atoll Formation, 27–29, 102, 117, 213n.21
Sunda Shelf, 131, 220n.34
sunken continents. *See* submergence, submerged continents
supercontinent, 13–14, 17–18, 31, 63, 65, 68, 209n.10. *See also* Pangea
swell (crustal), 23, 36, 92, 144, 211n.42

Taiwan, 23, 26, 73, 75, 139, 211nn.41, 45, 219n.21, 220nn.24, 33, 232n.41
Tangaloa, 106, 144
Tasman Rise, 19–21
Tasman Sea, 56–57, 215n.71, 216n.73
Terra Australis, 111–113, 115, 130, 223n.32, 227n.20
terrane, 17–19, 56, 64, 66, 68, 123, 217n.9
Tetragnatha, 58

Theosophy, 117, 119–120, 196, 228n.53
tidal wave, 148, 155, 157. *See also* tsunami
Tierra del Fuego, 111
Timor, 137
Tokelau, 183
Tonga, 36, 68–70, 105–106, 110, 142–145, 169, 185–186, 222n.53, 225n.99, 232n.56, 238n.13; 'Eua Island, 68–70; Fonuafo'ou Island, 142, 232n.56; Niuafo'ou Island, 145; Tafahi Island, 212n.2; Tongatapu Island, 22, 36, 142, 185–186
Tonga-Kermadec Trench, 22–23, 25, 36–37, 69
Tongareva. *See* Cook Islands, Penrhyn
trench (ocean-floor), 7, 14–17, 36–37, 148–151, 157–172
tsunami, 4, 44–47, 134–140, 154–160, 194, 222n.6, 231n.29, 232n.46, 235n.73; Aitape, 149–151; deposits, 45, 47, 214n.44; megatsunami, 4, 193, 214n.43
Tuanaki, 4, 174–175, 205–206, 236nn.104, 106, 240n.6
turbidite, 62–63, 217n.1
Tuvalu, 183, 237n.2, 237n.6

uplift, 18, 20–27, 52, 69, 160, 210n.15, 211nn.41, 47, 214n.44, 234n.42; coseismic, 23, 26, 30–31, 139–140, 156, 232n.41

Vanuatu, 22, 26, 36, 157–163, 188–192, 211n.41, 212n.3, 232n.66, 233n.21, 234n.42; Ambae (Aoba) Island, 89, 157, 161–163, 190, 238n.28;

Ambrym Island, 157, 163, 190, 192; Aneityum Island, 138; Aniwa Island, 138; Efate Island, 157; Epi Island, 157–158; Espiritu Santo (Santo) Island, 22, 36, 157; Karua Island, 142, 158, 160; Kuwae Island, 48, 50, 89, 142, 145, 158–160, 215n.49, 234n.39; Maewo Island, 161–162; Malakula Island, 22, 26, 36, 89, 157, 160–162; Malveveng Island, 89, 161–162; Pentecost Island, 161–162; Tolamp Island, 89, 161–162; Tongoa Island, 157–158, 160; Vanua Mamata Island, 89, 137, 162–163, 167, 190, 194
Victoria (Island), 89, 174–177, 236
vigia, 107, 225n.104
volcano: Hawaiian, 28–30, 38, 44, 133–136, 193, 210n.16, 212n.8, 231n.19; hot spot, 16, 32, 90; myths about, 91, 94, 103, 159; submarine (underwater), 15, 59, 140–143, 160, 232n.48
voyaging (ocean), 81, 90, 102, 105, 153–154, 209n.1, 221n.45, 227n.23

Wai'anae Slump, 44
West Wind Drift, 57, 216n.73
whales, 59, 69, 175, 178, 235n.78
Wrangellia Terrane, 56, 68

Yap-Palau Trench, 100–102
Yonaguni, 127, 230n.79
Younger Dryas, 75–76, 220n.25

ABOUT THE AUTHOR

Patrick Nunn holds a Personal Chair (Professor of Oceanic Geoscience) at the international University of the South Pacific in Suva, Fiji, where he has taught since 1985. He has also taught at universities in Australia, Canada, Japan, New Zealand, and the United States. He has been at the forefront of research into aspects of the geology, geography, and archaeology of the Pacific Islands region for more than twenty years, and he is the author of more than 170 professional papers and book chapters, in addition to several books. In March 2003 he was awarded the Gregory Medal of the Pacific Science Association for "outstanding service to science in the Pacific."